Solid Catalysts for the Upgrading of Renewable Sources

Solid Catalysts for the Upgrading of Renewable Sources

Special Issue Editors

Nicoletta Ravasio
Federica Zaccheria

MDPI • Basel • Beijing • Wuhan • Barcelona • Belgrade

MDPI

Special Issue Editors

Nicoletta Ravasio
CNR Istituto di Scienze e Tecnologie Molecolari
Italy

Federica Zaccheria
CNR Istituto di Scienze e Tecnologie Molecolari
Italy

Editorial Office
MDPI
St. Alban-Anlage 66
4052 Basel, Switzerland

This is a reprint of articles from the Special Issue published online in the open access journal *Catalysts* (ISSN 2073-4344) from 2018 to 2019 (available at: https://www.mdpi.com/journal/catalysts/special_issues/solid_catalysts)

For citation purposes, cite each article independently as indicated on the article page online and as indicated below:

LastName, A.A.; LastName, B.B.; LastName, C.C. Article Title. *Journal Name* **Year**, *Article Number*, Page Range.

ISBN 978-3-03897-572-4 (Pbk)
ISBN 978-3-03897-573-1 (PDF)

Contents

About the Special Issue Editors

Nicoletta Ravasio, Ph.D., Received her degree in Chemistry from University of Milano in 1982 and her PhD in Chemistry from the University of Bari (Italy) in 1987. She has been a Senior research Fellow at ISTM-CNR, Milano since 2001. Her activity is mainly devoted to the use of heterogeneous catalysis in organic synthesis and in renewable raw materials transformation, with special emphasis on vegetable oils and terpenes. In particular, she developed several selective hydrogenation and dehydrogenation processes based on supported Cu catalysts thanks to a particular preparation method that allows one to obtain very small Cu crystallites. Such nanoparticles can also show acidic properties and this dual nature of the metal particle can be exploited for the set-up of bifunctional processes to produce fine chemicals or biofuels, reducing the number of synthetic steps. She also investigates the use of amorphous solid acids, particularly showing Lewis character, for the sustainable synthesis of fine chemicals or oleochemicals.

Federica Zaccheria, Ph.D., Received her degree in Organic Chemistry in 1998 from the University of Milan and her PhD in Industrial Chemistry in 2002. She is currently a Research Scientist at the Institute of Molecular Science and Technology of CNR in Milan. Presently, her main research topic is heterogeneous catalysis applied to the synthesis of fine chemicals and to renewable materials selective transformations. Research activity has been mainly focused on the study of heterogeneous nontoxic and non-noble catalysts as substitutes of traditional stoichiometric reagents for organic synthesis and on the development of solid catalysts for the upgrade of renewable sources, such as vegetable oils and cellulose. Now, a great part of the work is also devoted to the use of agro-industrial wastes and by-products for the preparation of chemicals and materials.

Preface to "Solid Catalysts for the Upgrading of Renewable Sources"

The use of solid catalysts for the upgrading of renewable sources gives the opportunity to combine the two main cores of green chemistry, that is, on the one hand, the setting up of sustainable processes and, on the other, the use of biomass-derived materials. Solid catalysts have taken on a leading role in traditional petrochemical processes and could therefore also represent a key tool in new biorefinery-driven technologies.

This Special Issue covers topics related to the preparation and use of heterogeneous catalytic systems for the transformation of renewable sources, as well as of materials deriving from agro-industrial wastes and by-products. The valorization of rest raw materials represents a crucial challenge in the roadmap to a circular economy. At the same time, the ever-increasing importance of bioproducts, due to the acceptance and request of consumers, makes the upgrading of biomass into chemicals and materials not only an environmental issue but also an economical advantage.

In this Special Issue, we invite the main groups involved in heterogeneous catalysis applied to renewable materials to contribute original papers, mini reviews or commentaries in order to give an overview of the state-of-the-art in this field and an interpretation of the open challenges and opportunities. The main focus is devoted to the transformation and upgrading of:

1. Lignocellulosic materials;

2. Vegetable oils;

3. Terpenes;

4. Agro-industrial wastes and by-products.

Nicoletta Ravasio, Federica Zaccheria
Special Issue Editors

Editorial

Solid Catalysts for the Upgrading of Renewable Sources

Federica Zaccheria and Nicoletta Ravasio *

CNR-ISTM, via C.Golgi 19, 20133 Milano, Italy; f.zaccheria@istm.cnr.it
* Correspondence: nicoletta.ravasio@istm.cnr.it; Tel.: +39-02-50314382

Received: 9 January 2019; Accepted: 11 January 2019; Published: 15 January 2019

The use of renewable resources as raw materials for the chemical industry is mandatory in the transition roadmap toward the Bioeconomy. However, this is a challenge for the setup of catalytic processes based on heterogeneous catalysts.

First of all, when using biorenewables (particularly sugars) as starting materials, the process has to be designed in the condensed phase, as these kinds of molecules have little-to-no volatility and water is the solvent of choice with most bio-based systems. Moreover, many reactions designed to produce chemicals will also create water. This is the case for both etherification and esterification, which are widely used to produce fuel components and additives. For these reasons, the hydrothermal stability of the catalyst is one of the main problems when dealing with renewables.

Another issue is due to the highly oxygenated nature of plant-derived raw materials and platform molecules. This makes oxygen removal reactions such as dehydration, hydrogenolysis, hydrogenation, decarbonylation, or decarboxylation almost ubiquitous in biomass valorization pathways. Therefore, there is a need for robust hydrogenation or hydrogen transfer catalysts and also water-resistant acidic catalysts, and possibly for bifunctional materials where different active sites are present.

These challenges will be adressed in this special issue of *Catalysts* through several examples. A review article focused on the state of the art in the liquid phase depolymerization of lignin via catalytic transfer hydrogenolysis/hydrogenation reactions will open this interesting and current collection of papers [1]. Lignin is one of main structural components of lignocellulosic materials, and is widely available as a by-product in the pulp and paper industry and in the process of second generation bioethanol production. It could be a source of very valuable aromatic compounds if an effective method of depolymerization was available. It should be remembered that a shortage of aromatics, which are among the main building blocks in the chemical industry, is expected due to the shift from conventional fossil fuels to shale oil. This makes alternative routes to aromatics of particular interest. The review will also discuss the effect of lignin origin, as it is known that there are significant differences between hardwood, softwood, and straw lignins. The hydrogenolysis of dimethyl adipate to 1,6-hexandiol and the hydrogenolysis of xylitol in water to ethylene glycol, propylene glycol, and glycerol are the subjects of two other papers [2,3]. In particular, the hydrothermal stability of the catalyst used in the latter reaction was studied and improved by decreasing the amounts of aggressive by-products. Transfer hydrogenation is also one of the steps involved in the one-pot conversion of ethyl levulinate into gamma-valerolactone (GVL) [4]. In this reaction, a solid catalyst with both acidic and basic sites showing high thermal and chemical stability was successfully used. GVL is one of the most promising platform molecules we can obtain from biomass, as it can be upgraded to various chemicals and fuels, such as polymers, fuel additives, and jet fuel.

A second review deals with a class of hybrid materials that can act as bifunctional catalysts in biomass conversion due to their particular structure, namely Metal Organic Framework (MOF) [5]. The structures of MOF show coordinatively unsaturated (open) sites, with Lewis acidity in inorganic nodes (metal ions) of the networks. These Lewis acids are of paramount importance for cascade processes in catalytic biomass upgrades such as depolymerization, dehydration, and isomerization.

Catalysts **2019**, *9*, 88

The third review paper deals with catalysis processes for the synthesis of terpene-derived amines. Besides cellulose, hemicellulose, lignin and vegetable oils, mono- and sesqui-terpenes are one of the major classes of chemicals we can obtain from biomass. They are the main constituents of turpentines, obtained through the distillation of resins from trees, particularly coniferous trees, but also of essential oils. They can be used as raw materials for the synthesis of several products, including fuels, fine chemicals, and agro-chemicals. Moreover, in the last years they have attracted considerable interest as renewable resources for rubber and polymerization chemistry. Particularly relevant reactions in the field of terpenes are C–N bond formation ones. The review [6] reports on different strategies, namely reductive amination of carbonylic terpenes, hydroaminomethylation, hydroamination of double C=C bonds, hydrogen-borrowing methodology for amination of alcohols, and C–H amination of terpenes. The following paper [7] deals with an imino-Diels Alder reaction allowing one to produce tricyclic octahydroacridines in one pot and one step, starting from citronellal and aromatic amines and using a clay as the catalyst. Finally, the preparation of bio-derived carbon-derived materials to be used as hydrothermally stable catalysts for biomass transformation will be considered.

The fourth review [8] will compare two main methods for biochar synthesis (namely conventional pyrolysis and hydrothermal carbonization (HTC)) and the features of biochar with respect to other carbonaceous materials. Moreover, it will describe char modification strategies and some applications in the field of biofuels. The last paper [9] will describe a particular method to obtain a carbon-based material from sugars and taurine, allowing one to directly introduce strongly acidic groups on the surface.

This collection shows how numerous and multifaceted the research topics related to the exploitation of biomass are. Not only should catalysts comply with some particular stability requirements, but many processes should be re-thought to face the challenges of a new raw materials pool.

References

1. Margellou, A.; Triantafyllidis, K.S. Catalytic Transfer Hydrogenolysis Reactions for Lignin Valorization to Fuels and Chemicals. *Catalysts* **2019**, *9*, 43. [CrossRef]
2. Kikhtyanin, O.; Pospelova, V.; Aubrecht, J.; Lhotka, M.; Kubička, D. Effect of Calcination Atmosphere and Temperature on the Hydrogenolysis Activity and Selectivity of Copper-Zinc Catalysts. *Catalysts* **2018**, *8*, 446. [CrossRef]
3. Rivière, M.; Perret, N.; Delcroix, D.; Cabiac, A.; Pinel, C.; Besson, M. Ru-(Mn-M)OX Solid Base Catalysts for the Upgrading of Xylitol to Glycols in Water. *Catalysts* **2018**, *8*, 331. [CrossRef]
4. Wu, W.; Li, Y.; Li, H.; Zhao, W.; Yang, S. Acid–Base Bifunctional Hf Nanohybrids Enable High Selectivity in the Catalytic Conversion of Ethyl Levulinate to γ-Valerolactone. *Catalysts* **2018**, *8*, 264. [CrossRef]
5. Isaeva, V.I.; Nefedov, O.M.; Kustov, L.M. Metal—Organic Frameworks-Based Catalysts for Biomass Processing. *Catalysts* **2018**, *8*, 368. [CrossRef]
6. Simakova, I.L.; Simakov, A.V.; Murzin, D.Y. Valorization of Biomass Derived Terpene Compounds by Catalytic Amination. *Catalysts* **2018**, *8*, 365. [CrossRef]
7. Zaccheria, F.; Santoro, F.; Iftitah, E.D.; Ravasio, N. Brønsted and Lewis Solid Acid Catalysts in the Valorization of Citronellal. *Catalysts* **2018**, *8*, 410. [CrossRef]
8. Cheng, F.; Li, X. Preparation and Application of Biochar-Based Catalysts for Biofuel Production. *Catalysts* **2018**, *8*, 346. [CrossRef]
9. Ji, H.; Fu, J.; Wang, T. Pyrolyzing Renewable Sugar and Taurine on the Surface of Multi-Walled Carbon Nanotubes as Heterogeneous Catalysts for Hydroxymethylfurfural Production. *Catalysts* **2018**, *8*, 517. [CrossRef]

catalysts

MDPI

Review

Catalytic Transfer Hydrogenolysis Reactions for Lignin Valorization to Fuels and Chemicals

Antigoni Margellou [1] and Konstantinos S. Triantafyllidis [1,2,*

[1] Department of Chemistry, Aristotle University of Thessaloniki, 54124 Thessaloniki, Greece;
   amargel@chem.auth.gr
[2] Chemical Process and Energy Resources Institute, Centre for Research and Technology Hellas,
   57001 Thessaloniki, Greece
* Correspondence: ktrianta@chem.auth.gr

Received: 31 October 2018; Accepted: 10 December 2018; Published: 4 January 2019

Abstract: Lignocellulosic biomass is an abundant renewable source of chemicals and fuels. Lignin, one of biomass main structural components being widely available as by-product in the pulp and paper industry and in the process of second generation bioethanol, can provide phenolic and aromatic compounds that can be utilized for the manufacture of a wide variety of polymers, fuels, and other high added value products. The effective depolymerisation of lignin into its primary building blocks remains a challenge with regard to conversion degree and monomers selectivity and stability. This review article focuses on the state of the art in the liquid phase reductive depolymerisation of lignin under relatively mild conditions via catalytic hydrogenolysis/hydrogenation reactions, discussing the effect of lignin type/origin, hydrogen donor solvents, and related transfer hydrogenation or reforming pathways, catalysts, and reaction conditions.

Keywords: lignin; catalytic transfer hydrogenation; hydrogenolysis; liquid phase reductive depolymerization; hydrogen donors; phenolic and aromatic compounds

1. Introduction

The projected depletion of fossil fuels and the deterioration of environment by their intensive use has fostered research and development efforts towards utilization of alternative sources of energy. Biomass from non-edible crops and agriculture/forestry wastes or by-products is considered as a promising feedstock for the replacement of petroleum, coal, and natural gas in the production of chemicals and fuels. The EU has set the target of 10% substitution of conventional fuels by biomass-derived fuels (biofuels) by 2020, and USA of 20% substitution by 2030 [1–3].

Lignocellulosic biomass consists mainly of cellulose, hemicellulose and lignin, all of which can be converted into a wide variety of platform chemicals that can be further transformed to fuels, engineering polymers, pharmaceuticals, cosmetics, etc. (Figure 1). Cellulose is a linear polymer consisting of glucose molecules linked with β-1,4-glycosidic bonds and hemicellulose is branched polysaccharide composed of C_5 and C_6 sugars [4]. Lignin is an amorphous polymer with p-coumaryl, coniferyl, and sinapyl alcohols being its primary building units. Lignocellulosic biomass can be derived from hardwoods (beech, birch, poplar, etc.), softwoods (pine, spruce, cedar, etc.), grasses (switchgrass, miscanthus, etc.), as well as various agricultural byproducts/wastes (straws, husks, bagasse, etc.). The percentage of cellulose, hemicellulose and lignin in lignocellulosic biomass depends on the nature of the source as well as on the type of the individual member, i.e., hardwood vs. softwood and poplar vs. beech within hardwoods. A number of pretreatment methods have been proposed for the selective isolation of each biomass component. These include physical methods such milling [5–7], sometimes combined with H_2SO_4, chemical methods such as acid (H_2SO_4, HCl, H_3PO_4), alkaline (NaOH), organosolv, ozone and oxidative treatment [6,8–11] and physicochemical such as ammonia fiber, SO$_2$ and steam explosion [6,12–14], wet oxidation [15] and hydrothermal methods [16,17]. The isolated fractions in the form of carbohydrate or phenolic biopolymers of varying molecular weight, functionality, particle size and other physicochemical characteristics, can be utilized as such in polymer composites, pharmaceutical formulations, etc. [18–20] Furthermore, the downstream selective depolymerization of these biopolymers to their primary building units, i.e., glucose, xylose, alkoxy-phenols, etc., and their consequent transformation to a wide variety of platform chemicals and eventually to final products, may offer even higher value to biomass valorization, via the "biorefinery" concept. Pyrolysis and hydrogenolysis/hydrogenation [21–24] represent probably the most studied thermochemical biomass (or its components) depolymerization processes towards the production of valuable compounds with a potential in fuels, chemicals and polymers industry [23,25,26].

The aim of this review is to focus on the heterogeneous catalytic transfer hydrogenation reactions for the depolymerization of various types of lignins, including technical lignins deriving from established industrial processes, i.e., kraft, soda or lignosulphonate lignin from the pulp and paper or related industries, as well as enzymatic/acid hydrolysis and organosolv lignins as part of the 2nd generation bioethanol production process.

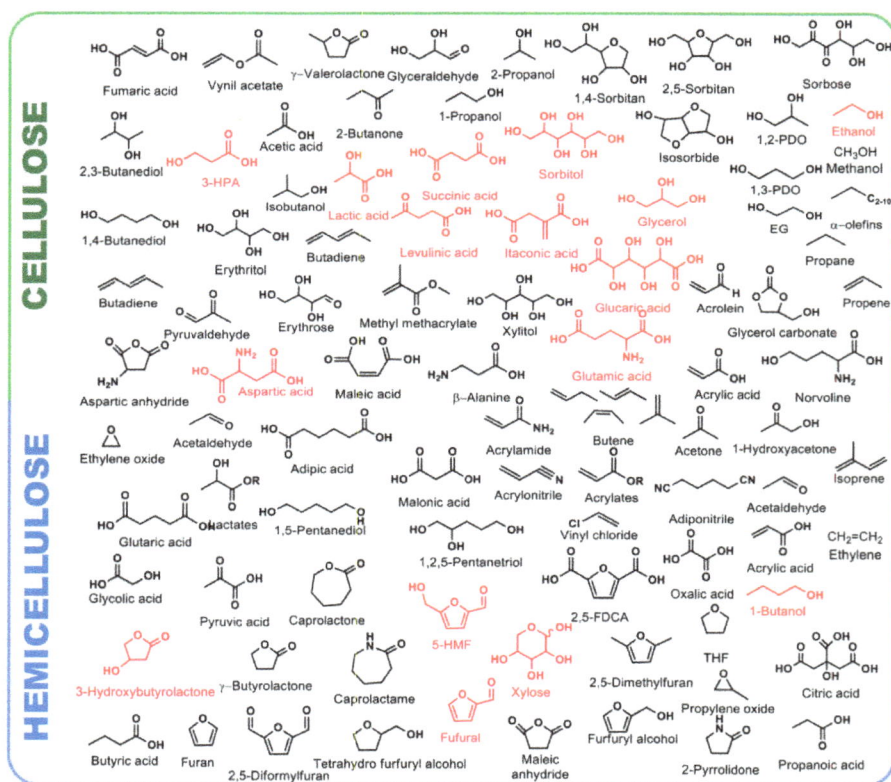

Figure 1. *Cont.*

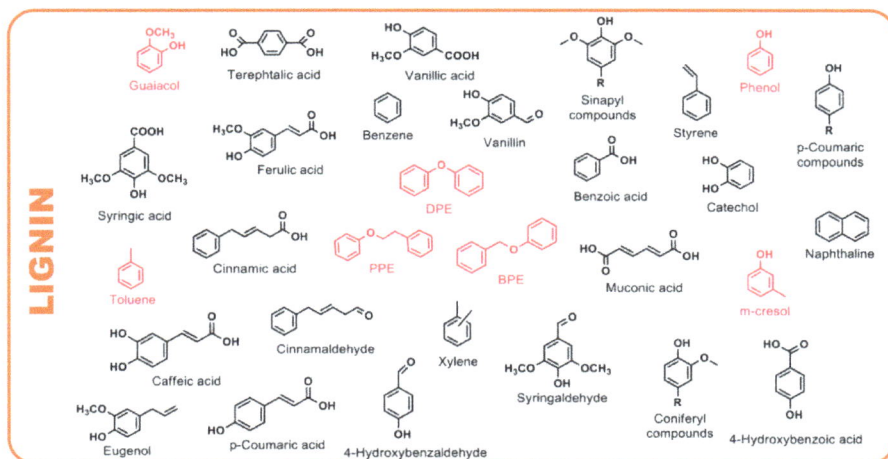

Figure 1. Chemicals derived by the valorization of lignocellulosic biomass. Reproduced from reference [27] with permission from MDPI.

2. Lignin Chemistry

2.1. Lignin Structure and Isolation

Lignin is an amorphous polymer formed by the polymerization of p-coumaryl, coniferyl and sinapyl alcohols via the phenylpropanoid pathway [28]. The structures of the three monolignols, being phenylpropene rings with one (coniferyl), two (sinapyl) or no (p-coumaryl) methoxy-substituents, are shown in Figure 2. Coniferyl alcohol (G units) is the main building block of softwood lignins with up to ca. 90% content, whereas hardwood lignins contain, in addition to coniferyl units, increased amounts of sinapyl alcohol (S units), reaching 50–75%. Grass lignin is also composed of coniferyl and sinapyl alcohol units, exhibiting also traces of p-coumaryl alcohol (H units) [29,30].

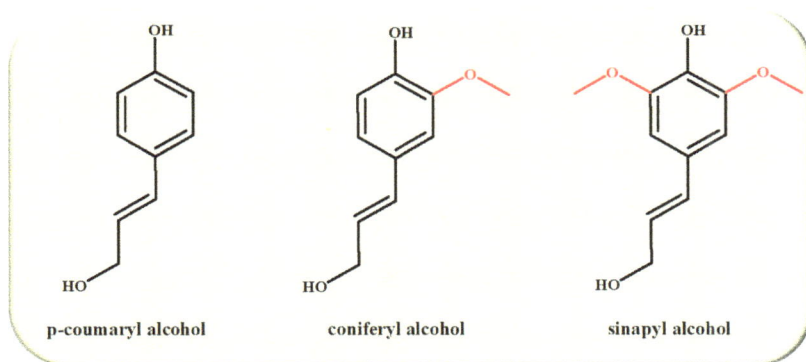

Figure 2. Building blocks of lignin.

The building blocks of lignin are linked via ether or carbon-carbon bonds formed between the aliphatic chain of monolignols and the aromatic moieties. The most dominant linkage is the β-O-4 aryl ether between the β-carbon of the aliphatic chain and the O-atom from the aromatic moiety, with

45–50% abundance in softwood and 60–62% in hardwood [29,31,32]. Other linkages appearing in lignin are β-β (resinol), β-5 (phenylcoumaran), β-1 (spirodienone), α-O-4, 4-O-5 (diaryl ether), α-O-γ, 5-5 (bisphenyl) and dibenzodioxocin [29,31–33]. Representative schematic representations of softwood and hardwood lignin structures, as well as the dominant linkages, are shown in Figure 3.

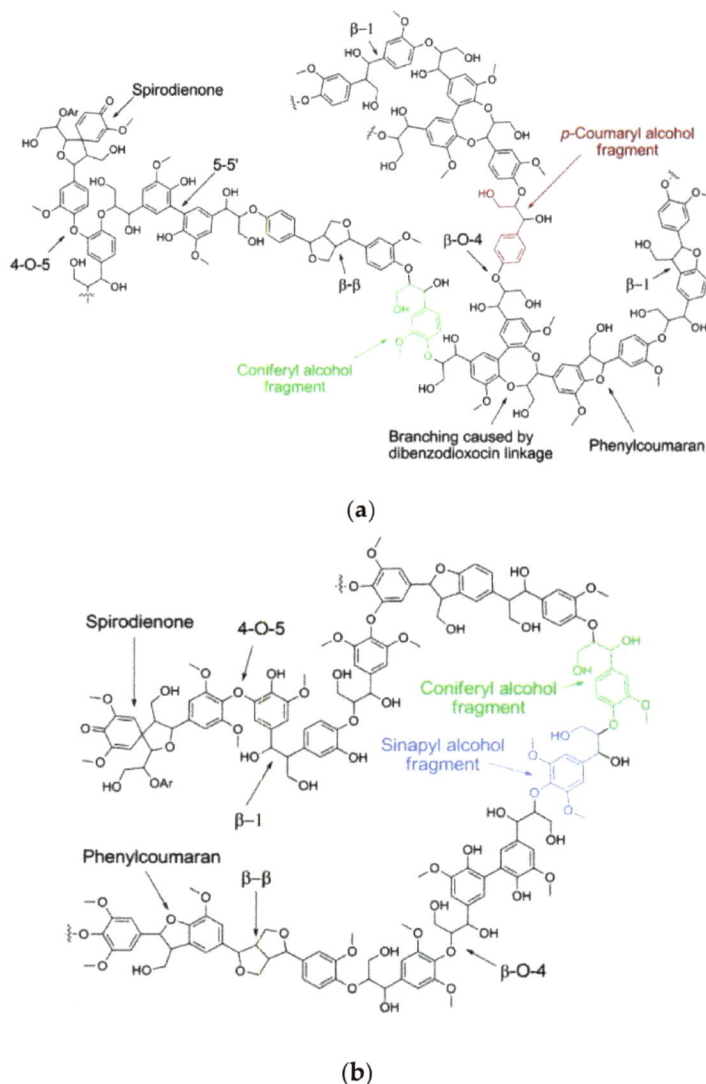

Figure 3. Schematic representations of (**a**) softwood and (**b**) hardwood lignin structures. Reprinted with permission from reference [3]. Copyright 2010, American Chemical Society.

The methods of lignin isolation can be classified into two categories based on the solubilization of lignin, as reported by Kim and co-workers [34]: the first category includes the methods in which lignin is isolated as insoluble residue after the solubilization of cellulose and hemicellulose while in the second category lignin is isolated in the process solution leaving cellulose and hemicellulose in the insoluble solids. Each isolation process may result in varying lignin yields with different molecular weight and other properties, and possible contaminations, as can be observed in Table 1.

Table 1. Major lignin isolation processes and the properties of the obtained lignin [4].

Process	Agent	T (°C)	MW (Da)	Polydispersity	Contamination
Kraft	NaOH + Na$_2$S	170	1000–3000	2.5–3.5	Sulfur
Soda	NaOH + anthraquinone	140–170	1000–3000	2.5–3.5	Sulfur
Sulfite	sulfite salts	140–170	1000–50,000	4.2–7.0	Sulfur
Organosolv	organic solvents	180–200	500–5000	1.5	Sulfur free

Kraft lignin is produced by the treatment of wood feedstock with NaOH and Na$_2$S at 170 °C for 2 h [3,33,35]. During kraft pulping, the hydroxide and hydrosulfide anions react with lignin, causing its depolymerizaiton into smaller water/alkali soluble fragments [31]. Besides the depolymerization via the cleavage of aryl ether bonds, introduction of thiol group, stilbene and carbohydrate linkages can occur [33,35]. Additionally, the isolated lignin is contaminated with carbohydrates from hemicellulose and a small amount of sulfur [4]. Kraft pulping is the dominant process and constitutes about 85% of total lignin production and is recognized as by-product in paper/pulp industry [36]. Similar to the Kraft process, soda pulping is more often used for the fractionation of non-woody biomass e.g grass, straw and sugarcane bagasse in the presence of NaOH or NaOH-anthraquinone at 140–170 °C [4,33]. Lignin is partially depolymerized during soda pulping via the cleavage of α- and β-aryl ether bonds, first in phenolic units and finally in non-phenolic units [35]. The resulting lignin is considered to be free of impurities compared to the Kraft lignin.

Another industrial process for the isolation of lignin is the sulfite pulping where the lignocellulosic biomass is digested at 140–170 °C with an aqueous solution of a sulfite or bisulfate salt of Na$^+$, NH$_4^+$, Mg or Ca [35]. This process can be carried out in the whole range of pH scale by selecting the appropriate salt. During the sulfite pulping, the linkages between the lignocellulosic compounds as wells as the ether bonds between lignin units can be cleaved by the nucleophilic attack of the sulfite anion [4,35]. As a consequence, sulfonation of the lignin aliphatic chain can occur.

The fractionation of lignocellulosic feedstocks via the organosolv process involves the treatment of biomass in organic solvents at the temperature range of 180–200 °C [4]. In this process a wide variety of organic compounds such as alcohols, ketones, acids, ethers and their mixtures with water have been used as solvents [37–41]. The fractionation can be improved by the addition of inorganic acids (H$_3$PO$_4$, HCl, H$_2$SO$_4$ [29,41,42]. Luterbacher et al. suggested the formaldehyde addition in the organosolv process for the stabilization of lignin during biomass pretreatment [43]. The subsequent hydrogenolysis of the extracted lignin resulted in 47–78% monomers, in contrast to the hydrogenolysis of lignin extracted in absence of formaldehyde which led to only 7–26% monomers. The organosolv pretreatment of biomass [11,44], as well as the recently reported hybrid steam explosion/organosolv process [45], have been proven beneficial for the enzymatic saccharification of the remaining cellulose, while at the same time achieving high yields of recovered lignin of relatively low molecular weight and high purity [11,33,45].

2.2. Lignin Valorization

The chemical structure and composition of lignin offer numerous exploitation opportunities towards the production of a vast variety of valuable products. For example, lignin itself can be used either directly without modification or after chemical modification in the polymer industry. One of the main applications of lignin is the substitution of phenol in the phenol-formaldehyde resins, without modifying the properties of the final product. Furthermore, lignin can be mixed with polymers such as polyolefins, polyesters and polyurethanes in the form of blends, copolymers and composites for the production of eco-friendly plastics with improved properties [19,46,47]. After chemical modifications lignin can be also added in epoxy resins. Another possible exploitation of lignin, is the thermochemical conversion to carbon functional materials [48] and chemicals for pharmaceutical applications [49].

In addition to utilizing lignin as such, the platform chemicals/monomers, i.e., phenolics, aromatics, alkanes that derive from various depolymerization processes may lead to the production of even higher added value fuels, chemicals and products, usually via more controlled selective catalytic reaction pathways and related processes. Of course, the economics and sustainability of the integrated technology and the final products depend greatly on the effectiveness of the initial depolymerization process. The main thermochemical processes for lignin depolymerization can be divided into three groups based on the temperature/energy requirements. i.e., pyrolysis and more specifically fast pyrolysis leading mainly to the production of bio-oil (relatively high temperature/energy, ca. 400–700 °C), hydrotreatment or hydroprocessing in the absence of solvents (moderate temperatures, ca. 350–450 °C) and liquid phase depolymerization comprising various acid/base and reductive/oxidative reactions (relatively low temperatures, ca. $\leq$ 400 °C) [29]. The "lignin-first" process is a relatively new strategy that applies directly on the lignocellulosic biomass and provides efficient lignin solubilization and depolymerization in a single step/reactor, as described below.

In fast pyrolysis, lignin is heated up to 400–700 °C under high heating/cooling rates in the absence of oxygen, with or without catalyst [34,50]. The main products of no-catalytic, thermal fast pyrolysis of lignin are bio-oil (containing substituted alkoxyphenols and few aromatics), char and gases (mainly CO, CO_2, CH_4). Despite being a high temperature/energy process that could lead to uncontrolled depolymerization and breaking of C-O and C-C bonds, in a recent work of Lazaridis et al. it has been shown on the basis of 2D HSQC NMR results that the composition profile in terms of G- or S-units of the parent lignin is "transferred" to the composition of the produced thermal pyrolysis lignin bio-oil [51]. On the other hand, the catalytic fast pyrolysis of lignin where the primary thermal pyrolysis vapors/products are in situ converted to less oxygenated products via dehydration, decarbonylation, dealkoxylation, cracking and aromatization reactions, may provide bio-oils with substantially altered composition, containing mainly alkyl-phenols, mono-aromatics (BTX) and naphthalenes, depending on the physicochemical characteristics of the catalysts [30,51,52]. Gasification is also an important thermochemical process, widely studied with biomass as feedstock, showing also potential for lignin valorization via synthesis gas production or hydrogen production and utilization [29]. The ratio of the produced gases (H_2, CO, CO_2 and CH_4) is dependent upon process parameters, i.e., temperature, pressure, presence of steam and oxygen, heating rate, and the elemental composition of feed lignin. Due to the sulfur content of technical lignins, the gasification process can also produce H_2S.

With regard to the liquid phase depolymerization processes, various catalytic reaction mechanisms have been proposed including acidic, alkaline, oxidative or reductive pathways. Lignin depolymerization under acidic conditions has been mainly studied by the use of metal salts (metal acetates, metal chlorides and metal triflates) with Lewis acid properties [53]. In supercritical water at 400 °C, the yield of products, composed mainly of oxygenated mono-aromatics, was in the range of 6.2–6.9 wt.% with metal (Fe, Cu, Co, Ni, Al; max. yield with $FeCl_2$) chlorides as acidic agent and 7.1–7.9 wt.% with metal (Fe, Cu, Co, Ni; max. yield with $Co(Ac)_2$) acetates. The conversion was increased when the solvent changed from water to ethanol. Formic acid has also been studied as acidic catalyst, in ethanol/water mixtures with relatively low yield of monomer phenolics [54], while H_2SO_4 was successfully used on hydrolysis lignin (Mw > 20,000 g/mol), yielding ~75 wt.% of depolymerized lignin with Mw of 1660 g/mol [55]. With regard to alkaline conditions, when NaOH and KOH were used as homogeneous base catalysts, up to ~20 wt.% yield of oil was obtained, consisting of monomeric phenolic compounds, such as catechol, cresols, syringol and guaiacol. However, the relatively low oil production was attributed to substantial repolymerization reactions [56] Hulterberg and co-workers have studied the base (NaOH)—catalyzed depolymerization of pine kraft lignin in a continuous flow reactor [57]. The optimum conditions for higher production of monomeric phenolic compounds, less char formation and partial deoxygenated dimeric/oligomeric fractions, were determined to be 240 °C for residence time of 2 min, using 5 wt.% lignin loading and NaOH/lignin ratio of ~1 (w/w).

The depolymerization of lignin under oxidative conditions has been studied by the use of H_2O_2, O_2 and nitrobenzene as oxidants and metal oxide catalysts (organometallic, single oxides and perovskites) at low temperatures. The oxidation resulted in the cleavage of lignin C-O and C-C bonds and the production of low molecular weight compounds mainly aldehydes, carboxylic acids and alcohols [30,47,58]. A well-known lignin oxidation process is the vanillin production from Borregaard Company via the catalytic oxidation of lignosulfonates with O_2 as oxidizing agent. A detailed description of various lignin depolymerization/valorization processes can be found in previous reviews of Zakzeski et al. [3], Pandey and Kim [34], Li et al. [30], Sun et al. [47], Xu et al. [59] and Schutyser et al. [58]. Apart from the thermochemical depolymerization processes, enzymatic deconstruction of lignin had been also proposed by the use of oxidative enzymes, mainly laccases and peroxidase, from fungi and bacteria [59–61].

A more detailed analysis and overview of the reductive depolymerization processes with emphasis on the use of hydrogen donor solvents and catalytic transfer hydrogenation/hydrogenolysis methods is presented in the next sections.

3. Reductive Depolymerization

In contrary to the oxidative depolymerization, reductive depolymerization is taking place in the presence of reducing agents and redox catalysts. Sels and co-workers have reported a categorization of reductive depolymerization process based on hydrogen source and reaction temperature [50,58]. When H_2 gas is used as the reducing agent, the process is called hydroprocessing and when hydrogen donor solvents are used, the process is usually called liquid phase reforming. On the other hand, there are many studies using hydrogen donor solvents which refer to transfer hydrogenation instead of reforming, without however elaborating on the possible reaction steps and mechanism. Further subcategories of hydroprocessing, in terms of the reaction temperature, are the mild (<320 °C) and the harsh hydroprocessing (>320 °C). Mild hydroprocessing is performed in liquid phase with solvents and catalysts leading to p-substituted methoxyphenols, while harsh hydroprocessing provides a wider spectrum of products including demethoxylated phenolic species, deoxygenated aromatics, alkanes, catechols and methoxy-phenols. Harsh hydroprocessing may also take place in the absence of solvents. In the solvent-free hydroprocessing of Kraft lignin by the use of NiMo/MgO-La$_2$O$_3$ at 350 °C, 4 h reaction time and 100 bar initial H_2 pressure, the conversion was 87% with the highest total monomer yield 26.4 wt.% which included 15.7 wt.% alkyl-phenolics [62]. Similar results were obtained for Alcell lignin by the use of supported noble metals at 400 °C, for 4 h reaction time and initial H_2 pressure of 100 bar, with Ru/TiO$_2$, exhibiting the highest catalytic activity providing bio-oil yield 78.3 wt.%, and 9.1 wt.% alkylphenolics, 2.5 wt.% aromatics, and 3.5 wt.% catechols [63]. The bifunctional hydroprocessing with metals supported on acidic materials has been also identified as a separate case, leading to alkane production via additional hydrolysis and dehydration reactions due to the acidic nature of the support, at temperatures below 320 °C [50].

3.1. Reductive Depolymerization of Lignin Model Compounds

Due to the complex nature of lignin, many studies were carried out with model compounds simulating the structure and the bonds of lignin, in order to elucidate the kinetics and pathways for lignin depolymerization. As mentioned in Section 2.1, the most abundant linkage in lignin polymer is the β-O-4 ether bond. The transformation of β-O-4 and 4-O-5 model compounds shown in Figure 4, is discussed below.

Zhu et al. studied the hydrogenolysis of nine compounds containing different functional groups, i.e., benzyl alcohol, aromatic methoxyl and phenolic hydroxyl groups, in methanol and formic acid, acting as hydrogen donors, and Pd/C as the catalyst [64]. In the compounds without a benzyl alcohol group, the β-O-4 linkages were cleaved directly and quickly, in contrast to the compounds that contained benzylic alcohol group, where additional reactions, such as the dehydrogenation of the benzyl alcohol group, hindered the cleavage of β-O-4 bonds to form aromatic monomer products.

The aromatic methoxyl and the phenolic hydroxyl groups had no impact on products distribution but the aromatic methoxyl group in both non-phenolic and phenolic compounds seemed to promote the cleavage of β-O-4 bonds, while the phenolic hydroxyl group had a small negative impact on the cleavage of these bonds.

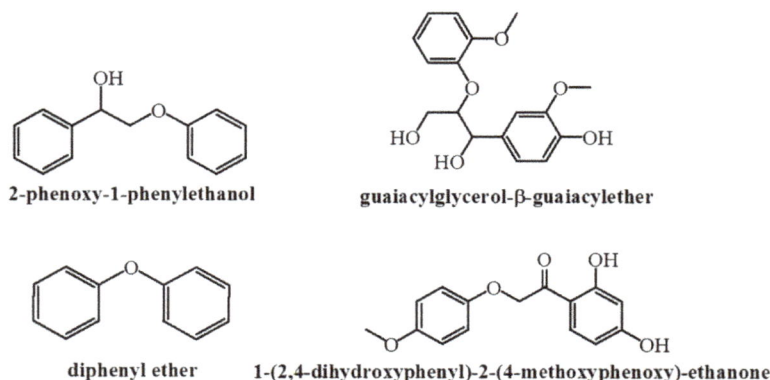

Figure 4. Lignin model compounds studied in the catalytic reductive depolymerization.

The group of Zhang et al. screened a wide range of monometallic catalysts (Ru, Rh, Pd, Pt, Ir, Ag, Au, Cu, Fe, Co, Ni, Re and Sn) in the hydrogenolysis of 2-phenoxy-1-phenylethanol (130 °C, 10 bar H_2) and Ni showed the highest selectivity towards monomers, i.e., 22% with 14% being cyclohexanol, but not the highest conversion (58%). The highest conversion (>99%) was observed by Ru and Rh [65]. The catalytic activity of nickel increased by the addition of Ru, Rh, Pd, Pt and Au leading to full conversion of 2-phenoxy-1-phenylethanol and the NiAu catalyst showed the highest monomer yield 71% (37% cyclohexanol). The optimum ratio Ni:Au was proved to be 7:3. The same group examined further the synergistic activity of NiM (M=Ru, Rh, Pd) and observed that $Ni_{85}Ru_{15}$ led to complete conversion of 2-phenoxy-1-phenylethanol with 58% monomer yield (32% cyclohexanol) at 130 °C, 10 bar H_2 [66].

The activity of Pd-Ni bimetallic nanoparticles supported on ZrO_2 was examined in the hydrotreatment of 2-phenoxy-1-phenylethanol with $NaBH_4$ or H_2 gas as hydrogen source at 80 °C. The optimum reaction system corresponded to Pd:Ni ratio of 1:8, with H_2 gas as hydrogen source at 80 °C for 12 h, yielding 100% formation of cyclohexanol [67]. The presence of $NaBH_4$ induced the initial formation of phenol which was further converted to cyclohexanol at increased $NaBH_4$ amount. The catalytic activity of the same bimetallic Pd-Ni catalysts was also tested in the hydrogenolysis of 2-phenethyl phenyl ether in isopropanol at 210 °C. Under inert atmosphere the conversion of the substrate reached 67% with 60% aromatic yield, whereas the addition of hydrogen gas increased the conversion to 75% with 70% aromatic yield [68].

The hydrogenolysis of 2-phenoxy-1-phenylethanol with HCOOH as hydrogen donor and carbon supported metal catalysts at 80 °C, resulted in the production of acetophenone and phenol [69]. The presence of equivalent amount of base (NH_3) promote the reaction and among the M/C (M=Pd, Rh, Ir, Re, Ni) catalysts tested, Pd/C exhibited the highest activity. Changing the hydrogen source from HCOOH to propanol or hydrogen gas, the reaction was not performed. On the other hand, the use of other amines, i.e., ethylamine, diethylamine and p-allylamine as bases instead of NH_3, can be successfully applied resulting in >95% conversion. In another work by the same group, the effect of other H-donor solvents was examined under redox-neutral reaction conditions and the $NaBH_4$ was proved to be the best H-donor with 100% substrate (2-phenoxy-1-phenylethanol) conversion [70].

The hydrogenolysis of 1-(2,4-dihydroxyphenyl)-2-(4-methoxyphenoxy)-ethanone (DHPMPE) model compound in water/ethanol and Pt/C as catalyst at 275 °C yielded quantitative amounts of 4-methoxypenol and a mixture of 6-hydroxy-3-coumaranone and 2,4 dihydroxyacetophenone [71]. The authors suggested that the reaction mechanism is based on β-O-4 bond cleavage and the cyclisation of 2,4 dihydroxyacetophenone (produced via hydrogenolysis of the initially formed pentacyclic ether) to 6-hydroxy-3-coumaranone favored by the presence of -OH group in α position of the aromatic ring. In the hydrogenolysis reaction of guaiacylglycerol-β-guaiacylether (GGGE) under the same conditions, the reaction products were 2-methoxyphenol and 4-propyl-2-methoxyphenol [71]. The reactant and product concentration profiles in the conversion of the DHPMPE and GGGE lignin model compounds are shown in Figure 5.

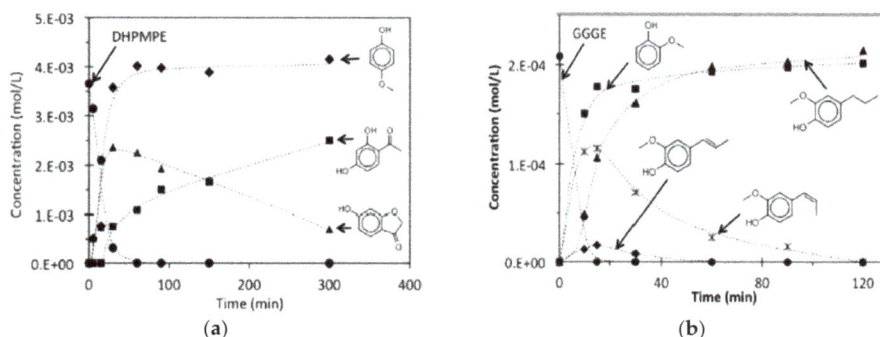

Figure 5. Reaction profiles of (**a**) 1-(2,4-dihydroxyphenyl)-2-(4-methoxyphenoxy)-ethanone (DHPMPE) and (**b**) guaiacylglycerol-β-guaiacylether (GGGE) lignin model compounds in 50 vol.% ethanol + 50 vol.% water, Pt/C, 275 °C, 80 bar. Reproduced from reference [71] with permission from Elsevier.

Rinaldi and co-workers examined the effect of solvent type as well as their hydrogen donor activity (protic with/without Lewis basicity and aprotic polar/non polar) in the hydrogenolysis reaction of diphenyl ether, a compound containing 4-O-5 ether bond [72]. The hydrogenolysis reaction was studied at 90 °C under hydrogen pressure (50 bar) by the use of Raney Ni catalyst. Among the protic solvents with Lewis basicity, 2-propanol exhibited the highest conversion 72.7% (83% monomers–40.5% cyclohexanol) and the use of Hex-F-2-PrOH as protic solvent without Lewis basicity led to full reactant conversion but with lower monomers yield (32.7% monomers–16.6% cyclohexanol). Among the aprotic solvents, 2-Me-THF led to 61.1% conversion (74.2% monomers–34.4% cyclohexanol) whereas full conversion was observed by methylcyclohexane (44.6% monomers–22.8% cyclohexanol), despite the fact that aprotic solvents are not hydrogen donor compounds. The ability of the above solvents to act as hydrogen donors was also tested in the absence of hydrogen gas. In the case of 2-propanol, the conversion decreased to 16.6% (100% monomers–49.8% benzene) while with the rest catalysts, the reaction was not performed.

Apart from β-O-4 model compounds, experiments with other compounds, simulating aryl ethers [71], carbon-carbon [71,73], and dibenzodioxocin [74] linkages, have also been carried out in order to define the pathways for the hydrogenolysis of native lignin.

3.2. Lignin-First Strategy

Lignin first strategy refers to the reductive catalytic fractionation of the whole biomass feedstock and not to the conversion of isolated lignins. Actually, this process is a combination of lignocellulosic biomass fractionation and simultaneous conversion of lignin to monomers [50]. The composition and the structure of the primary biomass influences significantly the obtained products. Klein et al. compared the extent of lignin depolymerization during the reaction of birch, poplar and eucalyptus wood in methanol under N_2 atmosphere (2 bars), at 200 °C and Ni/C as catalyst, and pointed out that birch wood (hardwood) is a better feedstock, enabling the formation of more lignin-derived products [75]. The preferred use of hardwood in this process is also remarked by Galkin et al. who found the following ranking considering the monophenol yield: birch > poplar > spruce > pine and attributed this effect to the relatively increased abundance of β-O-4 moieties in hardwoods (Figure 6) [76]. When hardwoods were compared with softwoods and grasses, hardwoods showed higher monomer yield and delignification amount, followed by grass and softwood [77]. Furthermore, the ratio between S- and G- units of lignin in biomass feedstock was also correlated with the respective ratio in the final products [77,78].

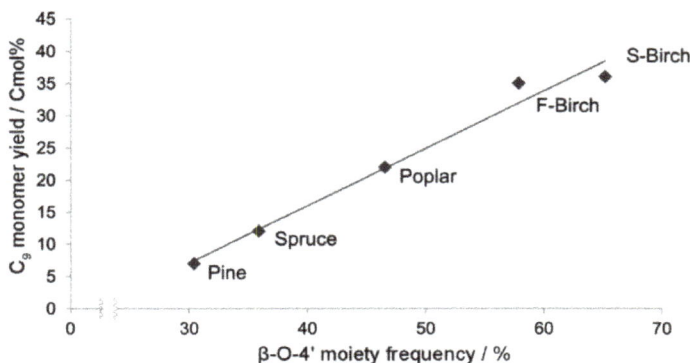

Figure 6. Yield of phenolic compounds produced by the lignin-first approach as a function of β-O-4′ moiety in the native lignin of different lignocellulose substrates. Reproduced from reference [76] with permission from John Wiley and Sons.

The catalytic conversion of woody feedstocks under H_2 gas pressure to lignin monomers is known from the early 1940s. The group of Hibbert studied the digestion of maple wood and woodmeal in the presence of 1,4 dioxane and Cu-CrO catalyst at 280 °C under H_2 pressure and found that the major products were 4-n-propylcyclohexanol and 3-(4-hydroxycyclohexyl)-propanol-1 [79]. Recently, the reductive catalytic fractionation of lignocellulosic feedstocks has been studied by the use of noble metals Pd [80–83], Pt [80], Ru [77], Rh [80] and transition metals, mainly Ni [75,84,85] as catalysts. A comparative study of the performance of Pd/C and Ru/C catalysts in the reductive processing of birch sawdust under H_2 pressure, showed that the two catalysts resulted in almost the same monomer yield and delignification efficiency but in completely different product selectivity and OH-content of lignin oil [86]. Pd/C favored the formation of para-propyl phenolics (75%) while Ru/C the formation of para-propanol phenolics (91%) and increased OH-content. Considering catalyst stability, recovery and reuse, Sels and co-workers studied the performance of Ni-Al_2O_3 pellets positioned in a basket within a stirred batch reactor [87].

As for the reaction medium, the most common solvents studied are H_2O [80,88], alcohols [75,77,83–85,88] and mixture of organic solvents with H_2O [80–82,85]. The polarity of the solvent was shown to influence delignification efficiency and formation of soluble mono-, di- and oligomer phenolics in the Pd/C-catalyzed reductive liquid processing of birch wood. The major phenolic monomers obtained were 4-propanolsyringol and 4-propanolguaiacol. The proposed ranking of solvents was: H_2O > MeOH ≈ EG > EtOH > 1-Pr-OH > 1-BuOH >THF >Dioxane>Hexane, although a too polar solvent like water caused significant solubilization of carbohydrates [88]. Song et al. studied the conversion of birch sawdust into 4-propylguaiacol and 4-propylsyringol with a range of alcohols over Ni/C catalyst and the highest conversion was 50% with >90% selectivity of the above products [78]. The authors proposed that lignin is initially fragmented into smaller lignin species with a molecular weight of m/z ca. 1100 to ca. 1600 via alcoholysis reaction, followed by hydrogenolysis of the fragments into the phenolic monomers. Considering the formic acid, there is no need to be added externally in the reaction, as it can be produced in situ from the acetyl groups in lignocellulosic biomass [4,82].

In order to promote the fractionation of lignocellulose, Hensen et al. added Brønsted acid co-catalysts (HCl, H_2SO_4, H_3PO_4 and CH_3COOH) for a possible replacement of the expensive $Al(OTf)_3$ in the process [89]. In the presence of Pd/C at 180 °C, the best co-catalyst was the HCl, resulting in 44 wt.% lignin monomers from oak sawdust, being similar to the performance of $Al(OTf)_3$ which provided 46 wt.% monomers. In a similar work, by Yan et al., H_3PO_4 enhanced the efficiency of lignin depolymerization from white birch wood sawdust with H_2 in dioxane/water using Pt/C as the catalyst, resulting in 46.4% monomer yield [80]. Among NaOH (alkaline), H_3PO_4 (acidic) and neutral conditions, H_3PO_4 resulted in enhanced delignification (85 % with NaOH vs. 96% with H_3PO_4) of poplar wood in methanol (MeOH) with Pd/C as the catalyst [90]. In the presence of H_3PO_4, the lignin-derived oil was characterized by a narrow molecular weight distribution and a monomer yield very close to theoretical whereas in the presence of NaOH, the monomer yield was lower due to repolymerization. More detailed description of the lignin-first process can be found in the reviews published by the groups of Samec [4], Barta [47] and Sels [58,91].

3.3. Depolymerization with External Hydrogen Source

Many groups examined the catalytic reductive depolymerization of various types of lignins utilizing external hydrogen source (H_2 gas). In these studies, the catalysts play a major role in the conversion and the selectivity of the final products. The most widely used catalysts are Pd/C [81,92–96], Ni/C [93–97], Pt/C [93–95,98], Pt/Al_2O_3 [99], Cu-based porous oxides [100,101], supported NiW and NiMo [102], Ru-based materials [93–96,103], bimetallic NiM (M = Ru, Rh, Pd) [66], Ni-Au [65,104], WP [105], Ru_xNi_{1-x}/SBA-15 [106], modified SBA-15 and Al-SBA-15 [107], MoO_x/CNT [108], $S_2O_8^{2-}$-KNO_3/TiO_2 [109]. Another significant parameter in the reductive depolymerization using molecular H_2 is the reaction medium. The most widely used solvents are water [104], alcohols [92–95,97,98,100–103,106,107] or mixtures of water with organic compounds [81,99,105] under sub or super- critical conditions. In most cases, the beneficial effect of a hydrogen donor solvent is aimed, as discussed in the following section. The conversion activity can be also enhanced by the addition of other compounds which may facilitate lignin depolymerization and formation of phenolic monomers, such as metal chlorides [92,93] and NaOH [104].

4. Catalytic Hydrogenolysis of Lignins Using Hydrogen Donors

4.1. Kraft Lignins

Ma et al. reported the complete ethanolysis of Kraft lignin by the use of α-MoC_{1-x}/AC (280 °C) and the main products were phenols, phenyl and C_6 alcohols and C_8-C_{10} esters [110]. The addition of H_2 gas in the reaction instead of the inert atmosphere had a negative impact on the formation of liquid products, increasing the alcohols and decreasing the ester yield. Comparing ethanol, methanol, isopropanol and water, ethanol was the most effective solvent providing more liquid products. In a following paper, the effect of catalyst properties was examined in the depolymerization of lignin under supercritical ethanol [111]. The activity sequence of the catalysts was: carbide > metal > nitride > oxide and the main products were esters, alcohols and aromatics with different ratios depended on the catalyst. Kraft lignin ethanolysis was also studied by the same group using alumina-supported molybdenum catalysts at 280 °C, reduced at different temperatures [112]. The depolymerization resulted in C_6 alcohols (mainly hexanol), C_8-C_{10} esters (2-hexanoic acid ethyl ester), monophenols (2-methoxy-4-methyl phenol), benzyl alcohols (O-methyl benzyl alcohol) and arenes (xylene). The increase in the reduction temperature from 500 to 750 °C resulted in a gradual increase of total production yield reaching 1390 mg/g lignin but further increase to 800 °C, decrease the yield. The same trend was exhibited in each group of products. The inferior activity of the material reduced at 750 °C was correlated with the metallic phase of Mo and the lower activity of the material reduced at 800 °C with the collapse of porous structure due to sintering phenomena. The effect of the reaction time was also studied. When the reaction time increased from 4 to 6 h the product yield increased dramatically from 142 to 1390 mg/g lignin but when the time was prolonged to 10 h, the product yield decreased.

The conversion of Kraft lignin into monomeric alkyl phenols was achieved over Cu/Mo-ZSM-5 catalysts at 220 °C using hydrogen produced by reforming and water gas shift reactions taking place in the water/methanol solvent system and in the presence of NaOH which was used for enhanced lignin solubility [113]. If no NaOH, methanol or water was used in the reaction, low conversions were observed, despite the fact that in the presence of methanol or water high selectivity to phenol was observed but low monomeric product yield. The optimum water: methanol ratios for the production of monomeric products were determined to be 1:1 and 3:1. The first ratio (1:1) led to 95.7 wt.% conversion and 70.3% selectivity for phenol, 3-methoxy and 2,5,6 trimethyl phenol. The molecular weight of EtOAc soluble products were 286.7 g/mol. In all reaction systems, almost no char was observed, with the exception of the experiment in the absence of NaOH which resulted in high char formation (20.4 wt.%). The proposed mechanism consisted of four steps and is shown in Figure 7.

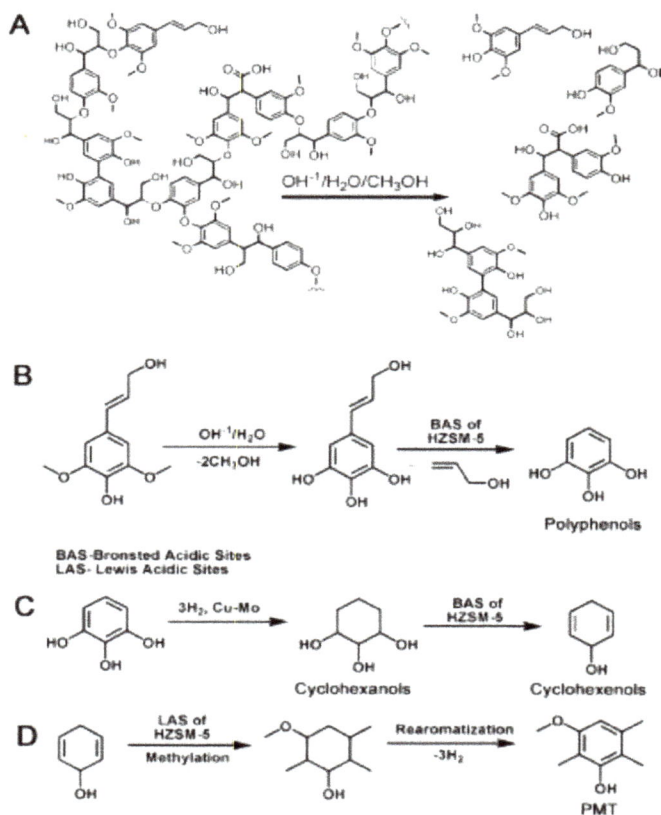

Figure 7. Possible mechanism for the selective formation of phenol, 3-methoxy, 2,5,6-trimethyl (PMT). Reproduced from reference [113] by permission of The Royal Society of Chemistry.

The reductive depolymerization of kraft lignin was also examined in a water–ethanol mixture 50/50 (v/v) with formic acid as an in-situ hydrogen source, by the use of Ni-based catalysts compared to 5% Ru/C [114]. The effectiveness of the formic acid is evident even in the absence of catalysts (89 wt.% yield of liquid depolymerized lignin at 200 °C). The catalysts resulted in decrease in the molecular weight of the products but also in an unfortunate increase of solid residue due to condensation reactions evoked by the acidic properties of the supports. 10% Ni/Zeolite led to a slight increase of the yield up to 93.5 wt.%, decrease of Mw to 3150 g/mol and 9.3 wt.% solid residue. The ability of formic acid to depolymerize Kraft lignin in the absence of any other catalyst was also reported by the same group in a previous publication [115]. Under the optimum operating conditions of ca. 300 °C, 1 h, 18.6 wt.% substrate concentration, 50/50 (v/v) water–ethanol medium containing formic acid (FA) with FA-to-lignin mass ratio of 0.7, lignin (Mw ~ 10,000 g/mol) was effectively de-polymerized towards a liquid product (DL, Mw 1270 g/mol) at a yield of ~90 wt.% and <1 wt.% yield of solid residue (SR). Higher acidity caused condensation of the intermediate products. Higher temperatures or prolonged reaction times resulted also in repolymerization. Formic acid as hydrogen source has been also used by Liguori and Barth for the depolymerization of Kraft lignin to phenols in water and with Pd-Nafion SAC-13 as catalyst at 300 °C [116]. The main products obtained were guaiacol, pyrocatechol and resorcinol. Nafion SAC-13 acted as a Brønsted acid, activating the lignin aryl ether sites and promoting the hydrogenolysis to phenols.

A combination of a supported metal catalyst (Ru/C) with various MgO based catalysts was studied in the depolymerization of kraft lignin in supercritical ethanol as hydrogen donor solvent [117]. In the absence of Ru/C, the most active catalyst proved to be MgO/ZrO$_2$ with the highest bio-oil yield (47.7 wt.%) consisting mainly of phenolic compounds, followed by MgO/C (43.3 wt.%) and MgO/Al$_2$O$_3$ (42.1 wt.%), as can be observed in Table 2. The number of base/acid sites of the MgO based catalysts was found to be correlated with the catalytic activity: MgO/ZrO$_2$ which possessed the highest number of base sites and the less acid sites, exhibited the highest catalytic activity. In contrast, the catalyst with the highest acidity, i.e., MgO/Al$_2$O$_3$, was the less active one. The addition of Ru/C led to higher bio-oil yield in the range of 70.9–82.7 wt.%, rich in higher alcohols and aliphatic esters, while the highest bio-oil yield (88.1 wt.%) and the lower solid residue (8.8 wt.%) was observed for Ru/C when used alone. The unexpected increase of catalytic bio-oil molecular weights, compared to the non-catalytic bio-oil, was attributed to the higher concentrations of heavy compounds. The addition of hydrogen gas in the experiment catalyzed by Ru/C + MgO/ZrO$_2$ did not improve the bio-oil yield (76.9 wt.%). Upon replacement of Kraft lignin with organosolv lignin, the Ru/C + MgO/ZrO$_2$ resulted in lower yield and lower molecular weight of bio-oil, with a higher yield of aromatic monomer. The higher depolymerization efficiency of organsolv lignin was attributed to the less condensed structure and the absence of catalyst-poising sulfur.

Table 2. Product yields, molecular weight averages of bio-oils, and monomer yields obtained from the depolymerization of kraft lignin over MgO loaded on different supports (i.e., carbon, Al_2O_3, and ZrO_2), Ru/C, and physical mixtures of Ru/C and each MgO catalyst at 350 °C after a reaction time of 60 min in the 40 mL batch reactor. Reproduced from reference [117] with permission from Elsevier.

Catalyst	Product Yield (wt.%)		Molecular Weight of Bio-Oil (g/mol)		Monomer Yield (wt.%)		
	Bio-Oil	SR	M_n (g/mol)	M_w (g/mol)	Total	Phenolic Monomers	Aliphatic Esters
No catalyst	36.2	47.3	347	737	1.75	1.70	0.04
MgO/C 10% [a]	43.3	37.3	370	834	2.10	1.83	0.26
MgO/Al$_2$O$_3$ 10%	42.1	37.5	383	844	1.98	1.76	0.23
MgO/ZrO$_2$ 10%	47.5	32.7	365	776	4.22	3.90	0.32
Ru/C 10%	88.1	8.8	398	906	5.05	4.54	0.51
Ru/C 10% + MgO/C 10%	79.2	10.0	371	944	5.73	5.15	0.58
Ru/C 10% + MgO/Al$_2$O$_3$ 10%	70.9	21.0	434	1064	4.52	3.76	0.77
Ru/C 10% + MgO/ZrO$_2$ 10%	82.7	11.3	323	832	6.10	5.16	0.94

[a] Percent value indicates the catalyst loading amount with respect to lignin weight.

Esposito et al. synthesized two different nickel-based materials, TiN-Ni and TiO_2-Ni and tested their activity in the hydrogenolysis of Kraft lignin in various alcohols, in a flow reactor system under relatively mild temperature and pressure conditions, considering mainly the effect of alcohols on lignin solubility, without discussing their potential hydrogen donating function [118]. Higher catalytic activity exhibited by the TiN-Ni, was attributed to the better dispersion of Ni in TiN phase as well as to the more favorable titanium oxidation state, i.e., being (III) in TiN compared to (IV) in TiO. Substituted phenols (3.2 wt.%) and aromatic fragments (60 wt.%) with small molecular weights were obtained for TiN-Ni.

Supercritical water/isopropanol systems were applied for the depolymerization of Kraft lignin over Fe on Rh/La_2O_3/CeO_2-ZrO_2 [119]. Different ratios of water and isopropanol were used to adjust the optimum in situ H_2 production, which was correlated with the hydrogen donating capability of water and isopropanol and the relative amount of Fe in the catalyst. Gradual increase in water content led to a gradual decrease in H_2 selectivity. Considering the products, the increase in water content, resulted in a progressive increase of aromatics and aliphatic acid/esters while a sharp decrease in hydrogenated cyclics was observed. The products distribution obtained at the different ratio of isopropanol/water can be seen in Figure 8. Apart from the liquid products, similar trends were observed in the gas products.

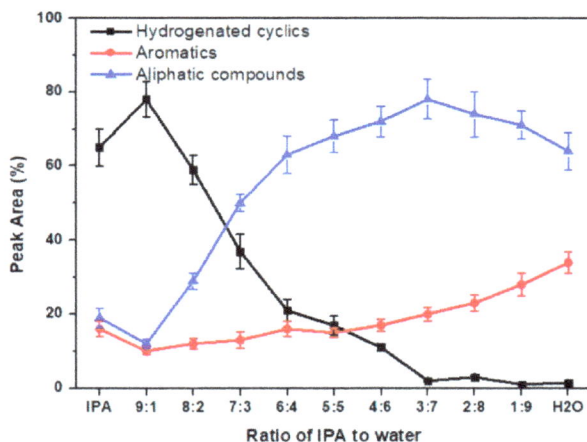

Figure 8. Qualitative liquid products distribution from Kraft lignin depolymerization under the different ratio of isopropanol to water (0.1 g kraft lignin, 40 wt.% Rh/La$_2$O$_3$/CeO$_2$-ZrO$_2$ catalyst, 373 °C reaction temperature, 2 h reaction time and 2 mmol Fe). Reproduced from reference [119] with permission from Elsevier.

Singh et al. studied the depolymerization of Kraft lignin in methanol at 220 °C by the use of homogeneous (NaOH) and heterogeneous catalysts (HZM-5 and iron turnings from lathe machining) [120]. Compared to the non-catalytic experiment, which resulted in 73.6 wt.% depolymerization yield, the homogeneous NaOH led to 68.5 wt.% yield, 5.1 wt.% monomeric compounds and 100–1000 g/mol molecular weight distribution. The heterogeneous HZSM-5 resulted in higher depolymerization yield of 85.1 wt.%, with lower monomers amount 4.2 wt.% and the same molecular weight distribution while the iron turnings resulted in lower depolymerization (44.4 wt.%) with 1.7 wt.% monomers and 100–2000 g/mol molecular weight distribution. In all cases, the main products were alkyl substituted phenols. It was suggested that methanol, in addition to being a solvent in the reaction, acted also as a hydrogen donor. The hydrogen released due to thermal reforming of methanol-induced hydrogenolysis of the ether linkages in lignin resulting in lignin depolymerization and demethoxylation.

4.2. Soda Lignins

Hensen et al. carried out a thorough investigation of the reductive depolymerization of soda lignin by the use of CuMgAlO$_x$ catalysts. In a first article, the group examined the influence of solvent, reaction time, and catalyst [121]. The most effective combination proved to be CuMgAlO$_x$ with ethanol at 300 °C, resulting in 17 wt.% monomers yield, comprising of aromatic products with small amounts of furans, hydrogenated substituted cyclic compounds and deoxygenated aromatics, as can be observed in Figure 9. The Cu content of 20 wt.% in the CuMgAlO$_x$ mixed oxide was found to induce the highest activity for Guerbet, esterification and alkylation reactions, resulting in higher monomers yield and low repolymerization [122]. Both the non-catalytic experiment and the use of methanol resulted in low monomer yields (5 and 6 wt.%) and in the latter case the products were mainly methylated phenols and guaiacol-type compounds. Copper in MgAlO$_x$ improved the monomer production and the formation of deoxygenated aromatics compared to the MgAlO$_x$, NiMgAlO$_x$, and PtMgAlO$_x$ catalysts. Considering the repolymerization via the phenolic hydroxyl groups, the authors suggested that ethanol not only acts as a hydrogen-donor solvent, but also as a capping agent and formaldehyde scavenger, as shown in Figure 10 [121,123]. In a following paper, they discussed the influence of reaction temperature on the products [124]. At lower temperatures in the range of 200–250 °C, recondensation reactions are dominant whereas at higher temperatures of 380–420 °C, the char formation due to carbonization played a major role. At the intermediate range of 300–340 °C, the depolymerization of lignin is enhanced, resulting in the formation of reactive phenolic intermediates which can be protected by alkylation, Guerbet and esterification reactions. CuO in the parent CuMgAlO$_x$ catalyst favors the alkylation reactions while its progressive reduction to Cu may lead to the undesirable increased hydrogenation of the aromatic ring.

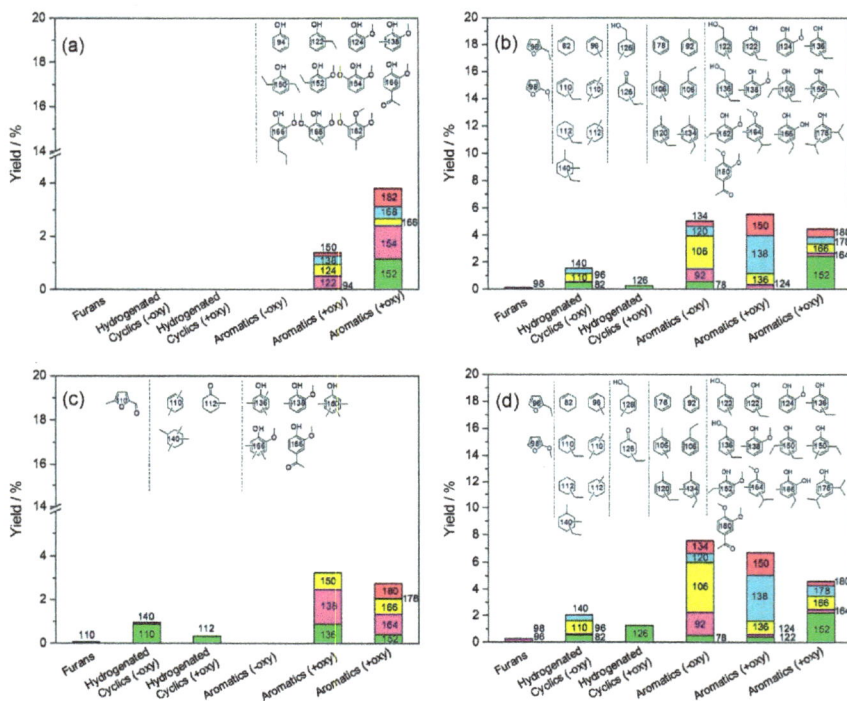

Figure 9. Product distribution following reaction: (**a**) blank reaction at 300 °C for 4 h in ethanol, (**b**) CuMgAlO$_x$ at 300 °C for 4 h in ethanol, (**c**) CuMgAlO$_x$ at 300 °C for 4 h in methanol, (**d**) CuMgAlO$_x$ at 300 °C for 8 h in ethanol. The numbers in the molecules correspond to the molecular weight, whereas colors are used to facilitate the reading of the figure. Reproduced from reference [121] with permission from John Wiley and Sons.

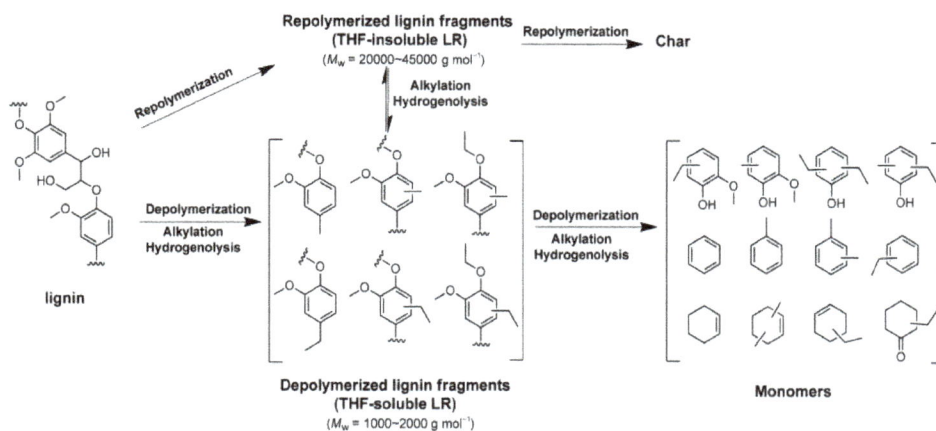

Figure 10. Possible reaction routes during lignin depolymerization in ethanol in the presence of catalyst. Reproduced from reference [121] with permission from John Wiley and sons.

Soda lignin depolymerization was performed over metals catalysts supported on ZSM-5 zeolite in supercritical ethanol at 440 °C [125]. When comparing the transition metals Ni, Co and Cu at 10 wt.% loading on ZSM-5 with Si/Al$_2$ of 200, the 10% Cu/ZSM-5(200) catalyst showed the highest yield of monoaromatic compounds (15.3 wt.%). Changing the metal loading from 10% to 5 and 30% a small decrease in monoaromatic compounds was observed. In order to find the optimum Si/Al$_2$ ratio of ZSM-5, Si/Al$_2$ was varied from 30 to 200. The highest yield 98.2 wt.% of monoaromatic compounds was obtained over 10 wt.% Cu/ZSM-5(30) due to the higher acid density. The beneficial effect of Cu in the depolymerization was confirmed by the experiment conducted in Cu-free ZSM-5 (30) which resulted in 89.4 wt.% monoaromatic compounds. The authors suggested that the in situ produced hydrogen atoms were adsorbed onto the surface of Cu leading to cleavage of ether bonds and thus promoting the depolymerization.

4.3. Alkali Lignins

In the hydrogenolysis of alkali lignin in supercritical ethanol, Zhou et al. found that CuNiAl-hydrotalcite was more active catalyst than Ni/ZSM-5 or Ru/C, resulting in 49.5 wt.% bio-oil yield at 290 °C [126]. The strong basic sites of hydrotalcite, compared to the acidic ZSM-5 and carbon, inhibited the recondensation of reactive compounds of bio-oil. The addition of phenol as co-solvent to ethanol (phenol/lignin=0.8), increased bio-oil yield to 72.3 wt.%. Phenol is suggested to promote the hydrogenolysis due to the enhanced solubilization of lignin and the capping agent action favoring the formation of mono-phenolics compounds and suppressing repolymerization. Higher amounts of phenol resulted in lower bio-oil yields and increased molecular weights, phenomena which attributed to secondary repolymerization reactions. Also, the addition of hydrogen gas did not enhanced the bio-oil yield (70.3 wt.%). The optimum temperature and time considering the bio-oil yield were determined to be 290 °C and 3 h.

The effective activity of nickel-based catalysts was also reported by Li et al., in the hydrogen transfer conversion of alkali lignin using isopropanol/water solvent [127]. Alkali lignin showed high conversion (93%) over Raney Ni catalysts at 180 °C, superior than with Pd/C catalyst which led to low conversion and liquefaction rates. Lower conversion was observed for Klason lignin due to its more condensed nature, attributed to the high acid concentrations in the Klason lignin preparation process.

The synergistic activity of formic acid and Pd/C was examined in the catalytic depolymerization of alkali lignin in subcritical water [128]. When the reaction was contacted without formic acid and

Pd/C at 265 °C for 1 h, the liquid products yield was 58.2 wt.% and the solid residue 30.6 wt.%. The addition of formic acid slightly increased the liquid products to 61.6 wt.% but extremely decreased the solid residue to 0.64 wt.%. The addition of Pd/C catalyst either in the presence of formic acid or not, resulted in lower liquid products (45.8 and 41.3 wt.%) and higher solid residues (16.3 and 54.4 wt.%). The products yields from lignin depolymerization in all reaction systems are shown in Table 3. The catalyst favored the conversion of formic acid and production of H_2 via reforming and water–gas shift reactions and promoted the repolymerization reactions. Significant differences are observed in the composition of liquid products. In the absence of formic acid and Pd/C, the main compound was guaiacol, while in the presence of formic acid or both formic acid and Pd/C, catechol was the main compound. Pd/C can catalyze the hydrogenolysis of the aryl–O ether bond resulting in significant yield of phenol and char formation.

Table 3. Product yields from lignin depolymerization with additives at 265 °C, 6.5 MPa from 1–6 h hold time. Reproduced from reference [128] by permission of The Royal Society of Chemistry.

Sample	Hold Time (h)	Liquid (wt.%)	Solid Residue (wt.%)	Gas (wt.%)	% Balance
Lignin	1.0	58.2	30.6	5.56	94.4
Lignin	3.0	51.6	35.9	9.12	96.6
Lignin	6.0	33.3	46.0	17.0	96.3
Lignin/FA	1.0	61.6	0.64	36.5	98.7
Lignin/FA	3.0	52.3	0.19	46.1	98.6
Lignin/FA	6.0	41.1	1.12	58.3	101
Lignin/Pd/C	1.0	41.3	54.4	0.84	96.5
Lignin/Pd/C	3.0	36.1	57.3	1.25	94.7
Lignin/Pd/C	6.0	35.6	59.1	2.25	97.0
Lignin/FA-Pd/C	1.0	45.8	16.3	38.7	101
Lignin/FA-Pd/C	3.0	38.5	18.3	49.8	107
Lignin/FA-Pd/C	6.0	31.6	22.9	53.4	108
Pd/C	6.0	-	98.5	-	98.5

4.4. Organosolv Lignins

Toledano et al. studied the hydrogenolysis of organosolv lignin from olive tree prunings under microwave irradiation, using a variety of hydrogen donor solvents without any H_2 addition [129,130]. In a first paper, they examined the activity of metallic (Ni, Ru, Pd, Pt) catalysts supported on Al-SBA-15 and found that 10% Ni/Al-SBA-15 with tetralin as solvent provided improved bio-oils with 17 wt.% yield, consisting of simple phenolics, including mesitol and syningaldehyde, as well as, a small amount of esters. The product distribution obtained over Ni, Ru, Pd, Pt catalysts supported on Al-SBA-15 can be seen in Table 4. The other metals exhibited lower activity with high remaining lignin due to repolymerization phenomena [129]. In a further work, the same group carried out detailed research on the solvent effect on the hydrogenolysis of organosolv lignin from olive tree prunings with 10% Ni/Al-SBA-15 at 150 °C [130]. The most effective solvent, as shown in Figure 11, was proved to be formic acid resulting in high bio-oil yield (28.89 wt.%), no biochar and a wide variety of phenolics compounds. Formic acid was noted to exhibit additional acidolytic properties for the depolymerization of lignin. Less efficient solvents were glycerol and tetralin with the later leading mainly to phthalates. The authors highlighted the unexpected behavior of isopropanol which in the case of the catalytic experiment, it was proven less efficient compared to the blank experiment due to the dehydrogenation of isopropanol to acetone over the Ni/Al-SBA-15 catalyst.

Table 4. Compounds identified in appreciable quantities in lignin bio-oils obtained by the hydrogenolysis of organosolv lignin from olive tree prunings under microwave irradiation using Ni, Pd, Pt, Ru catalysts supported on Al-SBA-15. Compounds yields expressed as mg of each compound per gram of lignin. Reproduced from ref. [129] with permission from Elsevier.

	Blank	Ni2% AlSBA	Ni5% AlSBA	Ni10% AlSBA	Pd2% AlSBA	Pt2% AlSBA	Ru2% AlSBA
Mesitol (**2**)	0.56	2.26	1.19	1.13	0.49	1.67	1.40
2,3,6-Trimehtylphenol (**3**)	0.19	0.76	0.22	0.28	0.18	0.17	0.45
6-Ethyl-*o*-cresol (**5**)	0.13	0.14	0.04	-	-	0.32	0.19
4-Ethyl-*m*-cresol (**6**)	-	0.11	0.04	-	-	0.08	-
Vanillin (**7**)	0.20	0.20	0.19	0.20	-	0.13	0.11
Diethylphthalate (**10**)	2.34	10.10	1.67	1.69	10.93	10.66	7.44
3,4-Dimethoxyphenol (**11**)	0.05	0.12	0.05	0.15	0.24	0.09	0.09
Syringaldehyde (**12**)	0.47	0.68	0.57	0.56	0.15	0.43	0.43
Butyl-octyl ester phthalic acid (**16**)	0.21	0.36	0.30	0.19	0.34	0.36	0.37

[a] Reaction conditions: 0.5 g lignin, 0.5 g catalyst, 12.5 mL tetralin, microwaves, 400 W, average temperature 140 °C, 30 min reaction.

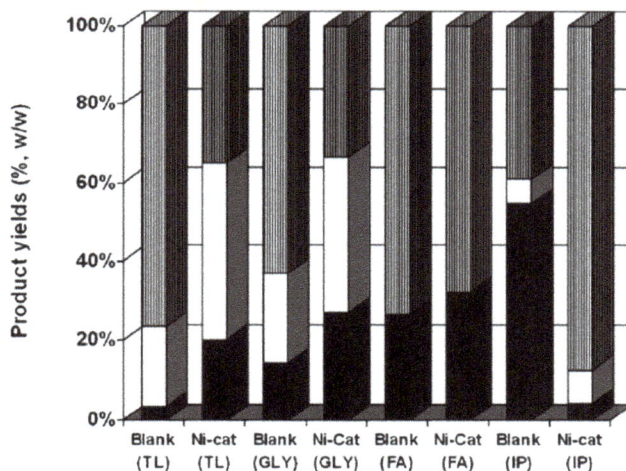

Figure 11. Products yields (%, w/w) referenced to initial lignin weight of bio-oil (black), biochar (white) and residual lignin (grey) derived by hydrogenolysis of lignin under mild, microwave-assisted, H_2 free conditions, using various solvents: tetralin (TL), glycerol (GLY), formic acid (FA) and isopropanol (IP). Reaction conditions: 0.5 g lignin, 0.5 g catalyst, 12.5 mL solvent, microwaves, Taverage = 423 K, t = 30 min. Reproduced from reference [130] with permission from John Wiley and sons.

Organosolv lignin from switchgrass had been successfully depolymerized in ethanol, 20 wt.% Pt/C and formic acid as hydrogen donor molecule at 350 °C for the production of phenol and substituted phenols with higher H/C and lower O/C molar ratios [131]. In a similar reaction system, organosolv lignin was depolymerized using isopropanol as solvent, Ru/C as catalyst and formic acid as hydrogen donor at 400 °C [132]. The catalytic experiments resulted in 71.2 wt.% lignin oil, negligible solids formation and significant amount of water (9.6%) due to the decomposition of formic acid. The activity of the catalyst was confirmed by the experiment conducted in the absence of Ru/C which led to only 18 wt.% conversion and a large amount of solids derived either from the unconverted lignin or repolymerization reactions. The effective conversion of lignin was further compared with the hydrotreatment experiment, conducted in the absence of solvent (only Ru/C + H_2 gas), exhibiting lower oil yield (63.1 wt.%) compared to the use of isopropanol as solvent. Considering the chemical composition of derived oils, the major products were ketones (methyl isobutylketone) followed by aromatics, catechols and alkylphenolics. In an attempt to improve the catalytic reaction system, methanol and ethanol were also tested as solvents. Methanol gave almost similar yield (68.4 wt.%) with isopropanol but higher amounts of alkylphenolics and aromatics whereas ethanol resulted in lower yield (63.4 wt.%).

The effect of biomass feedstock and isolation method of lignin on the depolymerization in supercritical ethanol and formic acid, in absence of catalyst, at 250–350 °C was also investigated [133]. The reaction was performed in lignin derived from oak (hardwood) and pine (softwood), isolated as ethanosolv, formasolv and Klason types. Regardless the isolation method, all lignins exhibited bio-oil and conversion yields above 90 wt.% with low solid residue <2.5 wt.% at 350 °C. At this temperature, the combination of ethanol with formic acid facilitated the hydrogen production which quenched the radicals, suppressing the repolymerization. For the hardwood type biomass, at lower reaction temperature (250 °C), the ethanosolv and the formasolv lignin exhibited lower bio-oil yield, 68.0 and 77.5 wt.%, respectively, while Klason lignin bio-oil yield dramatically decreased to 19.3 wt.% due to the abundance of C-C recalcitrant bonds formed in the Klason process. With regard to bio-oil composition, at 350 °C, linear and branched short-chain oxygenated species from the decomposition of ethanol and formic acid, monoaromatic species and long chain fatty acid alkyl esters from the esterification of woody biomass with ethanol were produced. At lower reaction temperature, where the deoxygenation/hydrogenation reactions are limited, carbonyl or double bond-containing monoaromatics were formed. With regard to softwood biomass, the ethanosolv and formasolv lignins exhibited 97.1 and 99.4 wt.% conversion and 88.1 and 90.7 wt.% bio-oil yield. Again, the Klason lignin depolymerization resulted in lower conversion (95.1 wt.%) and bio-oil yield (81.7 wt.%). At the low temperatures of 250–300 °C, the most important parameter was suggested to be the relative abundance of ether linkages in the lignin structure.

The catalytic activity of Cu based porous metal oxides towards the depolymerization of organosolv lignin extracted from candlenuts was examined in supercritical methanol at 310 °C [134]. Taking into consideration lignin conversion, the following rank was determined: $Cu_{20}PMO$ > $Cu_{20}PMO$ $Cu_{20}Cr_{20}PMO$ > $Cu_{20}La_{20}PMO$ (PMO stands for porous metal oxide). For the best catalyst $Cu_{20}PMO$, lignin conversion reached 48.3% after 1 h reaction. Due to the lower methanol reforming ability of lanthanum, the reaction was also carried out for higher times and the conversion reached 98% after 6 h. Under the same reaction conditions, the activity Cu-free porous metal oxides were also examined. [Mg/Al]PMO exhibited the highest lignin conversion of 74.6% for 5 h reaction. Comparison of CuPMOs to Cu-free analogs showed that Cu promotes higher yields of methanol-soluble products and suppresses re-condensation reactions. The $Cu_{20}La_{20}PMO$ variant was suggested as the most effective catalyst in terms of limiting over-reduction of aromatic intermediates due to the lower methanol-reforming of lanthanum, thus regulating the in situ production of hydrogen.

Organosolv lignin isolated from eucalyptus was depolymerized in water to syringol monomers over β-CaP_2O_6 and CoP_2O_6 [135]. Despite that eucalyptus is a hardwood feedstock, the use of phosphate catalysts selectively produced only syringol with yield of 8.47% over β-CaP_2O_6 and 6.67 % over CoP_2O_6. The hydrogenolysis of organosolv lignin from hybrid poplar in supercritical ethanol at 320 °C, by the use of amorphous B-containing FeNi alloyed catalysts, resulted in lignin depolymerization and the production of deoxygenated aliphatic side chains [136].

4.5. Lignosulfonate

Lignosulfonate depolymerization via hydrogen transfer reactions was investigated over Raney Ni at 200 °C and a solvent mixture composed of water, isopropanol, butanol and hexane, introducing the concept of an emulsion microreactor [137]. Isopropanol was considered to be the hydrogen donor solvent for the hydrogenolysis of lignolsulfonate while the combination of the above three solvents provided higher monomer yields, compared to the more classical water, alcohol or water-alcohol mixtures, as can be seen in Table 5. The higher yield of phenolic monomers was observed for the solvent mixture water: butanol: isopropanol: hexane = 1:1:3:0.1, with 4-ethyl guaiacol being the most abundant compound. The facile phase separation after the hydrogenolysis reaction was suggested as an important advantage of the described emulsion microreactor.

Table 5. Catalytic depolymerization of sodium lignosulfonate in emulsion microreactor. Reproduced from reference [137] with permission from Elsevier.

Solvent (mL)			Yield of Phenolic Monomers (mg·g^{-1})		
H$_2$O	*n*-BuOH	*i*-PrOH	4-Ethyl Guaiacol	Others	Total
25	/	/	18.8 ± 0.3	37.3 ± 0.5	56.1 ± 0.7
20	5.0	/	22.0 ± 0.4	39.8 ± 0.6	61.8 ± 0.8
15	10	/	23.5 ± 0.2	41.9 ± 0.5	65.4 ± 0.7
10	15	/	22.3 ± 0.3	40.2 ± 0.6	62.5 ± 0.8
20	/	5.0	21.3 ± 0.5	32.1 ± 0.6	53.4 ± 1.1
15	5.0	5.0	24.8 ± 0.4	50.2 ± 1.9	75.0 ± 2.2
10	5.0	10	33.8 ± 0.3	50.3 ± 2.5	84.1 ± 2.8
10	10	5.0	27.7 ± 0.1	52.8 ± 3.0	80.5 ± 3.0
5.0	15	5.0	27.2 ± 0.5	55.5 ± 2.4	82.7 ± 2.8
5.0	10	10	29.9 ± 0.6	65.5 ± 3.1	95.4 ± 3.5
5.0	5.0	15	39.3 ± 0.7	76.8 ± 2.3	116.1 ± 2.8

Condition: sodium lignosulfonate 0.5 g, Raney Ni 2.0 g, hexane 0.5 mL, 473 K for 2.0 h.

4.6. Enzymatic and Acid Hydrolysis Lignins

The role of formic acid in the reductive depolymerization of lignin has been studied in the work of Oregui-Bengoechea et al. [138]. In the hydrogenolysis of enzymatic hydrolysis eucalyptus lignin over NiMo/sulfated alumina and ethanol at 320 °C, the synergistic action of formic acid and catalyst was shown by the increased oil yield (38.4 wt.%) compared to the uncatalyzed experiment (23 wt.%) and the experiment where gaseous H$_2$ had been used instead of formic acid (19.7 wt.%). All the derived oils contained methoxy-, hydroxyl- and alkyl- substituted benzenes. The authors suggested that the formic acid is involved in lignin depolymerization via a formylation-elimination-hydrogenolysis mechanism, which includes also the catalytic decomposition of formic acid that provides molecular H$_2$ for the hydrogenolysis reaction (Figure 12). The important role of the solvent was discussed and ethanol proved to be more effective (oil yield 38.4 wt.%) than methanol (23.8 wt.%) or isopropanol (21.7 wt.%), as suggested in similar studies [130].

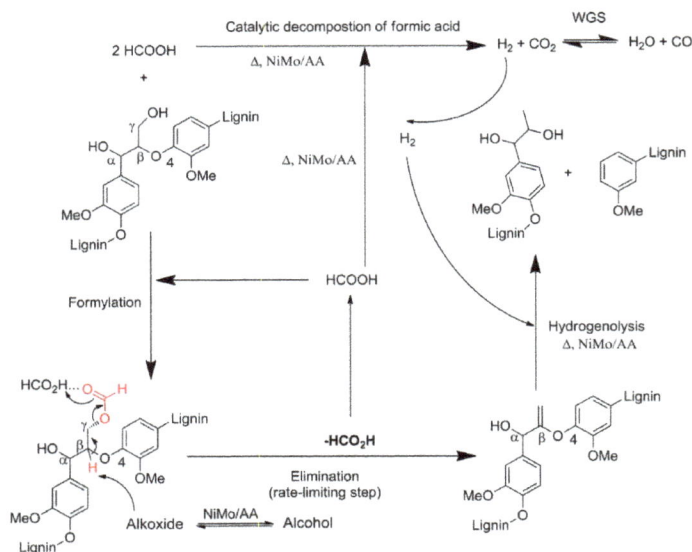

Figure 12. Proposed mechanism for formic acid-aided depolymerization of lignin through formylation, elimination, and hydrogenolysis. Reproduced from reference [138] with permission from John Wiley and sons.

Similar observations about the enhanced depolymerization in the presence of formic acid and ethanol had been reported also by Kristianto et al. for the concentrated acid hydrolysis lignin from fruit bunch palm oil [139]. From the catalytic activity screening of 5% Pd/C, 5% Ru/Al$_2$O$_3$, 5% Ru/C and 10% Ni/C at 300 °C it was shown that the most active catalyst was 5% Ru/C with 31 wt.% bio-oil yield and 47.1 wt.% solid residue. The addition of formic acid increased the bio-oil yield to 62.9 wt.% and decreased the solid residue to 19.8 wt.%, making formic acid better hydrogen donor than the external gas. The products obtained from the reaction were phenol and its derivatives whose amount increased with increase in formic acid/lignin ratio and reaction time. Small amounts of phenolic compounds with ethyl and ester groups were observed due to alkylation and esterification reactions of phenol intermediates with ethanol.

The synergistic effect of Raney Ni and zeolites catalysts had been investigated in the depolymerization of enzymatic hydrolysis lignin from bamboo residues [140]. In a methanol/water reaction mixture at 250 °C, when Raney Ni was combined with zeolite catalysts, the yield of mono-phenols significantly increased from 12.9 wt.% (Raney Ni) and 5 wt.% (zeolite) to 27.9 wt.% (HUSY and Raney Ni). The optimum ratio of zeolite: Raney Ni was determined to be 8:4. The authors proposed that Raney Ni could act as lignin cracking and methanol reforming catalyst and zeolite as Brønsted and/or Lewis solid acid essential for ether solvolysis and dehydration. Furthermore, zeolites can act as a blocking agent to prevent reactions between the original lignin and the unstable lignin fragments.

4.7. Lignins Extracted by Deep Eutectic Solvents

Das et al. studied the hydrogenolysis of deep eutectic solvent extracted lignin by the use of Ru/C, Pd/C and Pt/C at 270 °C with isopropanol as hydrogen donor solvent [141]. The most effective catalyst was Ru/C resulting in the highest oil yield 36.28 wt.% and the lowest char formation 46.43 wt.%. The main products were phenol, substituted phenols and long chain fatty acids derived by the reductive couple reaction of lignin. Increase in catalyst loading from 2 wt.% to 15 wt.% resulted in increase in oil yield but further increase (20 wt.%), decreased the oil yield due to side chain reactions, hindering the production of monomeric phenolic compounds. Higher reaction temperature (300 °C) did not enhance lignin depolymerization, resulting in decrease in oil yield and increase of gas products yield due to the further decomposition of lignin-derived compounds to gaseous products or condensation to char formation. Similar results was obtained by changing the reaction time. Gradual increase from 30 to 60 min, increased the oil yield but further increase to 180 min, resulted in decrease of oil yield and increase of gas products yield.

4.8. Selection and Design Criteria for an Effective Catalyst in Reductive Depolymerization of Lignin Using Hydrogen Donors

As discussed in the previous sections, the catalysts which have been widely studied in the reductive depolymerization of lignin by the use of hydrogen donors are based on noble (Pd, Ru, Pt) or transition metals (Ni, Cu) supported on carbon, zeolites and silica materials due to their known ability to catalyse the cleavage (hydrogenolysis) of C-O and C-C bonds, the hydrodeoxygenation of oxygenated compounds and the hydrogenation of aromatic double bonds. A summary of the most representative catalytic systems is shown in Table 6. However, the specific mechanisms of the in situ hydrogen production has been scarcely discussed with general reference to reforming of alcohols and related water gas shift reaction as well as decomposition of formic acid. Furthermore, while the terms "transfer hydrogenation" or "hydrogen transfer" have been used in some cases, there was no systematic effort to elucidate the reaction pathways involved in relation to the catalyst properties and the experimental conditions.

Table 6. Representative optimum results on catalytic hydrogenolysis of various lignins using small alcohols or formic acid as hydrogen donors.

Lignin	Catalyst & Reaction Conditions	Conversion/Yields [a] (wt.%)	Main Products	Reference
Kraft	α-MoC$_{1-x}$/AC, EtOH, 280 °C	100/-/-	Esters, alcohols, aromatics	[110]
Kraft	Cu/Mo-ZSM-5, H$_2$O:MeOH=1:1, 220 °C, NaOH	95.7/-/19.9	Phenolics	[113]
Kraft	10% Ni/Zeolite, H$_2$O:EtOH=1:1, HCOOH, 200 °C	93.5 [b]	Aromatics	[114]
Kraft	Pd-Nafion SAC-13, H$_2$O, HCOOH, 300 °C	-/-/10.5	Phenolics	[116]
Kraft	MgO/ZrO$_2$, EtOH, 350 °C	-/47.5/4.22	Phenolics	[117]
Kraft	Ru/C+ MgO/ZrO$_2$, EtOH, 350 °C	-/82.7/6.10	Phenolics, higher alcohols, aliphatic esters	[117]
Kraft	HZSM-5, MeOH, 220 °C	85.1/-/4.2	Phenolics	[120]
Kraft	Iron turnings, MeOH, 220 °C	44.4/-/1.7	Phenolics	[120]
Soda	CuMgAlO$_x$, EtOH, 300 °C	-/-/17	Aromatics, furans, hydrogenated cyclics	[121–124]
Soda	10 wt.% Cu/ZSM-5(30), EtOH, 440 °C	-/-/98.2	Aromatics	[125]
Alkali	CuNiAl-HT, EtOH, Phenol:lignin=0.8, 290 °C	-/72.3/-	-	[126]
Alkali	Raney Ni, i-PrOH/H$_2$O, 180 °C	93.5/-/-	Phenolics-aromatics	[127]
Alkali	Pd/C, H$_2$O, HCOOH, 265 °C	-/45.8/-	Phenolics	[128]
Organosolv	10% Ni/Al-SBA-15, Tetralin, 140 °C [c]	-/16.94/-	Phenolics, esters	[129]
Organosolv	10% Ni/Al-SBA-15, HCOOH, 150 °C [c]	-/28.89/-	Phenolics	[130]
Organosolv	20 wt.% Pt/C, EtOH, HCOOH, 350 °C	-/-/21	Phenolics	[131]
Organosolv	Ru/C, i-PrOH, HCOOH, 400 °C	-/71.2/-	ketones aromatics, phenolics	[132]
Organosolv	Cu$_{20}$PMO, MeOH, 310 °C	48.3/-/-	Phenolics, aromatics	[134]
Organosolv	FeNiB alloys, EtOH, 320 °C	-/-/-	Phenolics	[136]
Enzymatic hydrolysis	NiMo/sulfated alumina, EtOH, HCOOH, 320 °C	-/38.4/-	methoxy, hydroxyl, alkyl benzenes	[138]
Acid hydrolysis	5% Ru/C, EtOH, HCOOH, 300 °C	-/62.9/-	Phenolics, esters	[139]
Enzymatic hydrolysis	HUSY+Raney Ni, MeOH/H$_2$O, 250 °C	-/-/27.9	Phenolics	[140]
DES Extracted	Ru/C, i-PrOH, 270 °C	-/36.28/-	Phenolics, acids	[141]

[a] Conversion/Bio-oil/Monomers, [b] depolymerized lignin, [c] Microwave irradiation.

Supercritical alcohols can donate hydrogen in the form of molecular hydrogen, hydride, or protons, with the hydride deriving from α-hydrogen and the proton from the alcohol hydroxyl, forming simultaneously electron-deficient hydroxylalkylation species, alkoxide ions and aldehydes [142]. Similarly, decomposition of formic acid in supercritical water conditions leads to in situ H_2 formation. However, the relatively high temperatures, e.g., >280 °C, at which most of the lignin hydrogenolysis studies in supercritical solvents have been conducted, may also induce repolymerization-condensation reactions of the initially formed monomer phenolics, especially in the presence of acidic catalysts. Thus, optimization of the overall system, i.e., type of solvent/hydrogen donor—reaction temperature— catalyst properties, is required in order to achieve high yields of liquid products enriched in monomers.

With regard to the catalyst properties, three main interactive criteria should be considered: (i) effect on the in situ hydrogen production mechanism, (ii) high hydrogenation reactivity and facilitated activation of the lignin C-O or C-C bonds, and (iii) stabilization of reactive intermediates via alkylation or other reactions. The first criterion is related with the catalyst properties that should be tailored towards enhanced reforming and WGS reaction activity and/or transfer and stabilization via surface intermediates of hydride (H:) or protons (H⁺) by selecting appropriate metals and supports. In this latter case, which is less discussed in the literature, an effective catalyst or catalyst support surface would comprise of Lewis acid sites and Brønsted or Lewis basic sites that can attract H: and H⁺ from alcohols, respectively, thus initiating the steps of hydrogen transfer and lowering the overall reaction required temperature. Materials with such properties can be various transition metal oxides or mixed oxides, such as ZrO_2, TiO_2, MgO, etc and their modified/doped analogues, as well as metal-modified zeolites and other aluminosilicates. The surface acid-base properties of these materials can be tuned by selecting the appropriate composition. The second criterion refers to the intrinsic (de)hydrogenation and redox activity of noble or transition metals which includes the dissociative adsorption of molecular H_2, as two hydrogen atoms, which are available to participate in the various hydrogenation or hydrogenolysis pathways. The hydrogen atoms on the noble/transition metals may interact directly with the abundant ether bonds in the lignin fragments or with a double bond (at deeper hydrogenation conditions) or they can interact via spill-over phenomena with a nearby sorbed intermediate, such as an ether bond reacting with the Brønsted acid sites of the support of the hydrogenating metal. Thus, the synergistic action of the metal with the support could also lead to facile hydrogenation/hydrogenolysis reaction reducing further the required reaction temperature. The third criterion is related with the effect of the catalyst on the reactions occurring between the alcohol solvent and the formed reactive intermediates in lignin hydrogenolysis with the aim to inhibit their repolymerization, one representative example being that of favoring their alkylation with methyl or ethyl moieties (from methanol or ethanol respectively) in the presence of metal oxides such as SiO_2-Al_2O_3, CuO, TiO_2, etc. A balance between hydrogenation and alkylation may also be desirable, in order to limit repolymerization and not lose the aromatic nature of the obtained monomers, thus pointing to metal oxides (mainly being used as supports) that cannot be easily reduced in situ in the presence of hydrogen.

Along these lines and in addition to the more classical hydrogenation catalysts, e.g., noble or transition metals on carbon, zeolites, silica, etc., more "sophisticated" multifunctional catalytic formulation have been recently reported, as described in the sections above, including TiN-Ni and TiO_2-Ni, FeNiB alloys, Cu based materials such as $Cu_{20}La_{20}PMO$ (porous metal oxides) and $CuMgAlO_x$, Fe on $Rh/La_2O_3/CeO_2$-ZrO_2, β-CaP_2O_6 and CoP_2O_6, and others. With regard to catalyst recovery from batch reactor systems, new magneticcatalytic formulations with weakly acidic Brønsted-type centers, such as Fe_3O_4@SiO_2@Re and Co@Nb_2O_5@Fe_3O_4 which exhibited promising behavior in the reductive depolymerization of lignin using gaseous H_2, could be also effectively used in catalytic transfer hdyrogenolysis reactions [143,144].

5. Conclusions and Outlook

The valorization of lignin, being the most abundant natural phenolic/aromatic polymer, has tremendous potential provided that efficient depolymerization processes will be developed.

The reductive depolymerization via catalytic hydrogenation/hydrogenolysis reactions utilizing hydrogen donors offers a possible route with minimum requirements of molecular (gaseous) hydrogen. Small alcohols and water may act both as solvents and hydrogen donors, under sub- or supercritical conditions, providing the necessary hydrogen for breaking the primary ether bonds of the lignin macromolecule and leading to deeper C-C bond hydrogenolysis and hydrogenation of the derived alkoxy-phenols depending on process parameters and catalysts used. Methanol, ethanol, isopropanol or mixtures of these alcohols with water, can lead to high degrees of lignin depolymerization and monomers yield, with ethanol being the most promising candidate. Apart from the alcohol-based solvents, formic acid can also act as hydrogen donor, in cooperation with alcohols or water, thus leading to enhanced depolymerization and increased bio-oil production.

Although the selection of the most appropriate solvent is crucial for inducing transfer hydrogenolysis/hydrogenation reactions in lignin depolymerization, still the major role is being played by the catalyst type and its properties, especially when relatively low/moderate temperatures are aimed that do not favor undesirable hydrogenation of the aromatic ring and repolymerization/condensation reactions. As discussed in Section 4.8, an effective catalyst in reductive depolymerization of lignin using hydrogen donors should promote the in situ generation of hydrogen via alcohol reforming, WGS or dehydrogenation (via H: and H^+ abstraction) reactions or formic acid decomposition, should exhibit high hydrogenation reactivity and favor activation of the lignin aryl-ether bonds, and should be able to contribute to stabilization of reactive intermediates, for example, via alkylation reactions leading to increased bio-oil yield and minimum solid residues.

Finally, towards the potential upscaling of lignin depolymerization via catalytic transfer hydrogenolysis using hydrogen donors, continuous flow reactors and processes need to be developed by considering important parameters such as the effective solubilization of lignin as feedstock, increased catalyst reactivity to balance the relatively short contact times and low rate of catalyst deactivation.

The lignin bio-oils rich in alkoxy- or alkyl-phenols, derived either from reductive depolymerization or from thermal/catalytic fast pyrolysis, could be used without or after further upgrading (e.g., glycidilation) in the production of epoxy or phenol-formaldehyde resins and related polymers and composites, while specific components such as vanillin will find application in the food industry, cosmetics, and pharmaceuticals. Furthermore, the lignin bio-oil can be upgraded via hydrodeoxygation (HDO) reactions to hydrocarbon (alkanes) fuels.

Acknowledgments: We acknowledge support of this work by the project "INVALOR: Research Infrastructure for Waste Valorization and Sustainable Management" (MIS 5002495) which is implemented under the Action "Reinforcement of the Research and Innovation Infrastructure", funded by the Operational Programme "Competitiveness, Entrepreneurship and Innovation" (NSRF 2014–2020) and co-financed by Greece and the European Union (European Regional Development Fund). We would also like to acknowledge COST Action CA17128 (LignoCOST) for promoting exchange and dissemination of knowledge and expertise in the field of lignin valorization.

Conflicts of Interest: The authors declare no conflict of interest. The funders had no role in the design of the study; in the collection, analyses, or interpretation of data; in the writing of the manuscript, or in the decision to publish the results.

References

1. European Parliament and the Council, Directive 2009/28/EC. 2009. Available online: https://eur-lex.europa.eu/legal-content/EN/ALL/?uri=celex%3A32009L0028 (accessed on 21 December 2018).
2. Perlack, R.D.; Wright, L.L.; Turhollow, A.F.; Graham, R.L.; Stokes, B.J.; Erbach, D.C. *Biomass as Feedstock for a Bioenergy and Bioproducts Industry: The Technical Feasibility of a Billion-Ton Annual Supply*; U.S. Department of Energy: Oak Ridge National Laboratory, Oak Ridge, TN, USA, 2005.
3. Zakzeski, J.; Bruijnincx, P.C.; Jongerius, A.L.; Weckhuysen, B.M. The catalytic valorization of lignin for the production of renewable chemicals. *Chem. Rev.* **2010**, *110*, 3552–3599. [CrossRef] [PubMed]
4. Galkin, M.V.; Samec, J.S. Lignin Valorization through Catalytic Lignocellulose Fractionation: A Fundamental Platform for the Future Biorefinery. *ChemSusChem* **2016**, *9*, 1544–1558. [CrossRef] [PubMed]

5. Da Silva, A.S.; Inoue, H.; Endo, T.; Yano, S.; Bon, E.P. Milling pretreatment of sugarcane bagasse and straw for enzymatic hydrolysis and ethanol fermentation. *Bioresour. Technol.* **2010**, *101*, 7402–7409. [CrossRef]
6. Agbor, V.B.; Cicek, N.; Sparling, R.; Berlin, A.; Levin, D.B. Biomass pretreatment: Fundamentals toward application. *Biotechnol. Adv.* **2011**, *29*, 675–685. [CrossRef] [PubMed]
7. Barakat, A.; Mayer-Laigle, C.; Solhy, A.; Arancon, R.A.D.; de Vries, H.; Luque, R. Mechanical pretreatments of lignocellulosic biomass: Towards facile and environmentally sound technologies for biofuels production. *RSC Adv.* **2014**, *4*, 48109–48127. [CrossRef]
8. Gonzales, R.R.; Sivagurunathan, P.; Kim, S.H. Effect of severity on dilute acid pretreatment of lignocellulosic biomass and the following hydrogen fermentation. *Int. J. Hydrogen Energy* **2016**, *41*, 21678–21684. [CrossRef]
9. Xu, J.K.; Sun, R.C. Chapter 19—Recent Advances in Alkaline Pretreatment of Lignocellulosic Biomass. In *Biomass Fractionation Technologies for a Lignocellulosic Feedstock Based Biorefinery*; Mussatto, S.I., Ed.; Elsevier: Amsterdam, The Netherlands, 2016; pp. 431–459.
10. Garcia-Cubero, M.A.; Gonzalez-Benito, G.; Indacoechea, I.; Coca, M.; Bolado, S. Effect of ozonolysis pretreatment on enzymatic digestibility of wheat and rye straw. *Bioresour. Technol.* **2009**, *100*, 1608–1613. [CrossRef] [PubMed]
11. Nitsos, C.; Stoklosa, R.; Karnaouri, A.; Vörös, D.; Lange, H.; Hodge, D.; Crestini, C.; Rova, U.; Christakopoulos, P. Isolation and Characterization of Organosolv and Alkaline Lignins from Hardwood and Softwood Biomass. *ACS Sustain. Chem. Eng.* **2016**, *4*, 5181–5193. [CrossRef]
12. Dale, B.E.; Leong, C.K.; Pham, T.K.; Esquivel, V.M.; Rios, I.; Latimer, V.M. Hydrolysis of lignocellulosics at low enzyme levels: Application of the AFEX process. *Bioresour. Technol.* **1996**, *56*, 111–116. [CrossRef]
13. Duque, A.; Manzanares, P.; Ballesteros, I.; Ballesteros, M. Chapter 15—Steam Explosion as Lignocellulosic Biomass Pretreatment. In *Biomass Fractionation Technologies for a Lignocellulosic Feedstock Based Biorefinery*; Mussatto, S.I., Ed.; Elsevier: Amsterdam, The Netherlands, 2016; pp. 349–368.
14. Tang, Y.; Chandra, R.P.; Sokhansanj, S.; Saddler, J.N. Influence of steam explosion processes on the durability and enzymatic digestibility of wood pellets. *Fuel* **2018**, *211*, 87–94. [CrossRef]
15. Schmidt, A.S.; Thomsen, A.B. Optimization of wet oxidation pretreatment of wheat straw. *Bioresour. Technol.* **1998**, *64*, 139–151. [CrossRef]
16. Nitsos, C.K.; Matis, K.A.; Triantafyllidis, K.S. Optimization of hydrothermal pretreatment of lignocellulosic biomass in the bioethanol production process. *ChemSusChem* **2013**, *6*, 110–122. [CrossRef] [PubMed]
17. Hu, Z.; Foston, M.; Ragauskas, A.J. Comparative studies on hydrothermal pretreatment and enzymatic saccharification of leaves and internodes of alamo switchgrass. *Bioresour. Technol.* **2011**, *102*, 7224–7228. [CrossRef] [PubMed]
18. Oksman, K.; Aitomäki, Y.; Mathew, A.P.; Siqueira, G.; Zhou, Q.; Butylina, S.; Tanpichai, S.; Zhou, X.; Hooshmand, S. Review of the recent developments in cellulose nanocomposite processing. *Compos. Part A Appl. Sci. Manuf.* **2016**, *83*, 2–18. [CrossRef]
19. Sen, S.; Patil, S.; Argyropoulos, D.S. Thermal properties of lignin in copolymers, blends, and composites: A review. *Green Chem.* **2015**, *17*, 4862–4887. [CrossRef]
20. Liu, X.; Lin, Q.; Yan, Y.; Peng, F.; Sun, R.; Ren, J. Hemicellulose from Plant Biomass in Medical and Pharmaceutical Application: A Critical Review. *Curr. Med. Chem.* **2017**, *24*, 1–21. [CrossRef] [PubMed]
21. Espro, C.; Gumina, B.; Szumelda, T.; Paone, E.; Mauriello, F. Catalytic Transfer Hydrogenolysis as an Effective Tool for the Reductive Upgrading of Cellulose, Hemicellulose, Lignin, and Their Derived Molecules. *Catalysts* **2018**, *8*, 313. [CrossRef]
22. Huber, G.W.; Iborra, S.; Corma, A. Synthesis of transportation fuels from biomass: Chemistry, catalysts, and engineering. *Chem. Rev.* **2006**, *106*, 4044–4098. [CrossRef]
23. Zhou, C.H.; Xia, X.; Lin, C.X.; Tong, D.S.; Beltramini, J. Catalytic conversion of lignocellulosic biomass to fine chemicals and fuels. *Chem. Soc. Rev.* **2011**, *40*, 5588–5617. [CrossRef]
24. Lappas, A.A.; Kalogiannis, K.G.; Iliopoulou, E.F.; Triantafyllidis, K.S.; Stefanidis, S.D. Catalytic pyrolysis of biomass for transportation fuels. *Wires Energy Environ.* **2012**, *1*, 285–297. [CrossRef]
25. Isikgor, F.H.; Becer, C.R. Lignocellulosic biomass: A sustainable platform for the production of bio-based chemicals and polymers. *Polym. Chem.* **2015**, *6*, 4497–4559. [CrossRef]
26. Serrano-Ruiz, J.C.; Pineda, A.; Balu, A.M.; Luque, R.; Campelo, J.M.; Romero, A.A.; Ramos-Fernandez, J.M. Catalytic transformations of biomass-derived acids into advanced biofuels. *Catal. Today* **2012**, *195*, 162–168. [CrossRef]

27. Espro, C.; Gumina, B.; Paone, E.; Mauriello, F. Upgrading Lignocellulosic Biomasses: Hydrogenolysis of Platform Derived Molecules Promoted by Heterogeneous Pd-Fe Catalysts. *Catalysts* **2017**, *7*, 78. [CrossRef]

28. Rinaldi, R.; Jastrzebski, R.; Clough, M.T.; Ralph, J.; Kennema, M.; Bruijnincx, P.C.; Weckhuysen, B.M. Paving the Way for Lignin Valorisation: Recent Advances in Bioengineering, Biorefining and Catalysis. *Angew. Chem.* **2016**, *55*, 8164–8215. [CrossRef]

29. Azadi, P.; Inderwildi, O.R.; Farnood, R.; King, D.A. Liquid fuels, hydrogen and chemicals from lignin: A critical review. *Renew. Sustain. Energy Rev.* **2013**, *21*, 506–523. [CrossRef]

30. Li, C.; Zhao, X.; Wang, A.; Huber, G.W.; Zhang, T. Catalytic Transformation of Lignin for the Production of Chemicals and Fuels. *Chem. Rev.* **2015**, *115*, 11559–11624. [CrossRef] [PubMed]

31. Chakar, F.S.; Ragauskas, A.J. Review of current and future softwood kraft lignin process chemistry. *Ind. Crop. Prod.* **2004**, *20*, 131–141. [CrossRef]

32. Dorrestijn, E.; Laarhoven, L.J.J.; Arends, I.W.C.E.; Mulder, P. The occurrence and reactivity of phenoxyl linkages in lignin and low rank coal. *J. Anal. Appl. Pyrol.* **2000**, *54*, 153–192. [CrossRef]

33. Gillet, S.; Aguedo, M.; Petitjean, L.; Morais, A.R.C.; Lopes, A.M.D.; Lukasik, R.M.; Anastas, P.T. Lignin transformations for high value applications: Towards targeted modifications using green chemistry. *Green Chem.* **2017**, *19*, 4200–4233. [CrossRef]

34. Pandey, M.P.; Kim, C.S. Lignin Depolymerization and Conversion: A Review of Thermochemical Methods. *Chem. Eng. Technol.* **2011**, *34*, 29–41. [CrossRef]

35. Lora, J. Chapter 10—Industrial Commercial Lignins: Sources, Properties and Applications. In *Monomers, Polymers and Composites from Renewable Resources*; Belgacem, M.N., Gandini, A., Eds.; Elsevier: Amsterdam, The Netherlands, 2008; pp. 225–241. [CrossRef]

36. Tejado, A.; Pena, C.; Labidi, J.; Echeverria, J.M.; Mondragon, I. Physico-chemical characterization of lignins from different sources for use in phenol-formaldehyde resin synthesis. *Bioresour. Technol.* **2007**, *98*, 1655–1663. [CrossRef] [PubMed]

37. Zhou, Z.; Lei, F.; Li, P.; Jiang, J. Lignocellulosic biomass to biofuels and biochemicals: A comprehensive review with a focus on ethanol organosolv pretreatment technology. *Biotechnol. Bioeng.* **2018**, *115*, 2683–2702. [CrossRef]

38. Zhang, Z.Y.; Harrison, M.D.; Rackemann, D.W.; Doherty, W.O.S.; O'Hara, I.M. Organosolv pretreatment of plant biomass for enhanced enzymatic saccharification. *Green Chem.* **2016**, *18*, 360–381. [CrossRef]

39. Villaverde, J.; Ligero, P.; Vega, A. Formic and acetic acid as agents for a cleaner fractionation of Miscanthus x giganteus. *J. Clean. Prod.* **2010**, *18*, 395–401. [CrossRef]

40. Katahira, R.; Mittal, A.; McKinney, K.; Ciesielski, P.N.; Donohoe, B.S.; Black, S.K.; Johnson, D.K.; Biddy, M.J.; Beckham, G.T. Evaluation of Clean Fractionation Pretreatment for the Production of Renewable Fuels and Chemicals from Corn Stover. *ACS Sustain. Chem. Eng.* **2014**, *2*, 1364–1376. [CrossRef]

41. Smit, A.; Huijgen, W. Effective fractionation of lignocellulose in herbaceous biomass and hardwood using a mild acetone organosolv process. *Green Chem.* **2017**, *19*, 5505–5514. [CrossRef]

42. Pan, X.; Gilkes, N.; Kadla, J.; Pye, K.; Saka, S.; Gregg, D.; Ehara, K.; Xie, D.; Lam, D.; Saddler, J. Bioconversion of hybrid poplar to ethanol and co-products using an organosolv fractionation process: Optimization of process yields. *Biotechnol. Bioeng.* **2006**, *94*, 851–861. [CrossRef]

43. Shuai, L.; Amiri, M.T.; Questell-Santiago, Y.M.; Heroguel, F.; Li, Y.; Kim, H.; Meilan, R.; Chapple, C.; Ralph, J.; Luterbacher, J.S. Formaldehyde stabilization facilitates lignin monomer production during biomass depolymerization. *Science* **2016**, *354*, 329–333. [CrossRef]

44. Raghavendran, V.; Nitsos, C.; Matsakas, L.; Rova, U.; Christakopoulos, P.; Olsson, L. A comparative study of the enzymatic hydrolysis of batch organosolv-pretreated birch and spruce biomass. *AMB Express* **2018**, *8*, 114. [CrossRef]

45. Matsakas, L.; Nitsos, C.; Raghavendran, V.; Yakimenko, O.; Persson, G.; Olsson, E.; Rova, U.; Olsson, L.; Christakopoulos, P. A novel hybrid organosolv: Steam explosion method for the efficient fractionation and pretreatment of birch biomass. *Biotechnol. Biofuels* **2018**, *11*, 160. [CrossRef]

46. Calvo-Flores, F.G.; Dobado, J.A. Lignin as renewable raw material. *ChemSusChem* **2010**, *3*, 1227–1235. [CrossRef]

47. Sun, Z.; Fridrich, B.; de Santi, A.; Elangovan, S.; Barta, K. Bright Side of Lignin Depolymerization: Toward New Platform Chemicals. *Chem. Rev.* **2018**, *118*, 614–678. [CrossRef] [PubMed]

48. Liu, W.J.; Jiang, H.; Yu, H.Q. Thermochemical conversion of lignin to functional materials: A review and future directions. *Green Chem.* **2015**, *17*, 4888–4907. [CrossRef]

49. Bjørsvik, H.-R.; Liguori, L. Organic Processes to Pharmaceutical Chemicals Based on Fine Chemicals from Lignosulfonates. *Org. Process. Res. Dev.* **2002**, *6*, 279–290. [CrossRef]

50. Van den Bosch, S.; Koelewijn, S.F.; Renders, T.; Van den Bossche, G.; Vangeel, T.; Schutyser, W.; Sels, B.F. Catalytic Strategies Towards Lignin-Derived Chemicals. *Top. Curr. Chem.* **2018**, *376*, 36. [CrossRef] [PubMed]

51. Lazaridis, P.A.; Fotopoulos, A.P.; Karakoulia, S.A.; Triantafyllidis, K.S. Catalytic Fast Pyrolysis of Kraft Lignin With Conventional, Mesoporous and Nanosized ZSM-5 Zeolite for the Production of Alkyl-Phenols and Aromatics. *Front. Chem.* **2018**, *6*, 295. [CrossRef] [PubMed]

52. Custodis, V.B.F.; Karakoulia, S.A.; Triantafyllidis, K.S.; van Bokhoven, J.A. Catalytic Fast Pyrolysis of Lignin over High-Surface-Area Mesoporous Aluminosilicates: Effect of Porosity and Acidity. *ChemSusChem* **2016**, *9*, 1134–1145. [CrossRef]

53. Guvenatam, B.; Heeres, E.H.J.; Pidko, E.A.; Hensen, E.J.M. Lewis-acid catalyzed depolymerization of Protobind lignin in supercritical water and ethanol. *Catal. Today* **2016**, *259*, 460–466. [CrossRef]

54. Wang, H.; Tucker, M.; Ji, Y. Recent Development in Chemical Depolymerization of Lignin: A Review. *J. Appl. Chem.* **2013**, *2013*, 1–9. [CrossRef]

55. Mahmood, N.; Yuan, Z.; Schmidt, J.; Xu, C.C. Hydrolytic depolymerization of hydrolysis lignin: Effects of catalysts and solvents. *Bioresour. Technol.* **2015**, *190*, 416–419. [CrossRef]

56. Toledano, A.; Serrano, L.; Labidi, J. Organosolv lignin depolymerization with different base catalysts. *J. Chem. Technol. Biotechol.* **2012**, *87*, 1593–1599. [CrossRef]

57. Abdelaziz, O.Y.; Li, K.; Tunå, P.; Hulteberg, C.P. Continuous catalytic depolymerisation and conversion of industrial kraft lignin into low-molecular-weight aromatics. *Biomass Convers. Biorefinery* **2017**, *8*, 455–470. [CrossRef]

58. Schutyser, W.; Renders, T.; Van den Bosch, S.; Koelewijn, S.F.; Beckham, G.T.; Sels, B.F. Chemicals from lignin: An interplay of lignocellulose fractionation, depolymerisation, and upgrading. *Chem. Soc. Rev.* **2018**, *47*, 852–908. [CrossRef]

59. Xu, C.; Arancon, R.A.; Labidi, J.; Luque, R. Lignin depolymerisation strategies: Towards valuable chemicals and fuels. *Chem. Soc. Rev.* **2014**, *43*, 7485–7500. [CrossRef] [PubMed]

60. Bugg, T.D.; Ahmad, M.; Hardiman, E.M.; Rahmanpour, R. Pathways for degradation of lignin in bacteria and fungi. *Nat. Prod. Rep.* **2011**, *28*, 1883–1896. [CrossRef] [PubMed]

61. Sena-Martins, G.; Almeida-Vara, E.; Duarte, J.C. Eco-friendly new products from enzymatically modified industrial lignins. *Ind. Crops Prod.* **2008**, *27*, 189–195. [CrossRef]

62. Kumar, C.R.; Anand, N.; Kloekhorst, A.; Cannilla, C.; Bonura, G.; Frusteri, F.; Barta, K.; Heeres, H.J. Solvent free depolymerization of Kraft lignin to alkyl-phenolics using supported NiMo and CoMo catalysts. *Green Chem.* **2015**, *17*, 4921–4930. [CrossRef]

63. Kloekhorst, A.; Heeres, H.J. Catalytic Hydrotreatment of Alcell Lignin Using Supported Ru, Pd, and Cu Catalysts. *ACS Sustain. Chem. Eng.* **2015**, *3*, 1905–1914. [CrossRef]

64. Zhu, G.D.; Ouyang, X.P.; Jiang, L.F.; Zhu, Y.; Jin, D.X.; Pang, Y.X.; Qiu, X.Q. Effect of functional groups on hydrogenolysis of lignin model compounds. *Fuel Process. Technol.* **2016**, *154*, 132–138. [CrossRef]

65. Zhang, J.G.; Asakura, H.; van Rijn, J.; Yang, J.; Duchesne, P.; Zhang, B.; Chen, X.; Zhang, P.; Saeys, M.; Yan, N. Highly efficient, NiAu-catalyzed hydrogenolysis of lignin into phenolic chemicals. *Green Chem.* **2014**, *16*, 2432–2437. [CrossRef]

66. Zhang, J.; Teo, J.; Chen, X.; Asakura, H.; Tanaka, T.; Teramura, K.; Yan, N. A Series of NiM (M = Ru, Rh, and Pd) Bimetallic Catalysts for Effective Lignin Hydrogenolysis in Water. *ACS Catal.* **2014**, *4*, 1574–1583. [CrossRef]

67. Zhang, J.-W.; Cai, Y.; Lu, G.-P.; Cai, C. Facile and selective hydrogenolysis of β-O-4 linkages in lignin catalyzed by Pd–Ni bimetallic nanoparticles supported on ZrO₂. *Green Chem.* **2016**, *18*, 6229–6235. [CrossRef]

68. Mauriello, F.; Paone, E.; Pietropaolo, R.; Balu, A.M.; Luque, R. Catalytic Transfer Hydrogenolysis of Lignin-Derived Aromatic Ethers Promoted by Bimetallic Pd/Ni Systems. *ACS Sustain. Chem. Eng.* **2018**, *6*, 9269–9276. [CrossRef]

69. Galkin, M.V.; Sawadjoon, S.; Rohde, V.; Dawange, M.; Samec, J.S.M. Mild Heterogeneous Palladium-Catalyzed Cleavage of β-O-4′-Ether Linkages of Lignin Model Compounds and Native Lignin in Air. *ChemCatChem* **2014**, *6*, 179–184. [CrossRef]

70. Galkin, M.V.; Dahlstrand, C.; Samec, J.S. Mild and Robust Redox-Neutral Pd/C-Catalyzed Lignol beta-O-4' Bond Cleavage Through a Low-Energy-Barrier Pathway. *ChemSusChem* **2015**, *8*, 2187–2192. [CrossRef] [PubMed]

71. Besse, X.; Schuurman, Y.; Guilhaume, N. Reactivity of lignin model compounds through hydrogen transfer catalysis in ethanol/water mixtures. *Appl. Catal. B Environ.* **2017**, *209*, 265–272. [CrossRef]

72. Wang, X.; Rinaldi, R. Solvent effects on the hydrogenolysis of diphenyl ether with Raney nickel and their implications for the conversion of lignin. *ChemSusChem* **2012**, *5*, 1455–1466. [CrossRef]

73. Macala, G.S.; Matson, T.D.; Johnson, C.L.; Lewis, R.S.; Iretskii, A.V.; Ford, P.C. Hydrogen transfer from supercritical methanol over a solid base catalyst: A model for lignin depolymerization. *ChemSusChem* **2009**, *2*, 215–217. [CrossRef]

74. Subbotina, E.; Galkin, M.V.; Samec, J.S.M. Pd/C-Catalyzed Hydrogenolysis of Dibenzodioxocin Lignin Model Compounds Using Silanes and Water as Hydrogen Source. *ACS Sustain. Chem. Eng.* **2017**, *5*, 3726–3731. [CrossRef]

75. Klein, I.; Saha, B.; Abu-Omar, M.M. Lignin depolymerization over Ni/C catalyst in methanol, a continuation: Effect of substrate and catalyst loading. *Catal. Sci. Technol.* **2015**, *5*, 3242–3245. [CrossRef]

76. Galkin, M.V.; Smit, A.T.; Subbotina, E.; Artemenko, K.A.; Bergquist, J.; Huijgen, W.J.; Samec, J.S. Hydrogen-free catalytic fractionation of woody biomass. *ChemSusChem* **2016**, *9*, 3280–3287. [CrossRef] [PubMed]

77. Van den Bosch, S.; Schutyser, W.; Vanholme, R.; Driessen, T.; Koelewijn, S.F.; Renders, T.; De Meester, B.; Huijgen, W.J.J.; Dehaen, W.; Courtin, C.M.; et al. Reductive lignocellulose fractionation into soluble lignin-derived phenolic monomers and dimers and processable carbohydrate pulps. *Energy Environ. Sci.* **2015**, *8*, 1748–1763. [CrossRef]

78. Song, Q.; Wang, F.; Cai, J.; Wang, Y.; Zhang, J.; Yu, W.; Xu, J. Lignin depolymerization (LDP) in alcohol over nickel-based catalysts via a fragmentation–hydrogenolysis process. *Energy Environ. Sci.* **2013**, *6*, 994. [CrossRef]

79. Godard, H.P.; McCarthy, J.L.; Hibbert, H. Hydrogenation of wood. *J. Am. Chem. Soc.* **1940**, *62*, 988. [CrossRef]

80. Yan, N.; Zhao, C.; Dyson, P.J.; Wang, C.; Liu, L.T.; Kou, Y. Selective degradation of wood lignin over noble-metal catalysts in a two-step process. *ChemSusChem* **2008**, *1*, 626–629. [CrossRef] [PubMed]

81. Torr, K.M.; van de Pas, D.J.; Cazeils, E.; Suckling, I.D. Mild hydrogenolysis of in-situ and isolated Pinus radiata lignins. *Bioresour. Technol.* **2011**, *102*, 7608–7611. [CrossRef] [PubMed]

82. Galkin, M.V.; Samec, J.S. Selective route to 2-propenyl aryls directly from wood by a tandem organosolv and palladium-catalysed transfer hydrogenolysis. *ChemSusChem* **2014**, *7*, 2154–2158. [CrossRef] [PubMed]

83. Parsell, T.; Yohe, S.; Degenstein, J.; Jarrell, T.; Klein, I.; Gencer, E.; Hewetson, B.; Hurt, M.; Kim, J.I.; Choudhari, H.; et al. A synergistic biorefinery based on catalytic conversion of lignin prior to cellulose starting from lignocellulosic biomass. *Green Chem.* **2015**, *17*, 1492–1499. [CrossRef]

84. Luo, H.; Klein, I.M.; Jiang, Y.; Zhu, H.Y.; Liu, B.Y.; Kenttamaa, H.I.; Abu-Omar, M.M. Total Utilization of Miscanthus Biomass, Lignin and Carbohydrates, Using Earth Abundant Nickel Catalyst. *ACS Sustain. Chem. Eng.* **2016**, *4*, 2316–2322. [CrossRef]

85. Ferrini, P.; Rinaldi, R. Catalytic biorefining of plant biomass to non-pyrolytic lignin bio-oil and carbohydrates through hydrogen transfer reactions. *Angew. Chem.* **2014**, *53*, 8634–8639. [CrossRef]

86. Van den Bosch, S.; Schutyser, W.; Koelewijn, S.F.; Renders, T.; Courtin, C.M.; Sels, B.F. Tuning the lignin oil OH-content with Ru and Pd catalysts during lignin hydrogenolysis on birch wood. *Chem. Commun.* **2015**, *51*, 13158–13161. [CrossRef] [PubMed]

87. Van den Bosch, S.; Renders, T.; Kennis, S.; Koelewijn, S.F.; Van den Bossche, G.; Vangeel, T.; Deneyer, A.; Depuydt, D.; Courtin, C.M.; Thevelein, J.M.; et al. Integrating lignin valorization and bio-ethanol production: On the role of Ni-Al2O3 catalyst pellets during lignin-first fractionation. *Green Chem.* **2017**, *19*, 3313–3326. [CrossRef]

88. Schutyser, W.; Van den Bosch, S.; Renders, T.; De Boe, T.; Koelewijn, S.F.; Dewaele, A.; Ennaert, T.; Verkinderen, O.; Goderis, B.; Courtin, C.M.; et al. Influence of bio-based solvents on the catalytic reductive fractionation of birch wood. *Green Chem.* **2015**, *17*, 5035–5045. [CrossRef]

89. Huang, X.; Ouyang, X.; Hendriks, B.M.S.; Gonzalez, O.M.M.; Zhu, J.; Koranyi, T.I.; Boot, M.D.; Hensen, E.J.M. Selective production of mono-aromatics from lignocellulose over Pd/C catalyst: The influence of acid co-catalysts. *Faraday Discuss.* **2017**, *202*, 141–156. [CrossRef] [PubMed]

90. Renders, T.; Schutyser, W.; Van den Bosch, S.; Koelewijn, S.F.; Vangeel, T.; Courtin, C.M.; Sels, B.F. Influence of Acidic (H3PO4) and Alkaline (NaOH) Additives on the Catalytic Reductive Fractionation of Lignocellulose. *ACS Catal.* **2016**, *6*, 2055–2066. [CrossRef]

91. Renders, T.; Van den Bosch, S.; Koelewijn, S.F.; Schutyser, W.; Sels, B.F. Lignin-first biomass fractionation: The advent of active stabilisation strategies. *Energy Environ. Sci.* **2017**, *10*, 1551–1557. [CrossRef]

92. Shu, R.; Long, J.; Xu, Y.; Ma, L.; Zhang, Q.; Wang, T.; Wang, C.; Yuan, Z.; Wu, Q. Investigation on the structural effect of lignin during the hydrogenolysis process. *Bioresour. Technol.* **2016**, *200*, 14–22. [CrossRef]

93. Shu, R.Y.; Xu, Y.; Ma, L.L.; Zhang, Q.; Wang, T.J.; Chen, P.R.; Wu, Q.Y. Hydrogenolysis process for lignosulfonate depolymerization using synergistic catalysts of noble metal and metal chloride. *RSC Adv.* **2016**, *6*, 88788–88796. [CrossRef]

94. Kim, J.Y.; Park, J.; Hwang, H.; Kim, J.K.; Song, I.K.; Choi, J.W. Catalytic depolymerization of lignin macromolecule to alkylated phenols over various metal catalysts in supercritical tert-butanol. *J. Anal. Appl. Pyrol.* **2015**, *113*, 99–106. [CrossRef]

95. Kim, J.Y.; Park, J.; Kim, U.J.; Choi, J.W. Conversion of Lignin to Phenol-Rich Oil Fraction under Supercritical Alcohols in the Presence of Metal Catalysts. *Energy Fuels* **2015**, *29*, 5154–5163. [CrossRef]

96. Yuan, Z.S.; Tymchyshyn, M.; Xu, C.B. Reductive Depolymerization of Kraft and Organosolv Lignin in Supercritical Acetone for Chemicals and Materials. *ChemCatChem* **2016**, *8*, 1968–1976. [CrossRef]

97. Lama, S.M.G.; Pampel, J.; Fellinger, T.P.; Beskoski, V.P.; Slavkovic-Beskoski, L.; Antonietti, M.; Molinari, V. Efficiency of Ni Nanoparticles Supported on Hierarchical Porous Nitrogen-Doped Carbon for Hydrogenolysis of Kraft Lignin in Flow and Batch Systems. *ACS Sustain. Chem. Eng.* **2017**, *5*, 2415–2420. [CrossRef]

98. Park, J.; Oh, S.; Kim, J.Y.; Park, S.Y.; Song, I.K.; Choi, J.W. Comparison of degradation features of lignin to phenols over Pt catalysts prepared with various forms of carbon supports. *RSC Adv.* **2016**, *6*, 16917–16924. [CrossRef]

99. Bouxin, F.P.; McVeigh, A.; Tran, F.; Westwood, N.J.; Jarvis, M.C.; Jackson, S.D. Catalytic depolymerisation of isolated lignins to fine chemicals using a Pt/alumina catalyst: Part 1—Impact of the lignin structure. *Green Chem.* **2015**, *17*, 1235–1242. [CrossRef]

100. Barta, K.; Warner, G.R.; Beach, E.S.; Anastas, P.T. Depolymerization of organosolv lignin to aromatic compounds over Cu-doped porous metal oxides. *Green Chem.* **2014**, *16*, 191–196. [CrossRef]

101. Barta, K.; Matson, T.D.; Fettig, M.L.; Scott, S.L.; Iretskii, A.V.; Ford, P.C. Catalytic disassembly of an organosolv lignin via hydrogen transfer from supercritical methanol. *Green Chem.* **2010**, *12*, 1640–1647. [CrossRef]

102. Narani, A.; Chowdari, R.K.; Cannilla, C.; Bonura, G.; Frusteri, F.; Heeres, H.J.; Barta, K. Efficient catalytic hydrotreatment of Kraft lignin to alkylphenolics using supported NiW and NiMo catalysts in supercritical methanol. *Green Chem.* **2015**, *17*, 5046–5057. [CrossRef]

103. Verziu, M.; Tirsoaga, A.; Cojocaru, B.; Bucur, C.; Tudora, B.; Richel, A.; Aguedo, M.; Samikannu, A.; Mikkola, J.P. Hydrogenolysis of lignin over Ru-based catalysts: The role of the ruthenium in a lignin fragmentation process. *Mol. Catal.* **2018**, *450*, 65–76. [CrossRef]

104. Konnerth, H.; Zhang, J.G.; Ma, D.; Prechtl, M.H.G.; Yan, N. Base promoted hydrogenolysis of lignin model compounds and organosolv lignin over metal catalysts in water. *Chem. Eng. Sci.* **2015**, *123*, 155–163. [CrossRef]

105. Ma, X.L.; Tian, Y.; Hao, W.Y.; Ma, R.; Li, Y.D. Production of phenols from catalytic conversion of lignin over a tungsten phosphide catalyst. *Appl. Catal. A Gen.* **2014**, *481*, 64–70. [CrossRef]

106. Kim, J.Y.; Park, S.Y.; Choi, I.G.; Choi, J.W. Evaluation of RuxNi1-x/SBA-15 catalysts for depolymerization features of lignin macromolecule into monomeric phenols. *Chem. Eng. J.* **2018**, *336*, 640–648. [CrossRef]

107. Chen, P.; Zhang, Q.; Shu, R.; Xu, Y.; Ma, L.; Wang, T. Catalytic depolymerization of the hydrolyzed lignin over mesoporous catalysts. *Bioresour. Technol.* **2017**, *226*, 125–131. [CrossRef] [PubMed]

108. Xiao, L.P.; Wang, S.Z.; Li, H.L.; Li, Z.W.; Shi, Z.J.; Xiao, L.; Sun, R.G.; Fang, Y.M.; Song, G.Y. Catalytic Hydrogenolysis of Lignins into Phenolic Compounds over Carbon Nanotube Supported Molybdenum Oxide. *ACS Catal.* **2017**, *7*, 7535–7542. [CrossRef]

109. Wang, J.; Li, W.; Wang, H.; Ma, Q.; Li, S.; Chang, H.M.; Jameel, H. Liquefaction of kraft lignin by hydrocracking with simultaneous use of a novel dual acid-base catalyst and a hydrogenation catalyst. *Bioresour. Technol.* **2017**, *243*, 100–106. [CrossRef] [PubMed]

110. Ma, R.; Hao, W.; Ma, X.; Tian, Y.; Li, Y. Catalytic ethanolysis of Kraft lignin into high-value small-molecular chemicals over a nanostructured alpha-molybdenum carbide catalyst. *Angew. Chem. Int. Ed.* **2014**, *53*, 7310–7315. [CrossRef] [PubMed]

111. Ma, X.L.; Ma, R.; Hao, W.Y.; Chen, M.M.; Iran, F.; Cui, K.; Tian, Y.; Li, Y.D. Common Pathways in Ethanolysis of Kraft Lignin to Platform Chemicals over Molybdenum-Based Catalysts. *ACS Catal.* **2015**, *5*, 4803–4813. [CrossRef]

112. Ma, X.; Cui, K.; Hao, W.; Ma, R.; Tian, Y.; Li, Y. Alumina supported molybdenum catalyst for lignin valorization: Effect of reduction temperature. *Bioresour. Technol.* **2015**, *192*, 17–22. [CrossRef]

113. Singh, S.K.; Ekhe, J.D. Cu–Mo doped zeolite ZSM-5 catalyzed conversion of lignin to alkyl phenols with high selectivity. *Catal. Sci. Technol.* **2015**, *5*, 2117–2124. [CrossRef]

114. Huang, S.H.; Mahmood, N.; Zhang, Y.S.; Tymchyshyn, M.; Yuan, Z.S.; Xu, C.B. Reductive de-polymerization of kraft lignin with formic acid at low temperatures using inexpensive supported Ni-based catalysts. *Fuel* **2017**, *209*, 579–586. [CrossRef]

115. Huang, S.; Mahmood, N.; Tymchyshyn, M.; Yuan, Z.; Xu, C.C. Reductive de-polymerization of kraft lignin for chemicals and fuels using formic acid as an in-situ hydrogen source. *Bioresour. Technol.* **2014**, *171*, 95–102. [CrossRef]

116. Liguori, L.; Barth, T. Palladium-Nafion SAC-13 catalysed depolymerisation of lignin to phenols in formic acid and water. *J. Anal. Appl. Pyrol.* **2011**, *92*, 477–484. [CrossRef]

117. Limarta, S.O.; Ha, J.M.; Park, Y.K.; Lee, H.; Suh, D.J.; Jae, J. Efficient depolymerization of lignin in supercritical ethanol by a combination of metal and base catalysts. *J. Ind. Eng. Chem.* **2018**, *57*, 45–54. [CrossRef]

118. Molinari, V.; Clavel, G.; Graglia, M.; Antonietti, M.; Esposito, D. Mild Continuous Hydrogenolysis of Kraft Lignin over Titanium Nitride–Nickel Catalyst. *ACS Catal.* **2016**, *6*, 1663–1670. [CrossRef]

119. Luo, L.; Yang, J.; Yao, G.; Jin, F. Controlling the selectivity to chemicals from catalytic depolymerization of kraft lignin with in-situ H2. *Bioresour. Technol.* **2018**, *264*, 1–6. [CrossRef] [PubMed]

120. Singh, S.K.; Nandeshwar, K.; Ekhe, J.D. Thermochemical lignin depolymerization and conversion to aromatics in subcritical methanol: Effects of catalytic conditions. *New J. Chem.* **2016**, *40*, 3677–3685. [CrossRef]

121. Huang, X.; Koranyi, T.I.; Boot, M.D.; Hensen, E.J. Catalytic depolymerization of lignin in supercritical ethanol. *ChemSusChem* **2014**, *7*, 2276–2288. [CrossRef] [PubMed]

122. Huang, X.; Atay, C.; Korányi, T.I.; Boot, M.D.; Hensen, E.J.M. Role of Cu–Mg–Al Mixed Oxide Catalysts in Lignin Depolymerization in Supercritical Ethanol. *ACS Catal.* **2015**, *5*, 7359–7370. [CrossRef]

123. Huang, X.M.; Koranyi, T.I.; Boot, M.D.; Hensen, E.J.M. Ethanol as capping agent and formaldehyde scavenger for efficient depolymerization of lignin to aromatics. *Green Chem.* **2015**, *17*, 4941–4950. [CrossRef]

124. Huang, X.; Atay, C.; Zhu, J.; Palstra, S.W.L.; Koranyi, T.I.; Boot, M.D.; Hensen, E.J.M. Catalytic Depolymerization of Lignin and Woody Biomass in Supercritical Ethanol: Influence of Reaction Temperature and Feedstock. *ACS Sustain. Chem. Eng.* **2017**, *5*, 10864–10874. [CrossRef]

125. Jeong, S.; Yang, S.; Kim, D.H. Depolymerization of Protobind lignin to produce monoaromatic compounds over Cu/ZSM-5 catalyst in supercritical ethanol. *Mol. Catal.* **2017**, *442*, 140–146. [CrossRef]

126. Zhou, M.H.; Sharma, B.K.; Liu, P.; Ye, J.; Xu, J.M.; Jiang, J.C. Catalytic in Situ Hydrogenolysis of Lignin in Supercritical Ethanol: Effect of Phenol, Catalysts, and Reaction Temperature. *ACS Sustain. Chem. Eng.* **2018**, *6*, 6867–6875. [CrossRef]

127. Li, Z.; Bi, Z.; Yan, L. Two-step hydrogen transfer catalysis conversion of lignin to valuable small molecular compounds. *Green Process. Synth.* **2017**, *6*, 363–370. [CrossRef]

128. Onwudili, J.A.; Williams, P.T. Catalytic depolymerization of alkali lignin in subcritical water: Influence of formic acid and Pd/C catalyst on the yields of liquid monomeric aromatic products. *Green Chem.* **2014**, *16*, 4740–4748. [CrossRef]

129. Toledano, A.; Serrano, L.; Pineda, A.; Romero, A.A.; Luque, R.; Labidi, J. Microwave-assisted depolymerisation of organosolv lignin via mild hydrogen-free hydrogenolysis: Catalyst screening. *Appl. Catal. B Envrion.* **2014**, *145*, 43–55. [CrossRef]

130. Toledano, A.; Serrano, L.; Labidi, J.; Pineda, A.; Balu, A.M.; Luque, R. Heterogeneously Catalysed Mild Hydrogenolytic Depolymerisation of Lignin Under Microwave Irradiation with Hydrogen-Donating Solvents. *ChemCatChem* **2013**, *5*, 977–985. [CrossRef]

131. Xu, W.; Miller, S.J.; Agrawal, P.K.; Jones, C.W. Depolymerization and hydrodeoxygenation of switchgrass lignin with formic acid. *ChemSusChem* **2012**, *5*, 667–675. [CrossRef] [PubMed]

132. Kloekhorst, A.; Shen, Y.; Yie, Y.; Fang, M.; Heeres, H.J. Catalytic hydrodeoxygenation and hydrocracking of Alcell® lignin in alcohol/formic acid mixtures using a Ru/C catalyst. *Biomass Bioenergy* **2015**, *80*, 147–161. [CrossRef]

133. Park, J.; Riaz, A.; Insyani, R.; Kim, J. Understanding the relationship between the structure and depolymerization behavior of lignin. *Fuel* **2018**, *217*, 202–210. [CrossRef]

134. Warner, G.; Hansen, T.S.; Riisager, A.; Beach, E.S.; Barta, K.; Anastas, P.T. Depolymerization of organosolv lignin using doped porous metal oxides in supercritical methanol. *Bioresour. Technol.* **2014**, *161*, 78–83. [CrossRef]

135. Klamrassamee, T.; Laosiripojana, N.; Faungnawakij, K.; Moghaddam, L.; Zhang, Z.Y.; Doherty, W.O.S. Co- and Ca-phosphate-based catalysts for the depolymerization of organosolv eucalyptus lignin. *RSC Adv.* **2015**, *5*, 45618–45621. [CrossRef]

136. Regmi, Y.N.; Mann, J.K.; McBride, J.R.; Tao, J.M.; Barnes, C.E.; Labbe, N.; Chmely, S.C. Catalytic transfer hydrogenolysis of organosolv lignin using B-containing FeNi alloyed catalysts. *Catal. Today* **2018**, *302*, 190–195. [CrossRef]

137. Liu, S.; Lin, Z.; Cai, Z.; Long, J.; Li, Z.; Li, X. Selective depolymerization of lignosulfonate via hydrogen transfer enhanced in an emulsion microreactor. *Bioresour. Technol.* **2018**, *264*, 382–386. [CrossRef] [PubMed]

138. Oregui-Bengoechea, M.; Gandarias, I.; Arias, P.L.; Barth, T. Unraveling the Role of Formic Acid and the Type of Solvent in the Catalytic Conversion of Lignin: A Holistic Approach. *ChemSusChem* **2017**, *10*, 754–766. [CrossRef] [PubMed]

139. Kristianto, I.; Limarta, S.O.; Lee, H.; Ha, J.M.; Suh, D.J.; Jae, J. Effective depolymerization of concentrated acid hydrolysis lignin using a carbon-supported ruthenium catalyst in ethanol/formic acid media. *Bioresour. Technol.* **2017**, *234*, 424–431. [CrossRef] [PubMed]

140. Jiang, Y.T.; Li, Z.; Tang, X.; Sun, Y.; Zeng, X.H.; Liu, S.J.; Lin, L. Depolymerization of Cellulolytic Enzyme Lignin for the Production of Monomeric Phenols over Raney Ni and Acidic Zeolite Catalysts. *Energy Fuels* **2015**, *29*, 1662–1668. [CrossRef]

141. Das, L.; Li, M.; Stevens, J.; Li, W.Q.; Pu, Y.Q.; Ragauskas, A.J.; Shi, J. Characterization and Catalytic Transfer Hydrogenolysis of Deep Eutectic Solvent Extracted Sorghum Lignin to Phenolic Compounds. *ACS Sustain. Chem. Eng.* **2018**, *6*, 10408–10420. [CrossRef]

142. Seo, M.; Yoon, D.; Hwang, K.S.; Kang, J.W.; Kim, J. Supercritical alcohols as solvents and reducing agents for the synthesis of reduced graphene oxide. *Carbon* **2013**, *64*, 207–218. [CrossRef]

143. Tudorache, M.; Opris, C.; Cojocaru, B.; Apostol, N.G.; Tirsoaga, A.; Coman, S.M.; Parvulescu, V.I.; Duraki, B.; Krumeich, F.; van Bokhoven, J.A. Highly Efficient, Easily Recoverable, and Recyclable Re–SiO$_2$–Fe$_3$O$_4$ Catalyst for the Fragmentation of Lignin. *ACS Sustain. Chem. Eng.* **2018**, *6*, 9606–9618. [CrossRef]

144. Opris, C.; Cojocaru, B.; Gheorghe, N.; Tudorache, M.; Coman, S.M.; Parvulescu, V.I.; Duraki, B.; Krumeich, F.; van Bokhoven, J.A. Lignin fragmentation over magnetically recyclable composite Co@Nb2O5@Fe3O4 catalysts. *J. Catal.* **2016**, *339*, 209–227. [CrossRef]

catalysts MDPI

Article

Effect of Calcination Atmosphere and Temperature on the Hydrogenolysis Activity and Selectivity of Copper-Zinc Catalysts

Oleg Kikhtyanin [1], Violetta Pospelova [2], Jaroslav Aubrecht [2], Miloslav Lhotka [3] and David Kubička [1,2,*]

[1] Technopark Kralupy, VŠCHT Praha, Žižkova 7, 278 01 Kralupy nad Vltavou, Czech Republic; oleg.kikhtyanin@vscht.cz

[2] Department of Petroleum Technology and Alternative Fuels, VŠCHT Praha, Technická 5, 166 28 Praha 6—Dejvice, Czech Republic; violetta.pospelova@vscht.cz (V.P.); jaroslav.aubrecht@vscht.cz (J.A.)

[3] Department of Inorganic Technology, University of Chemistry and Technology Prague, Technická 5, 166 28 Praha 6—Dejvice, Czech Republic; miloslav.lhotka@vscht.cz

* Correspondence: david.kubicka@vscht.cz; Tel.:+420-220-446-106

Received: 23 September 2018; Accepted: 7 October 2018; Published: 11 October 2018

Abstract: A series of CuZn catalysts with a Cu/Zn ratio of 1.6 was prepared by the calcination of a single precursor, CuZn-P consisting of an equimolar mixture of aurichalcite and zincian malachite, in three different calcination atmospheres (air, nitrogen, and hydrogen) at three temperatures (220, 350, and 500 °C). All catalysts were characterized by XRD and N_2-physisorption to assess their phase composition, crystallite sizes and textural properties and tested in dimethyl adipate (DMA) hydrogenolysis in a batch reactor at 220 °C and 10 MPa H_2. The XRD examination of these catalysts proved that both parameters, calcination temperature and atmosphere, affected the resulting phase composition of the catalysts as well as their crystallite sizes. In an oxidizing atmosphere, CuO and ZnO in intimate contact prevailed whereas in inert or reducing atmosphere both oxides were accompanied by Cu_2O and Cu. The crystallite size of Cu_2O and Cu was larger than the size of CuO and ZnO thus indicating a less intimate contact between the Cu-phases and ZnO in catalysts calcined in nitrogen and hydrogen. Catalysts prepared by calcination at 220 °C and CuZn catalyst calcined in the air at 350 °C significantly outperformed the other catalysts in DMA hydrogenolysis with a 59–78% conversion due to the small crystallite size and intimate contact between the CuO and ZnO phases prior to catalyst reduction. Despite the low DMA conversion (<30%), transesterification products were the main reaction products with overall selectivities of >80% over the catalysts calcined in nitrogen or hydrogen at least at 350 °C. The obvious change in the preferred reaction pathway because of the atmosphere calcination and temperature shows that there are different active sites responsible for hydrogenolysis and transesterification and that their relative distribution has changed.

Keywords: hydrogenolysis; transesterification; CuZn catalysts; calcination atmosphere; calcination temperature

1. Introduction

Over the past few decades, increasing attention has been paid to green chemistry principles to achieve a higher sustainability of chemical technologies. One of these principles is based on the application of highly efficient catalysts in industrial processes. However, their efficiency alone does not guarantee the overall technological sustainability as some catalyst productions use or produce harmful materials. This is the case of the traditional, highly efficient Adkins catalysts containing chromium that have been used for the industrial hydrogenolysis of esters to alcohols since the 1950s [1–3]. Recently, copper-zinc catalysts have been shown to possess significant catalytic activity in the hydrogenolysis of biomass-derived esters [4–6] and could thus replace copper-chromium catalysts that do not meet the highest environmental standards.

Copper-zinc catalysts belong among widely studied catalysts due to a significant interest in the methanol synthesis from synthesis gas [7–9], as well as by the hydrogenation of carbon dioxide [10,11]. From a structural point of view, several studies have concluded that zinc oxide supports the adsorption of hydrogen on the copper surface and ensures the stabilization of the copper dispersion [7,12]. ZnO has also been suggested to act as a spacer between copper particles supporting high copper dispersion and thus maximizing the specific surface of copper [7,8,13]. This is of importance as Cu is the active component responsible for the hydrogenation and hydrogenolysis activity of catalysts. Nonetheless, the exact nature of the Cu active sites is still a matter of discussion even in the case of the most studied methanol synthesis [8] and it is not yet sufficiently described in hydrogenolysis catalysts. Recently, the formation of transesterification products during the hydrogenolysis of dimethyl adipate over a CuZnAl catalyst has been reported [6] and a reaction scheme has been proposed (Scheme 1). However, a clear understanding of the transesterification activity is still lacking. Moreover, due to the sensitivity of the copper-based catalysts to thermal treatment at elevated temperatures, the optimization and stabilization of the distribution of Cu active sites is challenging and studies dealing with the effect of calcination conditions are rather scarce [14].

Therefore, the present work is focused on a systematic study of the effect of three different calcination gas atmospheres, i.e., air, nitrogen, and hydrogen, on the structure and catalytic performance of catalysts obtained by the calcination of a precipitated CuZn catalyst precursor at three different temperatures (220, 350, and 500 °C) in these three calcination atmospheres. Particular attention is paid to the effect of the calcination on the relative activity of these CuZn catalysts in the hydrogenolysis and transesterification reactions in order to contribute to a better understanding of both catalytic functions.

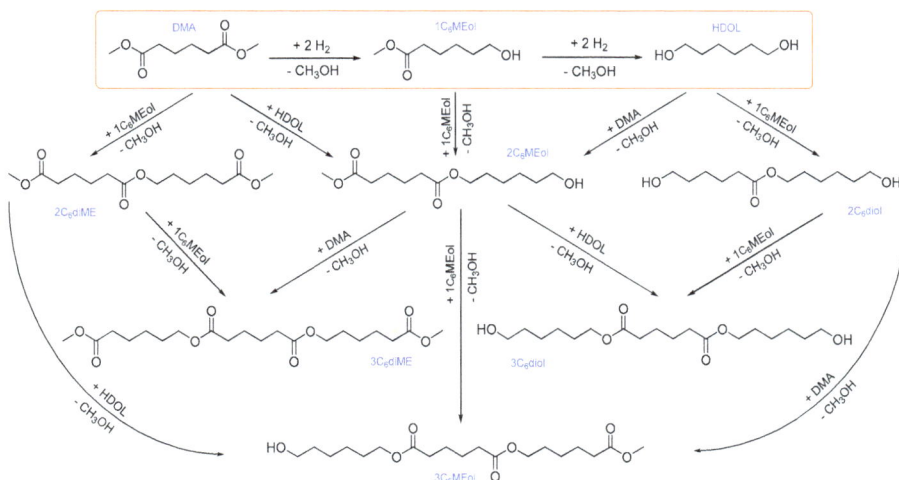

Scheme 1. The proposed reaction network explaining the formation of the transesterification by-products during the hydrogenolysis of dimethyl adipate [6]. DMA is dimethyl adipate, $1C_6$MEol is methyl 6-hydroxyhexanoate, HDOL is hexane-1,6-diol, $2C_6$diME is 6-methoxy-6-oxohexyl methyl adipate, $2C_6$MEol is 6-hydroxyhexyl methyl adipate, $2C_6$diol is 6-hydroxyhexyl 6-hydroxyhexanoate, $3C_6$diME is bis(6-methoxy-6-oxohexyl) adipate, $3C_6$MEol is 6-hydroxyhexyl (6-methoxy-6-oxohexyl) adipate, and $3C_6$diol is bis(6-hydroxyhexyl) adipate.

2. Results and Discussion

2.1. Catalyst Characterization

The Cu/Zn atomic ratio in the synthesized catalyst precursor determined by XRF was 1.6, i.e., very close to that expected from the chemical composition of the solution of copper and zinc nitrates used in the precipitation (1.5). Moreover, it was assumed that all catalyst samples prepared by different thermal treatments had the same chemical composition.

The phase composition of the catalyst precursor, as well as that of the final catalysts, was investigated by XRD. The XRD pattern of the as-prepared CuZn precursor (Figure 1) revealed that this sample was a mixture of two phases, zincian malachite ($(Cu_{0.8} Zn_{0.2})_2(OH)_2CO_3$; Ref. Code No. 01-079-7851) and aurichalcite ($Cu_2Zn_3(CO_3)_2(OH)_6$; Ref. Code No. 04-010-3227), found in almost equal amounts.

The XRD patterns of the catalysts prepared by calcination in static air at the three calcination temperatures (220, 350, and 500 °C) are compared with the XRD pattern of the catalyst precursor in Figure 1. When the catalyst precursor was treated at 220 °C, the phase composition did not change, although the intensity of the XRD reflexes decreased slightly. This can be attributed to the good thermal stability of zincian malachite and aurichalcite. It was reported that zincian malachite and aurichalcite are stable up to 270 °C and 360 °C, correspondingly [14,15]. When the calcination temperature was increased to 350 °C, the zincian malachite and aurichalcite crystalline domains present in the CuZn catalyst precursor decomposed completely and the CuO and ZnO crystalline particles were formed. A further increase in the calcination temperature to 500 °C did not change the phase composition of the catalyst anymore, i.e., the CuZn-500AS sample consisted exclusively of a mixture of CuO and ZnO. However, the intensity of the XRD reflexes notably increased, suggesting a growth of the crystallites size of both CuO and ZnO.

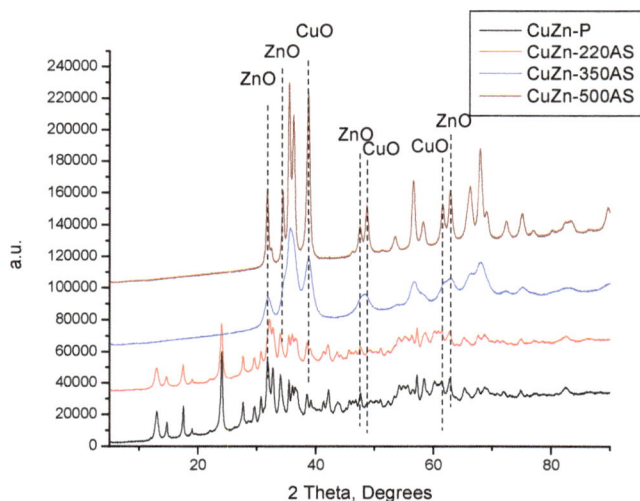

Figure 1. The XRD patterns of the catalysts prepared by the treatment of the CuZn precursor in static air at 220, 350, and 500 °C.

The XRD patterns of the catalysts prepared by calcination in nitrogen again at temperatures 220, 350, and 500 °C were compared with the XRD pattern of the catalyst precursor in Figure 2. Unlike in the case of treatment in static air at 220 °C, the thermal treatment of the CuZn-P in a nitrogen atmosphere at 220 °C (underflow conditions) resulted in a change in the phase composition of the prepared catalyst sample (CuZn-220N). Although the reflexes of zincian malachite and aurichalcite are still present (Figure 2), they are accompanied by the reflexes of CuO and ZnO. This evidences that the thermal decomposition of the initial crystalline phases has started under these calcination conditions. This difference can be explained by the different experimental setups, in particular, by the flow conditions or the partial local overheating of the precursor sample inside the autoclave (the sample is in contact with the autoclave wall while the thermocouple measuring the temperature is not). An increase in the treatment temperature to 350 °C resulted in a total disappearance of the reflexes of zincian malachite and aurichalcite accompanied by an increase in the intensity of the reflexes of CuO and ZnO. Additionally, the reflexes belonging to Cu_2O and Cu were identified in the XRD pattern of CuZn-350N as well (Figure 2). Because of the significant overlapping of the XRD reflexes of the different phases, only several characteristic reflexes are shown in Figure 2.

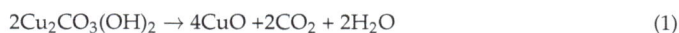

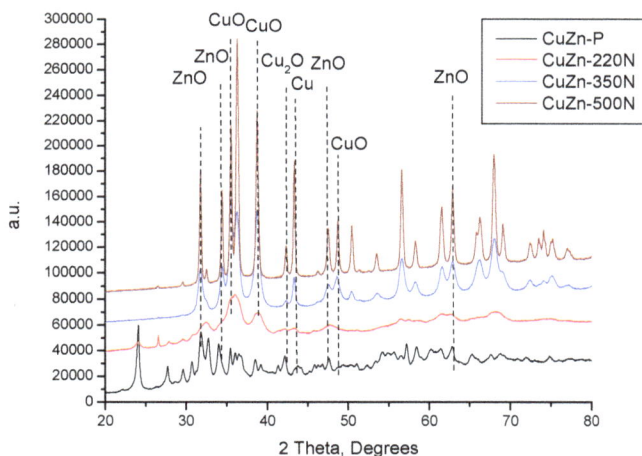

$$2Cu_2CO_3(OH)_2 \rightarrow 4CuO + 2CO_2 + 2H_2O \tag{1}$$

$$2CO_2 \rightarrow 2CO + O_2 \tag{2}$$

$$4CuO + 2CO \rightarrow 2Cu_2O + 2CO_2 \tag{3}$$

Figure 2. The XRD patterns of catalysts prepared by the treatment of the CuZn precursor in flowing nitrogen at 220, 350, and 500 °C.

The exclusive formation of CuO resulting from the calcination of malachite in air (Equation (1)) and the possible formation of Cu_2O during the calcination of malachite in an inert atmosphere was reported previously [15]. It was explained by CO_2 decomposition to CO and subsequent reduction of CuO by CO (Equations (2) and (3)) [16]. The decomposition of both zincian malachite, and aurichalcite in nitrogen, by analogy with malachite, can be expected to yield Cu_2O in addition to CuO. If CuO is reduced by CO originating from CO_2 to Cu_2O, the consecutive reduction step, i.e., the reduction of Cu_2O to Cu, is plausible as well. A further increase in the treatment temperature to 500 °C did not change the phase composition of the catalyst anymore, but the intensity of reflexes of all phases increased substantially due to the increased crystallite size of all phases. With a closer inspection of Figure 2, it can be established that the intensity of the reflexes corresponding to Cu_2O and Cu increased more than those belonging to CuO and ZnO (Figure 2).

Finally, the catalyst precursor was also treated in flowing hydrogen at 220, 350, and 500 °C. The resulting XRD patterns are shown in Figure 3 together with the XRD pattern of the catalyst precursor. In contrast to the treatment in air and nitrogen at 220 °C, there are no reflexes of the initial hydroxycarbonate phases in the XRD pattern of the CuZn-220H, i.e., a catalyst obtained by treatment in a hydrogen atmosphere at 220 °C. Instead, low-intensive reflexes of ZnO, CuO, Cu_2O, and Cu appeared (Figure 3). As the treatment temperature increased from 220 °C to 500 °C, the intensity of the reflexes of the oxidic Cu phases declined while that of the metallic copper substantially increased (Figure 3). Simultaneously, the intensity of the reflexes of ZnO also increased. This can be attributed to the increase in the crystallite size of ZnO and Cu (discussed below) and to the gradual reduction of the oxidic Cu phases to metallic Cu.

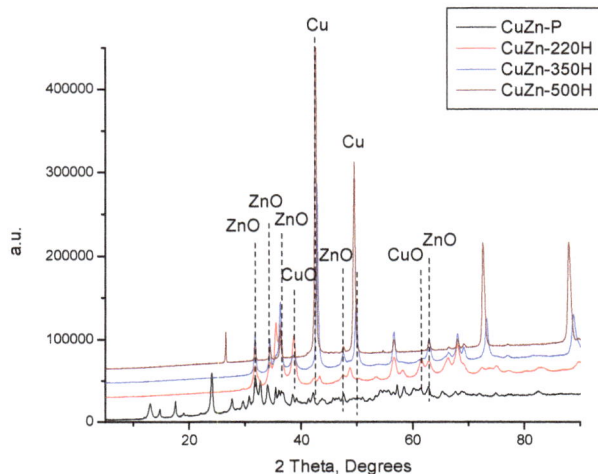

Figure 3. The XRD patterns of catalysts prepared by the treatment of the CuZn precursor in flowing hydrogen at 220, 350, and 500 °C.

The XRD results demonstrate that 220 °C is a sufficient temperature to completely decompose hydroxycarbonates (zincian malachite and aurichalcite) in a hydrogen atmosphere, but not in air or nitrogen atmospheres. The lowered thermal stability of both hydroxycarbonate phases may be a consequence of a direct reduction of the hydroxycarbonate Cu species in zincian malachite and/or aurichalcite structures according to the proposed overall reaction (Equation (4)). This conclusion is further corroborated by the observed partial destruction of the hydroxycarbonate phase under nitrogen at 220 °C. The partial destruction is caused by the transformation of the inert atmosphere into a partially reducing atmosphere due to the formation of CO during the treatment.

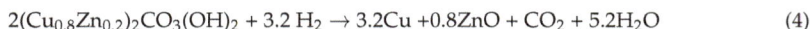

$$2(Cu_{0.8}Zn_{0.2})_2CO_3(OH)_2 + 3.2\ H_2 \rightarrow 3.2Cu + 0.8ZnO + CO_2 + 5.2H_2O \tag{4}$$

Nevertheless, Figure 3 also evidences that CuZn-220H still contained CuO in its composition after the calcination step (it can be considered as a reduction step at the same time), which is another distinctive feature of the experiment performed in hydrogen at 220 °C. This is rather unexpected since the reduction of CuO to metallic copper in CuO/γ-Al_2O_3 [17] or in CuO/ZnO [11] occurred at temperatures around 200 °C. On the other hand, the reduction of Cu^{2+} in bulky CuO was reported to occur at 300 °C [11]. It was proposed that the low-temperature reduction of CuO to metallic Cu took place for highly dispersed copper oxide species while the bulky CuO was reduced at higher temperatures [15,17]. The presence of oxidic Cu species in CuZn-220H allows for the assumption that the treatment conditions considerably influenced the properties (oxidation state, dispersion, etc.) of the Cu atoms in the resulting catalyst sample. Indeed, dedicated experiments showed (Figure A1) that the treatment of the CuZn-350AS sample (which is a mixture of ZnO and CuO phases; see Figure 1) in a hydrogen atmosphere at 220 °C resulted in the total reduction of copper oxide to metallic copper. Moreover, the exclusive presence of the reflexes of metallic copper, i.e., without any oxidic phases detected suggests that these metallic domains are not susceptible to an easy re-oxidation in the air. In contrast, the treatment of CuZn-P firstly in the air at 220 °C (yielding the CuZn-220AS sample) and then in hydrogen at the same temperature (Figure A1) resulted in the decomposition of zincian malachite and aurichalcite structures. The XRD pattern of CuZn-220AS-220H contained reflexes of CuO together with reflexes of metallic Cu at very low intensities. Thus, the existence of oxidic copper domains in the sample after a reductive treatment at 220 °C is undoubtedly a consequence of the sample preparation under specific conditions and their sequence.

The collected XRD patterns allowed for the calculation of the crystallite size of all phases detected in the catalysts using the Scherrer equation [18]. The sizes of the crystallites of zinc and copper compounds were calculated from the width of the corresponding diffraction lines that did not overlap with the lines belonging to other phases. As expected, the crystallite size depended on the pretreatment temperature and, more interestingly, on the calcination atmosphere (Table 1).

Table 1. The crystallite size of ZnO, CuO, Cu_2O, and Cu present in the catalysts prepared by calcination in different atmospheres at different temperatures.

Sample	Particle Size, nm (2 Theta, Degrees)			
	ZnO (31.7)	CuO (38.7)	Cu_2O (42.3)	Cu (43.3)
220AS	ND [1]	ND	ND	ND
220AS-220H [2]	7.3	7.3	8.7	10.2
350AS	5.1	7.7	ND	ND
350AS-220H [2]	12.5	ND	ND	18.1
500AS	15.8	11.9	ND	ND
220N	5.9	6.5	7.9	11.5
350N	11	10.3	21	22.2
500N	33.9	25.5	36	46.1
220H	9.8	9.2	11.2	18.1
350H	20	16.7	ND	23.4
500H	34.7	ND	ND	33.2

[1] ND stands for not detected; [2] After the reduction at 220 °C.

Due to the increasing pretreatment temperature in static air from 350 °C to 500 °C, the size of the ZnO and CuO crystallites increased from 5.1 nm and 7.7 nm to 15.8 and 11.9 nm, correspondingly. When the pretreatment was performed in a non-oxidizing atmosphere, either in nitrogen or in hydrogen, the size of the produced oxidic particles was larger compared to the corresponding experiments in static air (Table 1). Moreover, at a given temperature, the size of ZnO crystallites was larger in the catalysts prepared by calcination in nitrogen or hydrogen in comparison with those prepared in air (Table 1). This allows for the suggestion of a mutual stabilization of the CuO and ZnO domains if these are in intimate contact with each other. This conclusion is further supported by the larger size of the Cu crystallites obtained by direct reduction (calcination in hydrogen) than those obtained by calcination in air followed by reduction at 220 °C (Table 1). In addition, the size of metallic copper increased in comparison with the size of its oxidic precursors which indicates copper sintering during the reduction step. It is interesting to note that (i) the catalysts reduced after previous calcination in air, (ii) catalysts obtained by calcination in nitrogen (copper formed due to reduction by in-situ formed CO) and (iii) catalysts prepared by calcination in hydrogen (i.e., by direct reduction) exhibit a common linear dependency between the CuO crystallite size and the resulting Cu crystallite size (Figure 4). This might indicate that Cu is formed during the calcination of the hydroxycarbonate precursors either via CuO or that the formation of both crystallites is related in another way.

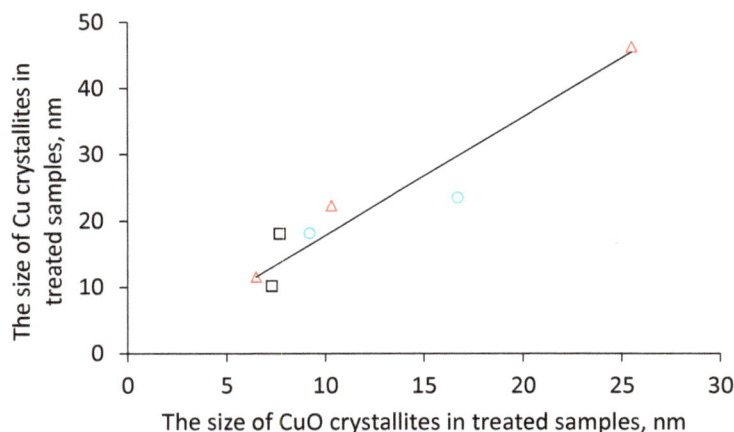

Figure 4. A relationship between the CuO and Cu crystallites size in catalyst samples prepared by calcination in air (□), nitrogen (△) and hydrogen ( ) at different temperatures.

The calcination atmosphere and temperature significantly affected the textural properties of the final catalysts. The results are summarized in Table 2. Among the samples calcined in air, CuZn-350AS possessed the best textural properties with respect to the BET surface, the external surface, pore volume, and average nanoparticle size. The CuZn-220AS had only slightly worse parameters compared to CuZn-350AS despite their substantially different phase composition. This could be attributed to the fact that CuZn-220AS was a mixture of hydroxycarbonates rather than of oxides. The increase in the calcination temperature to 500 °C resulted in a dramatic decrease in the textural properties of CuZn-500AS. On the other hand, the trend in the change in all characteristics observed for the samples prepared in nitrogen and hydrogen is more unambiguous. The BET surface, the external surface, and the pore volume substantially decreased while the average nanoparticle size, accordingly, increased with the increasing calcination temperature. It should also be noted that the samples prepared in hydrogen had worse adsorption characteristics, except for the micropore surface, compared with the similar samples prepared in nitrogen. Besides, Table 2 evidences that the pore diameter of differently prepared catalysts was approximately the same in the range of 13–39 nm and showed no dependence on either the atmosphere or calcination temperature.

Table 2. The textural properties of the CuZn catalysts prepared by the calcination of the CuZn precursor in air, nitrogen, and hydrogen at 220, 350, and 500 °C.

Sample	BET Surface, $m^2 \cdot g^{-1}$	t-Plot Surface, $m^2 \cdot g^{-1}$		BJH Pore Volume, $cm^3 \cdot g^{-1}$	BJH Pore Diameter, nm	Average Nanoparticle Size, nm
		Micropore	External			
CuZn-220AS	49	5.1	43.9	0.342	27	122
CuZn-350AS	62	1.3	61.1	0.463	25	96
CuZn-500AS	17	0.1	16.5	0.084	23	364
CuZn-220N	65	0	78.3	0.4	21	89
CuZn-350N	32	4.7	27.6	0.274	39	186
CuZn-500N	8	0.8	7.5	0.021	13	722
CuZn-220H	26	3.1	22.7	0.194	31	233
CuZn-350H	14	0.6	13.5	0.089	27	423
CuZn-500H	NM [1]					

[1] NM stands for not measured.

2.2. Activity and Selectivity of the Catalysts

Figure 5 depicts the dependency of the DMA conversion on the pretreatment conditions (i.e., different calcination atmosphere and temperature) applied to the synthesized CuZn-P precursor. CuZn-220AS prepared by a mild heating in static air possessed a reasonable activity in the reaction with DMA; the DMA conversion reached 64% after 2 h. When the calcination temperature was increased to 350 °C, the catalytic performance of the catalyst further improved; the DMA conversion amounted to 78% after 2 h (Figure 5). The observed increase is in agreement with the improvement in the textural properties of CuZn-220AS and CuZn-350 AS (Table 2). Additionally, the phase composition of the calcined catalysts after reduction has to be considered as well. The catalyst CuZn-220AS, if exposed to hydrogen at 220 °C for 2 h, still contained Cu in the oxidic state (Figure 4, sample CuZn-220AS-220H). On the other hand, the XRD pattern of this catalyst after the reaction (will be discussed further) revealed the presence of only metallic Cu clusters. This suggests the occurrence of a subsequent reduction of the oxidic Cu species during the reaction. In contrast, the catalyst CuZn-350AS exclusively contained metallic Cu clusters after the reduction step, so no additional activation during the reaction was necessary. This difference in the state of Cu in the catalysts, although inexplicable at the moment, could explain the difference in the catalytic performance of CuZn-220AS and CuZn-350AS together with their different textural properties. A further increase in the calcination temperature in air to 500 °C resulted in a decrease in the activity of the catalyst CuZn-500AS with the DMA conversion of 43% after 2 h (Figure 5).

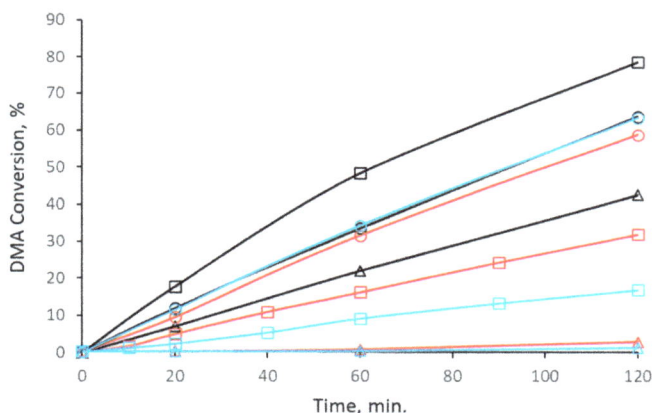

Figure 5. The DMA conversion observed on the CuZn catalysts in dependence on the temperature of their treatment in air (black line), N_2 (red line), and H_2 (blue line). Samples calcined at T = 220 °C (○), 350 °C (□), 500 °C (Δ).

The observed decrease in conversion can be attributed to the increase in the size of the copper crystallites determined from the XRD patterns (Table 1). The increase in the calcination temperature caused an increase in the size of copper crystallites leading to a decrease in the number of available catalytically active centers. A similar dependence of the catalytic performance of Cu-containing catalysts on the particle size of copper being influenced by the calcination temperature was also reported for the hydrogenolysis of ethyl acetate to ethanol [4]. In contrast, Peng Yuan et al. [5] observed an increase in the DMA conversion over a Cu–Zn–Al catalyst as the calcination temperature was increased from 400 °C to 500 °C.

The catalysts prepared by calcination in nitrogen and hydrogen also showed a trend of decreasing the DMA conversion with the increasing calcination temperature. For CuZn catalysts prepared by calcination in nitrogen, the DMA dropped significantly (from 59 to 3%) when the calcination temperature was increased from 220 °C to 500 °C (Figure 5). The effect of the hydrogen calcination atmosphere was even more pronounced. The DMA conversion decreased from 63% to a negligible 1% when the calcination temperature was increased from 220 °C to 500 °C. The physicochemical characterization of the catalysts allows the attributing of this behavior to the increase in the Cu crystallite size with increasing calcination temperature and to the worse textural properties of catalysts prepared by calcination at high temperatures. The dependence of the conversion on crystallite size is obvious when the DMA conversion is plotted as a function of CuO or Cu crystallites (Figures 6 and A2).

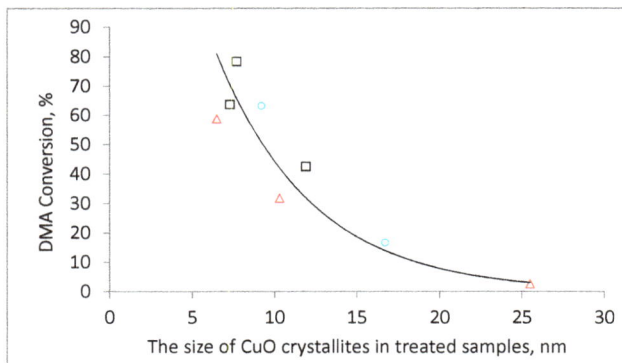

Figure 6. A correlation between the DMA conversion (after 120 min. of the reaction) and the size of CuO crystallites in the prepared CuZn catalysts. □: calcined in air, △: calcined in nitrogen, ○: calcined in hydrogen.

The observed correlation between the catalytic performance of CuZn catalysts prepared in the present study and the size of the Cu domains can be expected since the performance of the Cu-containing catalysts in different reactions has been repeatedly shown to depend on copper particle size [5,11,17]. Nevertheless, the treatment of the starting precursor in air, on the one hand, and in nitrogen or hydrogen, on the other hand, introduces some unique (previously not described) features to the performance of the CuZn hydrogenolysis catalyst. Regardless of the atmosphere used, all catalysts which were thermally treated at 220 °C exhibited approximately the same catalytic activity. The DMA conversion was in the range 59–64% (Figure 5) despite their phase composition after calcination differing significantly (Table 1). Moreover, the comparison of catalytic results with the physicochemical properties of the prepared catalysts exhibits no clear correlation between the catalytic performance of the catalysts and their textural properties.

On the other hand, catalytic results suggest that the high DMA conversion over catalysts prepared at 220 °C is observed provided that the samples possess small-size copper crystallites (Table 1). An increase in the calcination temperature to 350 °C resulted in an increase in the DMA conversion to 78% for CuZn-350AS while the activity of CuZn-350N and CuZn350H markedly decreased. According to XRD, CuZn-350AS represents a mixture of ZnO and CuO whereas the other samples contain notable amounts of Cu_2O and Cu besides CuO and ZnO. The crystallite size of Cu_2O and Cu is significantly larger than that of CuO and ZnO and they are plausibly present as separate phases that are not in intimate contact with the ZnO phase. This difference in the phase composition of the samples allows for the suggestion that an intimate contact between the CuO and ZnO crystallites because of the thermal treatment could be an important factor that stabilizes the small size of the Cu domains. Consequently, a high activity of CuZn catalysts is achieved. When calcining in air, the increase in the calcination temperature from 220 °C to 350 °C resulted in a complete destruction of the zincian malachite and aurichalcite structures. This led to an increased number of CuO and ZnO crystallites, thus, promoting the probability of close contact between the two phases. Additionally, the very small size of ZnO crystallites (Table 1) could also contribute to the high dispersion of the CuO domains. In contrast, the presence of the Cu_2O and Cu crystallites in samples prepared in N_2 and H_2 reduced the probability of the necessary interaction between CuO and ZnO. As a result, the number of potentially active sites formed on the interface between the CuO and ZnO domains after the reduction differs for catalysts prepared by the calcination in different atmospheres. Additionally, the existence of the said interfaces contributes to the stabilization of the Cu crystallites in ZnO surroundings. As a result, CuZn-350AS has a higher BET surface area, external surface, and pore volume with lower crystallites size values than CuZn-350N and CuZn-350H. As can be deduced from Table 1, the absence of intimate CuO-ZnO contact resulted in a more facile sintering of both the Cu and Zn domains at elevated temperatures. Consequently, the DMA conversion over CuZn-350AS was the highest among all catalysts (78%), while the DMA conversion observed over CuZn-350N and CuZn-350H was much lower, i.e., 32% and 17%, respectively. The significantly lower conversion over CuZn-350H in comparison with CuZn-350N is the consequence of a more extensive reduction of CuZn-350H during calcination and, hence, the lower probability to stabilize the CuO domains by intimate contact with ZnO. Consequently, the CuO and ZnO crystallites are approximately 17–20 nm large in CuZn-350H whereas only 10–11 nm in CuZn-350N (Table 1).

As reported previously [6], CuZn catalysts catalyze not only hydrogenolysis, but higher-molecular-weight products originating from transesterification reactions between DMA, 1,6-hexanediol (1,6-HDO), and partially hydrogenolyzed DMA (1HMEol) are also formed. These transesterification products consist of products with two C_6 backbones in their molecules (denoted H2) or with three C_6 backbones in their molecules (denoted H3). Thus, the selectivity to and yields of different products allow for obtaining additional important information on the performance of catalysts particularly with respect to the origins of the transesterification activity of the CuZn catalysts. Figure 7 depicts the selectivity towards reaction products formed by DMA hydrogenolysis over catalysts prepared by the thermal activation of CuZn-P in air. Regardless, the calcination temperature selectivity to the reaction products observed over the three CuZn-*T*AS catalysts followed the same trend as a function of DMA conversion. To maintain readability of the graph, the selectivities over all CuZn-*T*AS catalysts were not distinguished according to the calcination temperature (Figure 7).

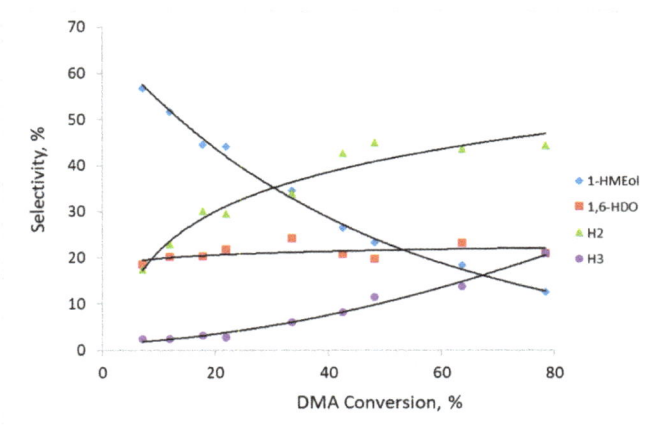

Figure 7. The change of selectivity to the reaction products in dependence on the DMA conversion observed over the catalysts prepared by the calcination of the precursor, CuZn-P, in the air at 220, 350, and 500 °C. 1HMEol is methyl 6-hydroxyhexanoate; 1,6-HDO is hexane-1,6-diol; H2 is the sum of 6-methoxy-6-oxohexyl methyl adipate, 6-hydroxyhexyl methyl adipate, and 6-hydroxyhexyl 6-hydroxyhexanoate; and H3 is the sum of bis(6-methoxy-6-oxohexyl) adipate, 6-hydroxyhexyl (6-methoxy-6-oxohexyl) adipate, and bis(6-hydroxyhexyl) adipate.

The composition of the main reaction products and their tendency to change with the increasing DMA conversion is consistent with the results from our previous work [6]. 1HMEol is evidently a primary product formed by the hydrogenolysis of only one ester group of the two present in the DMA. Its selectivity approached 100% at DMA conversion close to zero. Subsequently, 1HMEol can be transformed either by the hydrogenolysis of the second ester group yielding 1,6-HDO or by participating in transesterification reactions yielding bulky H2 and H3 molecules (for structures and chemical names, see Scheme 1). The hydrogenolysis of the second ester group in the DMA, i.e., the ester group in 1HMEol, also proceeded very fast on the catalysts prepared by calcination in air. Consequently, the selectivity to 1,6-HDO reached 20% at a DMA conversion as low as 12%. However, once 1HMEol and 1,6-HDO were formed, the transesterification reactions involving DMA, 1HMEol, and/or 1,6-HDO became possible and resulted in a formation of bulky compounds with either two or three "hexane backbones" in their molecule, denoted as H2 and H3 (for structures, see Scheme 1). During the reaction, the formation of the heavy H2 and H3 compounds became prevalent. Therefore, at DMA conversion of about 80%, the selectivity to H2 and H3 attained 44% and 21%, respectively. In addition, other reaction products, including cyclic ones (due to intramolecular transesterification), have also been identified among the reaction products. Their amount was small and they were excluded from further consideration. It can be inferred that the transesterification reactions are preferred to the second ester group hydrogenolysis under the studied reaction conditions as the selectivity to 1,6-HDO increased at a much more moderate rate with the increasing DMA conversion than the selectivity to the H2 and H3 products.

It has been reported that the hydrogenolysis of both ester functions in the DMA molecule does not occur at the same rate [5,19]. Hence, a ratio between the selectivities to HDO and the sum of the H2 and H3 products can help in evaluating the contribution of different reaction routes to the total composition of the reaction products. The ratio S_{HDO}/S_{H2+H3} observed for CuZn catalysts calcined in air varied only slightly for the three calcination temperatures; it was in the range of 0.61–0.94 at a DMA conversion ≈10%. This shows that the ratio between the active sites responsible for hydrogenolysis and transesterification, respectively, is approximately the same in these catalysts and does not change significantly due to different calcination temperature in air.

In the case of catalysts prepared by calcination in nitrogen and hydrogen at 220 °C (CuZn-220N and CuZn-220H, respectively), the maximum yield of 1HMEol reached ca. 10–12% at a DMA conversion of ca. 40%. The same maximum yield of 1HMEol was also obtained for all three catalysts prepared by calcination in air (Figure 8(1A,2A,3A)), suggesting that all catalysts prepared by calcination at 220 °C have a similar distribution of the Cu active sites regardless of the calcination atmosphere. Consequently, the S_{HDO}/S_{H2+H3} calculated for CuZn-220N and CuZn-220H at DMA conversion $\approx$10% was 0.77 and 0.7, correspondingly, which is within the range of values observed for the catalysts prepared by the calcination in air. Similar to the air-calcined CuZn catalysts, the selectivity to H2 and H3 increased markedly with the increase in the DMA conversion for CuZn-220N and CuZn-220H catalysts, implying the secondary nature of H2 and H3 formation.

A different selectivity character was observed in the case of samples prepared by the calcination of the CuZn-P precursor at 350 and 500 °C in nitrogen or hydrogen. The yields of the 1HMEol were minor—the maximum yield of 1HMEol was achieved over CuZn-350N and amounted to 2% at a DMA conversion of 30%. This is significantly less than the samples calcined at 220 °C in nitrogen, hydrogen, or in air (ca. 10–12% yield at a 30% DMA conversion). The 1,6-HDO yields were virtually 0%, which is again dramatically lower than the other catalysts that yielded 4–6% HDO at comparable conversions (Figure 8(1B,2B,3B)). Surprisingly, the esters (H2 and H3 products) requiring 1HMEol and/or HDO in addition to DMA were the prevailing reaction products. At a given conversion, their yield significantly exceeded the yield of esters formed over the CuZn catalysts calcined at 220 °C (in all atmospheres) and all catalysts calcined in air (i.e., calcined at 220, 350, or 500 °C); see Figure 8(1C,2C,3C,1D,2D,3D). It can be inferred that the calcination temperature in a non-oxidizing atmosphere had a severe impact on the hydrogenolysis activity of the CuZn catalysts, as the S_{HDO}/S_{H2+H3} dropped to values as low as 0.002 for catalysts calcined at 350 °C in N_2 or H_2. The results, thus, provide strong evidence that the hydrogenolysis of the methyl ester function and the transesterification reaction leading to the formation of bulky molecules, H2 and H3, occur on different active sites. Obviously, both the calcination atmosphere and temperature strongly affect the performance of the CuZn catalysts. The identification (and possibly also quantification) of the active sites, however, remains a task for forthcoming studies.

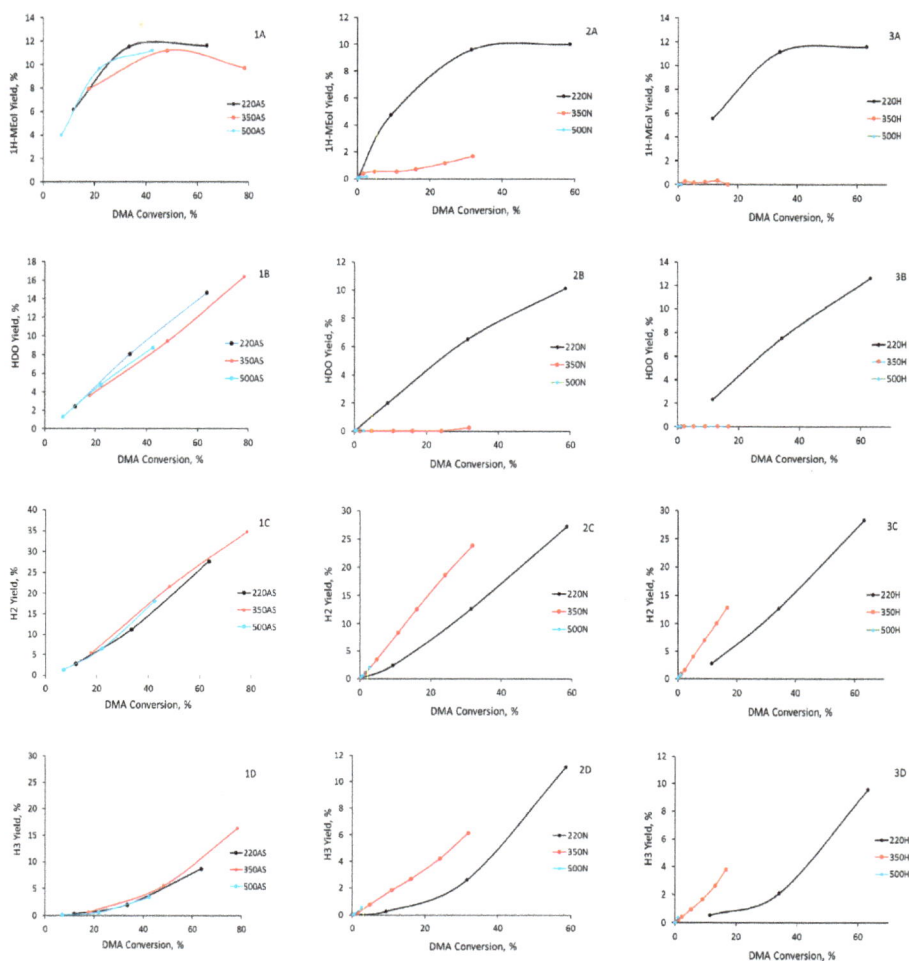

Figure 8. The change of the yield of reaction products (**A**: 1HMEol, **B**: HDO, **C**: H2, **D**: H3) with a dependence on the DMA conversion observed over the samples prepared by the thermal treatment of CuZn-P in air (**1**), nitrogen (**2**) and hydrogen (**3**).

3. Materials and Methods

A CuZn precursor with the copper/zinc molar ratio of 1.6 was prepared by a co-precipitation method based on the neutralization of a starting acidic solution with a basic precipitating agent. For the synthesis of the precursor, a glass beaker was filled with 200 mL of distilled water and heated to T = 60 °C under stirring at 600 RPM. An aqueous solution (0.5 M) containing $Cu(NO_3)_2 \cdot 3H_2O$ (99.0%, Penta, s.r.o., Prague, Czech Republic), $Zn(NO_3)_2 \cdot 6H_2O$ (99.6%, Lach:Ner, s.r.o., Neratovice, Czech Republic), and an aqueous solution (0.5 M) of Na_2CO_3 (99.4%, Lach:Ner, s.r.o., Neratovice, Czech Republic) used as the precipitant were dosed simultaneously dropwise to keep the pH value in the beaker in the range of 7 ± 0.1. The temperature and stirring rate were kept constant at 60 °C and 600 RPM, correspondingly, during the precipitation. The resulting suspension was aged under stirring for 90 min which caused an increase in the pH value to 7.6–7.7. The prepared precipitate was then filtered using a vacuum pump, washed with plenty of distilled water, and finally dried at 70 °C overnight.

To prepare a series of catalysts for the investigations, the starting CuZn precursor denoted as CuZn-P (fine powder, 1 g used for each experiment) was thermally treated either in an oven in static open air or in a Parr autoclave in a flow of either nitrogen (99.99% SIAD Czech, s.r.o., Prague, Czech Republic) or hydrogen (99.9% SIAD Czech, s.r.o, Prague, Czech Republic). The thermal activation of the precursor in each of the gas atmospheres was performed at 220, 350, and 500 °C for 2 h. The prepared samples were denoted as CuZn-TAS, CuZn-TN, and CuZn-TH, where T stands for the temperature of the heat treatment (calcination) while AS, N, and H stand for static air, flowing nitrogen and hydrogen used as a gas atmosphere, respectively. For example, CuZn-350AS denotes a catalyst prepared by the calcination of the precursor, CuZn-P, at 350 °C in an air atmosphere.

The copper and zinc content in the as-prepared CuZn-P sample were determined by XRF using a spectrometer ARL 9400 XP equipped with a rhodium lamp (Thermo ARL, Ecublens, Switzerland). The phase composition of the CuZn catalysts and the size of the crystallites of the relevant phases present in the catalysts were determined by X-Ray diffraction using a diffractometer PANanalytical X'Pert3 Powder and Cu Kα radiation (PANanalytical, Eindhoven, The Netherlands). The XRD patterns were recorded in a range of $2\theta = 5$–$90°$. The particle sizes were calculated using Scherrer's equation. Reflections at $2\theta = 31.8°$, $36.2°$, and $38.6°$ were used for the particle size calculations of the ZnO, Cu, and CuO crystallites, respectively. Equilibrium adsorption isotherms of nitrogen were measured at 77 K using a static volumetric adsorption system (TriFlex analyzer, Micromeritics, Norcross, GA, USA). The samples were degassed at 473 K (12 h) prior to an N_2 adsorption analysis, in order to obtain a clean surface. The adsorption isotherms were fitted using the Brunauer-Emmett-Teller (BET) method for the specific surface area, the t-plot method for the external specific surface area, and the BJH method for the distribution of mesopores.

The hydrogenolysis of dimethyl adipate (>99%, Sigma-Aldrich, s.r.o., Prague, Czech Republic) was carried out in a Parr SS autoclave with a reactor volume of 300 mL using the prepared CuZn catalysts. Prior to the catalytic tests, all catalysts (1 g) were reduced in situ in the autoclave at 220 °C using flowing hydrogen (H_2 99.9%, SIAD Czech, s.r.o., Prague, Czech Republic) at atmospheric pressure for 1 h. After the catalyst reduction, the temperature in the autoclave was decreased to below 100 °C and the reactor was loaded with 50 mL of dimethyl adipate (DMA) by a peristaltic pump (i.e., the autoclave with the reduced catalyst was not opened to avoid reoxidation of the catalyst). After that, the temperature in the autoclave was gradually increased to $T = 210$ °C. Then, the hydrogen pressure was increased to 10 MPa and the catalytic experiment was initiated by starting the stirring of the reaction mixture (DMA with a catalyst) at 600 RPM. Liquid reaction products were periodically withdrawn from the autoclave, centrifuged, and analyzed by a GC (Agilent 6890N, HPST, s.r.o., Prague, Czech Republic) equipped with a FID and non-polar column (ULTRA 1, 0.32 mm internal diameter, 20 m length). The DMA conversion was calculated according to Equation (5). Due to the absence of cracking reactions, the C6 backbone present in the DMA molecule, as well as in the reaction products, was used to calculate the product selectivity (Equation (6)). Methanol (also a reaction product) was excluded from these calculations.

$$conversion_{DMA} = \frac{n_{DMA,i} - n_{DMA,t}}{n_{DMA,i}} \tag{5}$$

where n_{DMA} is the number of DMA moles in the reaction mixture, where i stands for the initial number of DMA moles and t stands for the number of DMA moles at the sampling time.

$$S_{x,t} = \frac{n_{x,t} - n_{x,i}}{\sum n_{products\ 1C6\ backbone} + 2 \sum n_{products\ 2C6\ backbones} + 3 \sum n_{products\ 3C6\ backbones}} \tag{6}$$

where $S_{x,t}$ is the selectivity to product x at time t of the experiment, n_x is the number of moles of product x in the starting/initial mixture (index i) or at the time of sampling (index t), $n_{products\ yC6\ backbone}$ refers to the number of product moles having one ($y = 1$), two ($y = 2$) or three ($y = 3$) C_6 backbones in their molecules.

4. Conclusions

The results from the present study show that starting from the same CuZn hydroxycarbonate precursor, it is possible to prepare samples greatly varying in physicochemical properties and catalytic performance. This can be achieved by the thermal treatment of the precursor using different atmospheres (air, nitrogen, and hydrogen) and calcination temperatures (220, 350, and 500 °C). Both calcination parameters affected the resulting phase composition of the catalysts as well as their crystallite size. In an oxidizing atmosphere, a mixture of CuO and ZnO was produced with increasing crystallite size as the calcination temperature grew. In contrast, in inert or reducing atmospheres, both oxides were accompanied by Cu_2O and Cu. The crystallite sizes of Cu_2O and Cu were larger than the sizes of CuO and ZnO, thus indicating a less intimate contact between the Cu-phases and ZnO in catalysts calcined in nitrogen and hydrogen. All three catalysts calcined in different atmospheres at $T = 220$ °C demonstrated good activity, resulting in a DMA conversion of 59–63%. The increase in the calcination temperature to 350 °C resulted in a growth of the DMA conversion to 78%, observed for the sample prepared in the oxidizing atmosphere. This is explained by the small crystallite size and intimate contact between the CuO and ZnO phases prior to the catalyst reduction. In contrast, the DMA conversion notably decreased over the samples prepared by the calcination in nitrogen or hydrogen at higher temperatures. This is explained by the decreased number of potentially active sites formed on the interfaces between the CuO and ZnO domains after the reduction and the lower stabilization of Cu crystallites in the ZnO surrounding. As a result, the DMA conversion decreased to a negligible 1% observed for samples calcined in N_2 and H_2 at 500 °C.

Calcination parameters used for the catalyst preparation greatly influenced not only DMA conversion but also product selectivity. The reaction over the air-calcined samples yielded 1,6-hexanediol with selectivity exceeding 20% at DMA conversion >10% independent of the calcination temperature. In contrast, the catalysts calcined in nitrogen or hydrogen at 350 and 500 °C exhibited selectivity to 1,6-hexanediol below 2% at DMA conversion ranging from 1% to 30%. Transesterification products (demoted H2 and H3) were the main reaction products with overall selectivities of >80% in these cases. The observed change in the catalytic performance of samples prepared by calcination in N_2 and H_2 allowed for suggesting that different active sites are responsible for the hydrogenolysis and transesterification reaction pathways and that the relative distribution of these sites has changed in dependence on the calcination procedure applied to the starting CuZn precursor.

Author Contributions: Conceptualization, Methodology and Writing—original draft preparation, O.K. and D.K.; Investigation—experiments and analyses, O.K, J.A., V.P. and M.L.; Writing—review and editing, J.A., V.P. and D.K.

Funding: This research was funded by the Czech Science Foundation (project No. GA17-05704S).

Acknowledgments: This work was realized within the Operational Programme Prague—Competitiveness (CZ.2.16/3.1.00/24501) and "National Program of Sustainability" (NPU I LO1613) MSMT-43760/2015.

Conflicts of Interest: The authors declare no conflict of interest.

Appendix A

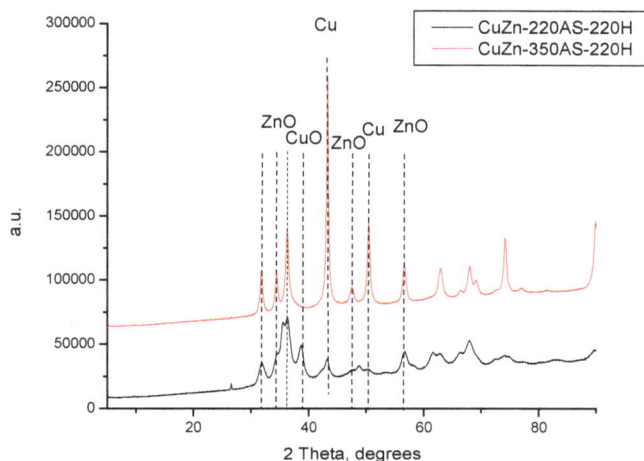

Figure A1. The XRD patterns of catalyst samples prepared by the treatment of the CuZn precursor in static air at 220 and 350 °C followed by the treatment in a hydrogen at T = 220 °C.

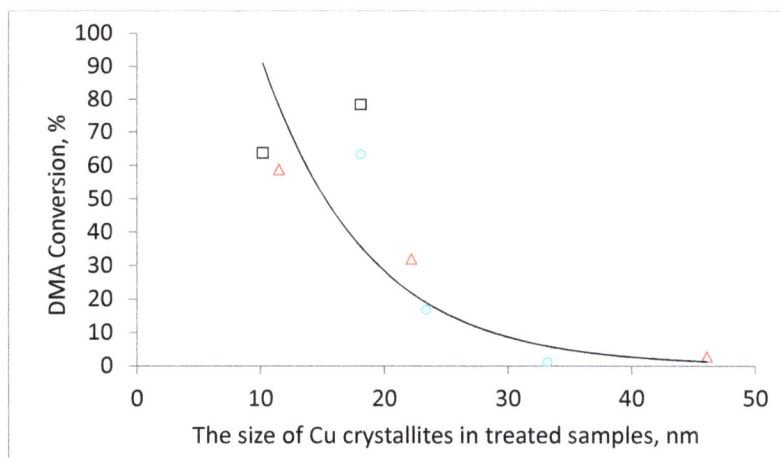

Figure A2. A correlation between the DMA conversion and the size of the Cu crystallites in the prepared CuZn catalysts. □: calcined in air, △: calcined in nitrogen,   : calcined in hydrogen.

References

1. Adkins, H.; Folkers, K. The catalytic hydrogenation of esters to alcohols. *J. Am. Chem. Soc.* **1931**, *53*, 1095–1097. [CrossRef]
2. Turek, T.; Trimm, D.L. The catalytic hydrogenolysis of esters to alcohols. *Catal. Rev.-Sci. Eng.* **1994**, *36*, 645–683. [CrossRef]
3. Adkins, H. Catalytic hydrogenation of esters to alcohols. *Org. React.* **1954**, *8*, 1–27.
4. Zhu, Y.-M.; Shi, L. Zn promoted Cu–Al catalyst for hydrogenation of ethyl acetate to alcohol. *J. Ind. Eng. Chem.* **2014**, *20*, 2341–2347. [CrossRef]

5. Yuan, P.; Liu, Z.; Hu, T.; Sun, H.; Liu, S. Highly efficient Cu–Zn–Al catalyst for the hydrogenation of dimethyl adipate to 1,6-hexanediol: Influence of calcination temperature. *React. Kinet. Mech. Catal.* **2010**, *100*, 427–439. [CrossRef]

6. Kubička, D.; Aubrecht, J.; Pospelova, V.; Tomášek, J.; Šimáček, P.; Kikhtyanin, O. On the importance of transesterification by-products during hydrogenolysis of dimethyl adipate to hexanediol. *Catal. Commun.* **2018**, *111*, 16–20. [CrossRef]

7. Zander, S.; Kunkes, E.L.; Schuster, M.E.; Schumann, J.; Weinberg, G.; Teschner, D.; Jacobsen, N.; Schloegl, R.; Behrens, M. The Role of the Oxide Component in the Development of Copper Composite Catalysts for Methanol Synthesis. *Angew. Chem. Int. Ed.* **2013**, *52*, 6536–6540. [CrossRef] [PubMed]

8. Behrens, M.; Studt, F.; Kasatkin, I.; Kuehl, S.; Haevecker, M.; Abild-Pedersen, F.; Zander, S.; Girgsdies, F.; Kurr, P.; Kniep, B.-L.; et al. The Active Site of Methanol Synthesis over $Cu/ZnO/Al_2O_3$ Industrial Catalysts. *Science* **2012**, *336*, 893–897. [CrossRef] [PubMed]

9. Behrens, M.; Kissner, S.; Girsgdies, F.; Kasatkin, I.; Hermerschmidt, F.; Mette, K.; Ruland, H.; Muhler, M.; Schloegl, R. Knowledge-based development of a nitrate-free synthesis route for Cu/ZnO methanol synthesis catalysts via formate precursors. *Chem. Commun.* **2011**, *47*, 1701–1703. [CrossRef] [PubMed]

10. Fujita, S.-I.; Moribe, S.; Kanamori, Y.; Kakudate, M.; Takezawa, N. Preparation of a coprecipitated Cu/ZnO catalyst for the methanol synthesis from CO_2—Effects of the calcination and reduction conditions on the catalytic performance. *Appl. Catal. A* **2001**, *207*, 121–128. [CrossRef]

11. Natesakhawat, S.; Lekse, J.W.; Baltrus, J.P.; Ohodnicki, P.R.; Howard, B.H.; Deng, X.; Matranga, C. Active Sites and Structure-activity Relationships of Copper based Catalysts for Carbon Dioxide Hydrogenation to Methanol. *ACS Catal.* **2012**, *2*, 1667–1676. [CrossRef]

12. Wang, S.; Liu, H. Selective hydrogenolysis of glycerol to propylene glycol on Cu–ZnO catalysts. *Catal. Lett.* **2007**, *117*, 62–67. [CrossRef]

13. Hu, Q.; Fan, G.; Yang, L.; Li, F. Aluminum-doped zirconia-supported copper nanocatalysts: Surface synergistic catalytic effects in the gas-phase hydrogenation of esters. *Chemcatchem* **2014**, *6*, 3501–3510. [CrossRef]

14. Schumann, J.; Tarasov, A.; Thomas, N.; Schlögl, R.; Behrens, M. Cu,Zn-based catalysts for methanol synthesis: On the effect of calcination conditions and the part of residual carbonates. *Appl. Catal. A* **2016**, *516*, 117–126. [CrossRef]

15. Millar, G.J.; Holm, I.H.; Uwins, P.J.R.; Drennan, J. Characterization of precursors to methanol synthesis catalysts Cu/ZnO system. *J. Chem. Soc. Faraday Trans.* **1998**, *94*, 593–600. [CrossRef]

16. Seciuin, M.K. Thermogravimetric and differential thermal analysis of malachite and azurite in inert atmospheres and in air. *Can. Mineral.* **1975**, *13*, 127–132.

17. Qi, G.-X.; Zheng, X.-M.; Fei, J.-H.; Hou, Z.-Y. A novel catalyst for DME synthesis from CO hydrogenation 1. Activity, structure and surface properties. *J. Mol. Catal. A Chem.* **2001**, *176*, 195–203. [CrossRef]

18. King, H.P.; Alexander, L.E. *X-ray Diffraction Procedures*, 2nd ed.; Wiley: New York, NY, USA, 1973.

19. Santos, S.M.; Silva, A.M.; Jordão, E.; Fraga, M.A. Hydrogenation of dimethyl adipate over bimetallic catalysts. *Catal. Commun.* **2004**, *5*, 377–381. [CrossRef]

catalysts

MDPI

Article

Ru-(Mn-M)O$_X$ Solid Base Catalysts for the Upgrading of Xylitol to Glycols in Water

Maxime Rivière [1], Noémie Perret [1], Damien Delcroix [2], Amandine Cabiac [2], Catherine Pinel [1] and Michèle Besson [1,*]

[1] Institut de recherches sur la catalyse et l'environnement de Lyon, University of Lyon, Univ. Claude Bernard Lyon 1, CNRS, IRCELYON, UMR5256, 2 Avenue Albert Einstein, 69626 Villeurbanne, France; mriviere@outlook.fr (M.R.); noemie.perret@ircelyon.univ-lyon1.fr (N.P.); catherine.pinel@ircelyon.univ-lyon1.fr (C.P.)

[2] IFP Energies Nouvelles, Rond-Point de l'Echangeur de Solaize, BP 3, 69360 Solaize, France; damien.delcroix@ifpen.fr (D.D.); amandine.cabiac@ifpen.fr (A.C.)

* Correspondence: michele.besson@ircelyon.univ-lyon1.fr; Tel.: +33-(0)472-445-358

Received: 26 July 2018; Accepted: 11 August 2018; Published: 14 August 2018

Abstract: A series of Ru-(Mn-M)O$_X$ catalysts (M: Al, Ti, Zr, Zn) prepared by co-precipitation were investigated in the hydrogenolysis of xylitol in water to ethylene glycol, propylene glycol and glycerol at 200 °C and 60 bar of H$_2$. The catalyst promoted with Al, Ru-(Mn-Al)O$_X$, showed superior activity (57 h^{-1}) and a high global selectivity to glycols and glycerol of 58% at 80% xylitol conversion. In comparison, the catalyst prepared by loading Ru on (Mn-Al)O$_X$, Ru/(Mn-Al)O$_X$ was more active (111 h^{-1}) but less selective (37%) than Ru-(Mn-Al)O$_X$. Characterization of these catalysts by XRD, BET, CO$_2$-TPD, NH$_3$-TPD and TEM showed that Ru/(Mn-Al)O$_X$ contained highly dispersed and uniformly distributed Ru particles and fewer basic sites, which favored decarbonylation, epimerization and cascade decarbonylation reactions instead of retro-aldol reactions producing glycols. The hydrothermal stability of Ru-(Mn-Al)O$_X$ was improved by decreasing the xylitol/catalyst ratio, which decreased the formation of carboxylic acids and enabled recycling of the catalyst, with a very low deactivation.

Keywords: hydrogenolysis; ethylene glycol; propylene glycol; xylitol; solid base catalyst; aqueous phase; alditol

1. Introduction

Ethylene glycol (EG) and propylene glycol (PG) are employed as raw or starting materials in various useful industrial applications, such as heat-transfer fluids and cosmetics, in the pharmaceutics and packaging sectors [1,2]. The global production of these glycols is currently based on the hydration of petroleum-based ethylene and propylene via epoxides intermediates [3], with 25 million metric tons being produced in 2017. Given recent strategies to diversify carbon sources, the use of bio-alditols, including xylitol and sorbitol, which can be obtained from inedible cellulose and hemicelluloses, appears to be a promising complementary process. Indeed, there are several advantages including the high C/O molar ratio of carbohydrates, the smaller environmental footprint, and their abundance [4–6].

In the last few decades, it has been reported that the upgrading of alditols to glycols can be performed in the aqueous phase under harsh conditions with temperatures of 160–230 °C and H$_2$ pressure (40–120 bar). The selective hydrogenolysis is challenging since parallel and consecutive reactions can occur and lead to a complex mixture of products [7–10]. The initial dehydrogenation of xylitol leads to an aldose (preferential route) or a ketose (minor route), according to the different rates of dehydrogenation of the primary or secondary hydroxyl groups. Then, C–C cleavage forms C$_3$ and C$_2$ intermediates by the retro-aldol reaction of the aldose or ketose catalyzed by base promotors, such as $Ca(OH)_2$. They are then hydrogenated to glycols or glycerol. Therefore, the (de)hydrogenation reaction is catalyzed by a heterogeneous metal-based catalyst and the retro-aldol step by an alkaline base. The decarbonylation reaction may also be catalyzed by the metallic sites and forms C_4 alditols (i.e., threitol) which can lead to the formation of butanediols (BDO) and CO via dehydroxylation and chain decarbonylation reactions [11]. In parallel, the epimerization reaction of xylitol produces other C_5 alditols (arabitol and adonitol) on metallic sites [12]. Moreover, lactate (LA) and 2-hydroxybutyrate (2-HBA) can be produced in the presence of a base by Cannizzaro-type reaction [9]. The main products observed in the hydrogenolysis of xylitol are reported in Figure 1. Thus, different strategies have been exploited in order to enhance the catalytic activity as well as glycols yield.

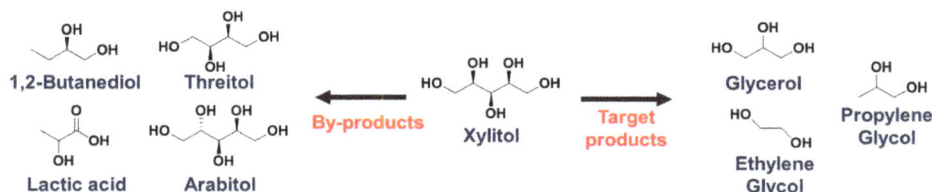

Figure 1. The main products observed during the hydrogenolysis reaction of xylitol.

Many metal-based catalysts, such as Cu, Ni, Ru, and Pt, have been studied for the hydrogenolysis of sugar alcohols to glycols, promoted or not by an alkali.

In the presence of $Ca(OH)_2$ base, 90%Cu-SiO_2 catalysts achieved glycols and glycerol yields of 50–60% in the aqueous phase at 200 °C under 60 bar of H_2. However, the catalytic activity was relatively low [13]. Ni-based catalysts, such as 6%Ni-NaY [14] and bimetallic 10%Ni-80%Cu-SiO_2 [15] catalysts displayed high selectivity to glycols and glycerol (up to 85%) at high conversion of the alditols in the presence of $Ca(OH)_2$. The Ru-based catalysts seem to be the most interesting candidates due to their high activity and their ageing resistance. Sun and Liu [9] reported a selectivity to glycols of 67% in the presence of 4%Ru/C catalyst in the hydrogenolysis of xylitol ($Ca(OH)_2$, 10 wt % xylitol in the aqueous phase, 200 °C and 40 bar H_2). Zhao et al. [16] observed a selectivity of 61% to glycols and glycerol over 3% Ru/carbon nanofibers. More recently, N-containing supports (amine-functionalized carbon nanotubes or covalent triazine framework) improved the performance of Ru-based catalysts in the presence of a base, in terms of activity [17] or yield of glycols and glycerol up to 85% [18].

Among strategies investigated to avoid both the addition of alkaline promoters and the formation of carboxylates, the use of bifunctional catalysts combining both metallic and basic sites has been studied. For instance, Ni-MgO catalysts [19] displayed a total selectivity to glycols and glycerol of 80% at 68% sorbitol conversion (in water, 200 °C, 40 bar H_2, 4 h). Some mixed oxide catalysts were also developed. Ni supported on fly ash (a waste mainly composed of SiO_2, Al_2O_3 and Fe_2O_3) showed a high global selectivity to glycols and glycerol (58% at 41% conversion) as well as high hydrothermal stability for four consecutive runs [20]. Catalysts supported on Ca-containing supports or directly on $Ca(OH)_2$ reached up to 60–70% selectivity to glycols at high conversion under base-free conditions. These include Cu-CaO-Al_2O_3 [21], Ni-Ru/$Ca(OH)_2$ (not stable because of the solubility of the support) [22], and more recently Ni-Ca(Sr)/hydroxyapatite catalysts [23]. Ni-Mg-Al-O hydrotalcite-like catalysts with different Mg/Al ratios were reported to be relatively stable during five runs with reasonable production of glycols [24]. In a previous study, we investigated the Mn promotion of a Ru/C catalyst for hydrogenolysis of xylitol in water [10]. We reported high activity (221 h^{-1}) and reasonable selectivity to glycols and glycerol (32%) over a 3%Ru/MnO(4.5%)/C catalyst. Moreover, we observed that the activity increased with the Mn-loading up to 384 h^{-1} and the selectivity to glycols and glycerol to 50% over 3.1%Ru/MnO(19.3%)/C catalyst. The selectivity over the same 3%Ru/MnO(4.5%)/C catalyst was then enhanced up to 70% in a water/alcohol mixture solvent, which avoided the competitive decarbonylation reaction on the Ru sites [25]. Nevertheless, the catalyst had a poor hydrothermal stability.

Hereby, we developed promising mixed-oxide catalysts Ru-(Mn-M)O_X catalysts (M: Al, Ti, Zr, Zn) synthetized by co-precipitation, calcination and reduction. We compared different methods of Ru incorporation, which influenced the activity and the product distribution. Finally, we demonstrated that the Ru-(Mn-Al)O_X catalyst stability could be improved by varying the operating conditions.

2. Results and Discussion

2.1. Catalyst Characterization

The series of Ru-(Mn-M)O_X catalysts were obtained by thermal decomposition of co-precipitated precursors and reduction. For comparison, a Ru/(Mn-Al)O_X was prepared by impregnation of a Ru salt on (Mn-Al)O_X mixed oxide. The catalyst synthesis methods are reported in the Experimental Section. The M metal loading, the specific surface areas, the Ru crystallite sizes, and the number of acidic or basic sites are shown in Table 1. The loading of Ru could not be measured experimentally by ICP-OES, since Ru could not be totally solubilized, either in the solution of H_2SO_4 and HNO_3, or in concentrated HF solution. However, in the filtrate recovered after impregnation of Ru precursor, no traces of Ru were detected. The actual M loadings (i.e., Al, Ti, Zn, Zr) were very different from the nominal ones; in particular up to 97% of Al was lost (Table 1, entry 1). This might originate from the too high pH applied during the synthesis of the material, which is outside the limits of precipitation of aluminum hydrate and therefore alkaline dissolution of amorphous Al(OH)$_3$ to soluble sodium aluminate occurs. ICP-OES analysis of the solids after each synthesis step demonstrated that M loss took place during the filtration step; the M precursor had not been entirely precipitated despite the basic pH (12), the temperature (80 °C) and the ageing time (20 h). Meanwhile, whatever the method, the experimental Mn loadings were close to the nominal loadings.

Table 1. Characterization of the catalysts based on Ru, Mn and M: Al, Ti, Zr, Zn.

Entry	Catalysts	Nominal Loading [wt %]			Experimental Loading [1] [wt %]			Δ [2] [%]	S_{BET} [3] [m^2 g^{-1}]	Crystallite Size of Ru [4] [nm]	Total Number of Sites [μmol g^{-1}]	
		Ru	Mn	M	Ru	Mn	M				Basic [5]	Acid [6]
1	Ru/(Mn-Al)O_X	3	56	13	n.m.	66.5	0.5	97	13	6	28 (2.2)	18 (1.3)
2	Ru-(Mn-Al)O_X	3	70	4	n.m.	68.4	1.4	60	26	29	44 (1.7)	49 (1.8)
3	Ru-(Mn-Ti)O_X	3	64	11	n.m.	55.9	7.6	32	47	29	127 (2.7)	66 (1.4)
4	Ru-(Mn-Zr)O_X	3	60	15	n.m.	56.7	10.0	29	61	27	123 (2.0)	81 (1.3)
5	Ru-(Mn-Zn)O_X	3	60	15	n.m.	54.7	13.4	10	37	13	123 (3.3)	83 (2.2)

[1] Determined by ICP-OES, [2] Δ: loss of M during the synthesis, [3] relative error: ±5%; [4] based on Rietveld refinement, [5] determined by CO$_2$-TPD, [6] determined by NH$_3$-TPD; numbers in brackets represent the acid or basic site density [μmol m^{-2}]; n.m.: not measurable.

The specific surface area of the mixed-oxide catalysts varied in a large range from 13 to 61 m^2 g^{-1}. The catalyst where the Ru was deposited on the mixed-oxide (Ru/(Mn-Al)O_X, Table 1, entry 1) possessed a lower surface area than the catalyst where the Ru precursor was co-precipitated (Ru-(Mn-Al)O_X, Table 1 entry 2) with values of 13 and 26 m^2 g^{-1}, respectively. In the series of Ru-(Mn-M)O_X catalysts, the surface specific area varied as follows: Ru-(Mn-Al)O_X < Ru-(Mn-Zn)O_X- < Ru-Mn-Ti)O_X- < Ru-(Mn-Zr)O_X with 26, 37, 47 and 61 m^2 g^{-1}, respectively. These values are lower than those reported in the literature for dried or calcined Mn-Cu-Al layered double hydroxides [26]. The lowest specific area is associated with the lowest M content in the mixed oxide catalysts. This is not surprising, since MnO$_X$ exhibits low surface area of 24 m^2 g^{-1} [10].

The XRD patterns of the Ru-(Mn-Al)O_X catalyst after each step of the synthesis highlighted the modification of the oxide species during the different thermal treatments, namely drying, calcination at 460 °C under air, and reduction at 450 °C under H$_2$ flow (Figure 2a). After drying, the only detectable crystalline phase was Mn$_3$O$_4$, whereas no peak corresponding to Al$_2$O$_3$ could be observed. The calcination step led to a mixture of Mn$_3$O$_4$ and Mn$_5$O$_8$, as reported in the literature under the same atmosphere and similar temperature [27]. The Mn_3O_4 species consists of tetrahedral Mn(II) and octahedral Mn(III) units, which stem from partial oxidation of Mn(II) under hydrothermal conditions, as shown on the Pourbaix diagram [28,29]. The Mn_5O_8 species is a metastable oxide of equal amounts of Mn(II) and Mn(IV) [30]. Meanwhile, no peak corresponding to RuO_2 was observed, suggesting an amorphous phase or a small crystallite size (<2 nm). After reduction treatment, only peaks attributed to MnO were detected. Ru-(Mn-Zr)O_X, Ru-(Mn-Zn)O_X and Ru-(Mn-Ti)O_X samples presented similar patterns (Figure 2b). Furthermore, despite the high M amount in some of the catalysts, no peak attributed to the M oxide was observed, except ZnO in the Ru-(Mn-Zn)O_X material. This observation suggests that TiO_2 and ZrO_2 were amorphous, unlike in a Cu-ZrO_2-MgO material obtained after similar thermal treatment [31]. It is worth noting that the pattern of Ru-(Mn-Al)O_X recorded with a higher time per step (Figure A1) exhibited peaks associated with Mn_3O_4 in the 2θ region of 28–38°. This phase might also be present in the other Ru-(Mn-M)O_X. The Ru^0 crystallites were detected in all XRD patterns with different sizes depending on the synthesis method (Table 1). The Ru-(Mn-M)O_X catalysts prepared by co-precipitation exhibited higher Ru mean crystallite sizes (13–29 nm) than the catalyst prepared by deposition of Ru over the (Mn-Al)O_X support (Ru/(Mn-Al)O_X, 6 nm). This is probably due to the calcination step. As reported in the literature for 4%Ru/Al_2O_3 catalyst [32], small RuO_2 crystallite sizes (<2 nm) are formed upon calcination, however, sintering of Ru particles occurs during reduction. On the other hand, the Ru/(Mn-Al)O_X catalyst underwent the reduction step without the initial calcination step leading to a small mean crystallite size of Ru.

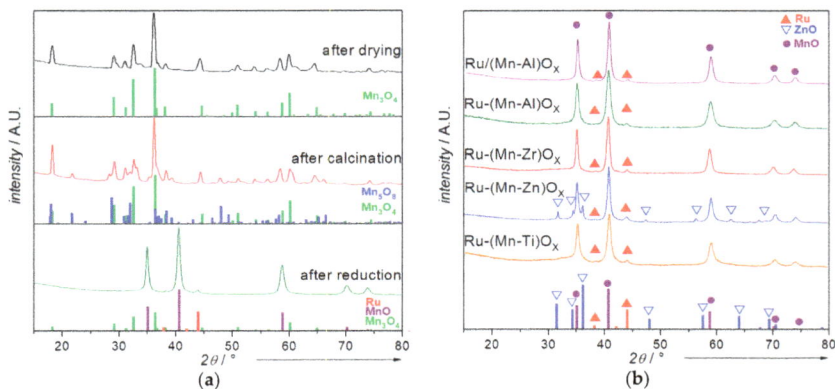

Figure 2. XRD patterns of: (**a**) Ru-(Mn-Al)O_X catalyst after the different thermal treatments; (**b**) Ru/(Mn-Al)O_X catalyst and Ru-(Mn-M)O_X catalysts (M: Al, Zr, Zn, Ti).

Morphology of Ru/(Mn-Al)O_X and Ru-(Mn-Al)O_X catalysts was analyzed by TEM. Images of Ru-(Mn-Al)O_X replica (Figure 3b) revealed a broad size distribution of large (50–100 nm) and small (<5 nm) Ru particles. The images of Ru/(Mn-Al)O_X replica (Figure 3a) showed a homogeneous size of small Ru particles (3–7 nm) gathered together as wormlike nanoparticles. These TEM observations are in accordance with the Ru crystallite size measured by XRD. Significant differences in the size and distribution of Ru particles may influence the activity and the product distribution in hydrogenolysis of xylitol.

The acid-base properties were investigated by CO_2 and NH_3 TPD. In the case of Al-containing materials, the result was dependent on the method of introduction of Ru. In CO_2-TPD profiles (Figure 4a), three main peaks at 175, 250 and 450 °C were detected and attributed to weak, moderate and strong basic sites, respectively. Ru/(Mn-Al)O_X catalyst contained mostly weak basic sites. The signal at 450 °C would be associated with CO_2 adsorbed on strong basic sites or on Ru particles, according to the literature [10,33]. The Ru-(Mn-M)O_X catalysts showed different desorption patterns that are a function of the nature of M. Catalysts Ru-(Mn-M)O_X with M: Zr, Zn, or Ti contained a majority of weak and moderate basic sites due to the presence of possible ZrO_2, ZnO and TiO_2 phases [34,35]. Stronger basic sites were observed on Ru-(Mn-Al)O_X catalysts, which is likely due to the higher content of Mn in this material. The total number of basic sites (in $\mu mol\ g^{-1}$) was as follows: Ru/(Mn-Al)O_X (28) < Ru-(Mn-Al)O_X (44) < Ru-(Mn-Zr)O_X, Ru-(Mn-Zn)O_X (123) $\approx$ Ru-(Mn-Ti)O_X (127), while the basic site density (in $\mu mol \cdot m^{-2}$) was as follows: Ru-(Mn-Al)O_X (1.7) < Ru-(Mn-Zr)O_X (2.0) < Ru/(Mn-Al)O_X (2.2) < Ru-(Mn-Ti)O_X (2.7) < Ru-(Mn-Zn)O_X (3.3) (Table 1). Since the M oxides have an amphoteric character, acid sites were also analyzed (Figure 4b). They are present in lower amounts than basic sites (18 to 83 $\mu mol\ g^{-1}$, i.e., acid density of 1.3–2.2 $\mu mol\ m^{-2}$).

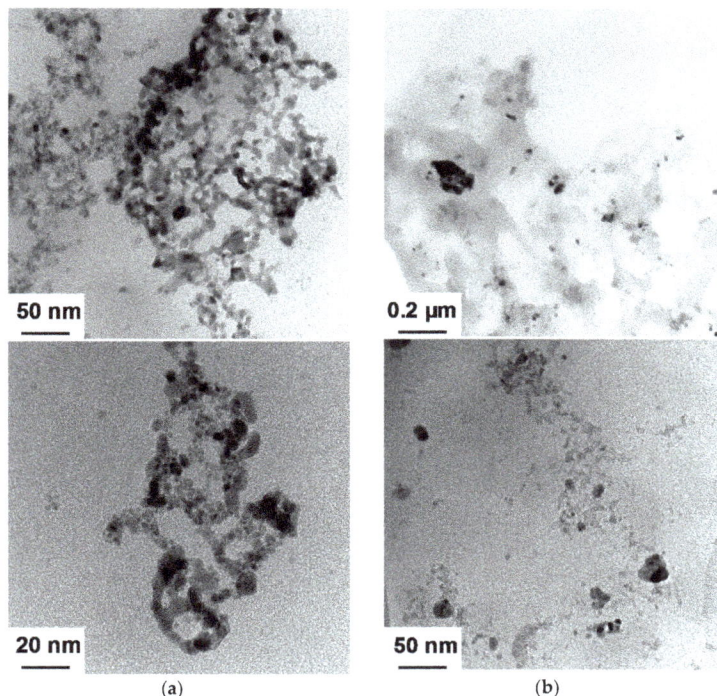

Figure 3. TEM images of replica: (**a**) Ru/(Mn-Al)O_X; (**b**) Ru-(Mn-Al)O_X catalysts.

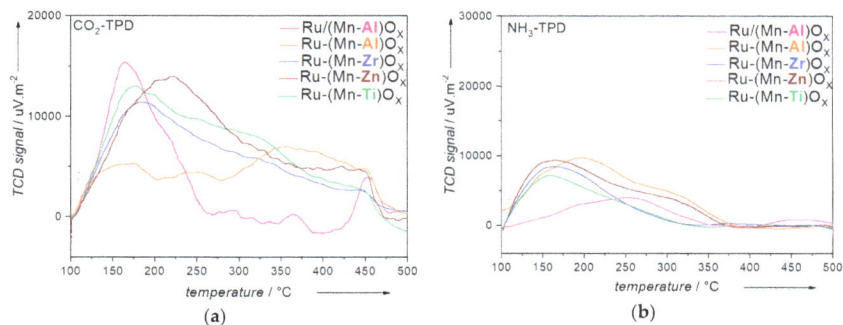

Figure 4. TPD profiles of mixed-oxide catalysts using as probe: (**a**) CO_2; (**b**) NH_3.

To summarize, the introduction of a M promoting element modified the specific surface area and total number of acid and basic sites, whereas the MnO_X phases (Mn_3O_4 and MnO) and the Ru crystallite size remained relatively the same. Meanwhile, the two methods of addition of Ru (via co-precipitation and deposition methods) showed that Ru/(Mn-Al)O_X contained small Ru particles, however, they were gathered while Ru-(Mn-Al)O_X contained small and large Ru particles with a heterogeneous distribution. These differences may influence considerably the catalytic performances in hydrogenolysis of xylitol.

2.2. Hydrogenolysis of Xylitol

The reactions were conducted in water at 200 °C and 60 bar H_2, without any alkali addition. The results of activity and selectivity to the main products are shown in Table 2. The selectivities are given at 80% conversion when the maximum concentration of the desired products was reached, as described previously [10,25]. The total carbon balance (comparison between TOC measured and TOC calculated from HPLC analysis) was around 90%.

Table 2. Activity and product distribution in the hydrogenolysis of xylitol over Ru-based mixed-oxide catalysts [1].

| Catalyst | Activity [h^{-1}] | Carbon Selectivity [2] [%] | | | | | | | CB$_T$ [4] [%] |
		EG	PG	GLY	LA	C$_4$ Products [3]	C$_5$ Alditols	Gas Phase Products	
Ru/(Mn-Al)O_X	111	12	17	8	4	11	6	32	90
Ru-(Mn-Al)O_X	57	21	28	9	3	5	2	20	87
Ru-(Mn-Zr)O_X	33	21	29	9	2	5	3	23	92
Ru-(Mn-Ti)O_X	33	22	29	8	2	7	3	26	99
Ru-(Mn-Zn)O_X	0	-	-	-	-	-	-	-	-

[1] Conditions: xylitol 10 wt % (15 g), 135 mL H_2O, 0.5 g catalyst (molar ratio xylitol/Ru: 764), 60 bar H_2, 200 °C; [2] determined at 80% conversion; [3] C_4 products: threitol, erythritol, butanediols; [4] CB$_T$: Total carbon balance.

The catalytic performance depended on the mode of introduction of Ru in the Al-promoted materials. Indeed, the activity was halved from 111 h^{-1} for Ru/(Mn-Al)O_X to 57 h^{-1} for Ru-(Mn-Al)O_X. In parallel, the selectivities towards glycols were enhanced in the presence of Ru-(Mn-Al)O_X (21% to EG and 28% to PG) vs. Ru/(Mn-Al)O_X (12% to EG and 17% to PG). The selectivity to lactic acid (LA) and to glycerol (GLY) remained the same. On the contrary, the cumulated selectivity to C_4 and C_5 alditols decreased from 17% to 7%. Moreover, the products transferred into the gas phase (determined as the difference of initial TOC and TOC measured in the liquid phase) were produced in lower amounts over Ru-(Mn-Al)O_X. This observation suggests that the side reactions of cascade decarbonylation leading to the degradation of glycols, particularly EG, were minimized [25]. This was confirmed by the calculation of the molar ratio of C_3 to C_2 products, which was 1.3 and 1.9 for the reactions in the presence of Ru-(Mn-Al)O_X and Ru/(Mn-Al)O_X, respectively, while the theoretical molar ratio is 1.0 for a totally selective hydrogenolysis reaction of xylitol to EG plus PG. Therefore, the Ru-(Mn-Al)O_X catalyst favored the retro-aldol reaction instead of the decarbonylation reactions. In contrast, Ru/(Mn-Al)O_X contributed to the decarbonylation reactions. Since the latter reactions take place on Ru particles, the larger particle sizes of Ru-(Mn-Al)O_X (as shown by TEM and XRD) may explain why side-reactions observed at 80% xylitol conversion are not significant, in addition to showing lower activity. Otherwise, the presence of a higher number of basic sites on Ru-(Mn-Al)O_X than on Ru/(Mn-Al)O_X favors the retro-aldol reaction to the detriment of epimerization or cascade decarbonylation reactions, which explains the decrease in catalytic activity.

The Ru-(Mn-Zn)O_X material was not active at all and no conversion was measured after a reaction time of 30 h. A comparable result was observed in our previous study using Ru/ZnO/C catalyst; the catalytic activity decreased from 220 to 13 h^{-1} in comparison with Ru/MnO/C catalyst [25]. Yet, previous studies have reported that the combination of Cu and ZnO was active and selective in the hydrogenolysis of alditols to glycols [21,36–38]. The activity of Ru-(Mn-Zr)O_X and Ru-(Mn-Ti)O_X catalysts was 33 h^{-1}, i.e., lower than the activity of Ru-(Mn-Al)O_X (57 h^{-1}). The product distribution was quite similar whatever the promoter. At 80% xylitol conversion, the cumulated selectivity to glycols and GLY was as high as 59% and the values of selectivity to C_4 and C_5 products were maintained in the range 7–10%. The selectivity to gaseous products increased slightly in the presence of Ru-(Mn-Zr)O_X and Ru-(Mn-Ti)O_X to 23% and 26%, respectively, vs. 20% over Ru-(Mn-Al)O_X catalyst. Thus, in general, despite the relatively high amount of M introduced (10 wt % Zr and 7.0 wt % Ti) and the various total number of acid and basic sites, the use of M promotor had no significant effect on the product distribution. This suggests that a low number of basic sites was sufficient to efficiently perform the retro-aldol reaction, whereas the acid sites favor the cascade decarbonylation to produce gaseous products. Moreover, the activity of Ru-(Mn-Al)O_X is higher due to the higher amount of Mn in the bulk. It has been previously shown that the amount of Mn introduced on the support before deposition of Ru accelerated the conversion of xylitol in base-free medium [10].

Compared to Ru-based catalysts described in the literature for hydrogenolysis of alditols in water under base-free conditions, the Ru-(Mn-Al)O_X catalyst is the most selective to glycols and GLY (58%). This selectivity was 41% from sorbitol over Ru/Ca(OH)$_2$ [22] and 32% from xylitol over Ru/MnO/C [10]. The selectivity of Ru-(Mn-Al)O_X is equal to that of a few Ni-based catalysts, such as Ni-(Mg-Al)O_X [24] or Ni/fly ash [20]. It is close to the one of the best catalytic systems, such as Ni-CaO/C [39] and Cu/CaO-Al$_2$O$_3$ [21], which displayed a selectivity to GLY and glycols of 69% and 73%, respectively; however, the latter Ni catalysts were less active in the hydrogenolysis under similar reaction conditions of alditols with values of 20 and 0.4 h^{-1} respectively, compared to Ru-(Mn-Al)O_X (57 h^{-1}).

The stability of the catalysts was investigated by analysis of the leaching of the catalyst in the final solution and of the used catalyst by XRD. No Ru leaching was observed (detection limit ICP-OES; <0.2 mg L^{-1}, <0.15 wt %), Zr and Ti leaching was less than 1 wt %, while 13–16 wt % of Al were leached. These observations are in line with the literature on the higher stability of TiO$_2$ and ZrO$_2$ than Al$_2$O$_3$ as support for Ru under hydrothermal conditions under H$_2$ atmosphere [40]. On the contrary, a Mn leaching as high as 42–48 wt % was observed for all catalysts. Although this value was still too high, the Mn leaching from Ru-(Mn-M)O_X catalysts was below the one from Ru/MnO/C of 63–70 wt % under the same reaction conditions [10]. The Mn leaching is associated with the production of carboxylic acids, such as lactic or 2-hydrobutyric acids, which were analyzed in the final reaction medium [10,33]. While Mn$_3$O$_4$ and MnO were observed in the fresh catalyst, only MnCO$_3$ was detected in the used catalyst. Similarly, calcium carbonate was detected over Ru/Ca(OH)$_2$ catalyst after hydrogenolysis of sorbitol under similar conditions [22].

2.3. Influence of Operating Conditions

The operating conditions (H$_2$ pressure, temperature) may have a significant influence on the product distribution, as shown over a Ru/C catalyst in hydrogenolysis of xylitol [9]. We investigated the H$_2$ pressure in order to decrease the carboxylic acids production and thus avoid the Mn leaching. Figure 5 shows the selectivity to the main products as a function of pressure over Ru-(Mn-Al)O_X catalyst.

An increase of H$_2$ pressure from 60 to 100 bar at 200 °C doubles the molar fraction of H$_2$ solubilized in aqueous solution [41]. However, catalytic activity increased only slightly from 52 to 59 h^{-1}, suggesting that the solubility of H$_2$ in water was not the limiting step. As for the product distribution, the cumulated selectivity to glycols decreased slightly from 48% to 43%. The selectivity to GLY increased gradually from 7% to 10%, and the selectivity to carboxylic acids (LA + 2–HBA) was reduced from 8.5% (7.5% LA; 1.0% 2–HBA) to 3.4% (2.7% LA; 0.7% 2–HBA). According to the scheme of the proposed reaction, GLY and LA come from the same intermediate, namely glyceraldehyde, via hydrogenation of the terminal carbonyl or via dehydration followed by Cannizzaro-like reaction; thereby, the pressure of H$_2$ favored GLY production at the expense of LA [9]. Nevertheless, despite the decrease in carboxylic acid production, the crystalline phase of the catalyst at the end of the reaction was $MnCO_3$, and Mn leaching still occurred.

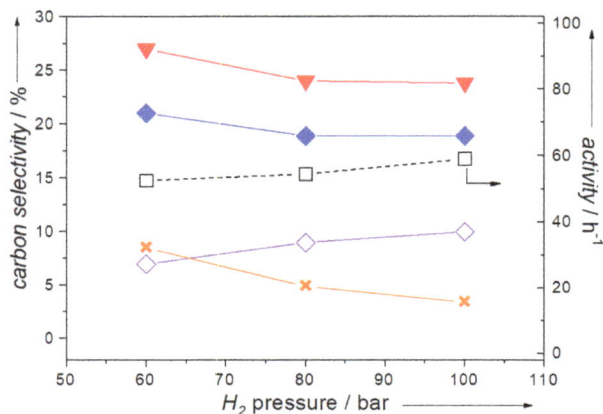

Figure 5. Influence of H_2 pressure on the activity and the selectivities to the main products in hydrogenolysis of xylitol in the presence of Ru-(Mn-Al)O$_X$. □ activity, ▼ PG, ◆ EG, × carboxylic acids (LA plus 2–HBA), ◇ GLY; Conditions: xylitol 10 wt % (15 g), 135 mL H_2O, 0.5 g Ru-(Mn-Al)O$_X$ (molar ratio xylitol/Ru: 764), 60–100 bar H_2, 200 °C.

In addition, the substrate to catalyst ratio was investigated at 200 °C and 60 bar H_2 by modifying the amount of catalyst. The mass ratio was varied from 30 to 3, i.e., a molar xylitol/Ru ratio from 764 to 77. The activity, the product distribution, and the Mn leaching at the end of reaction are shown in Figure 6.

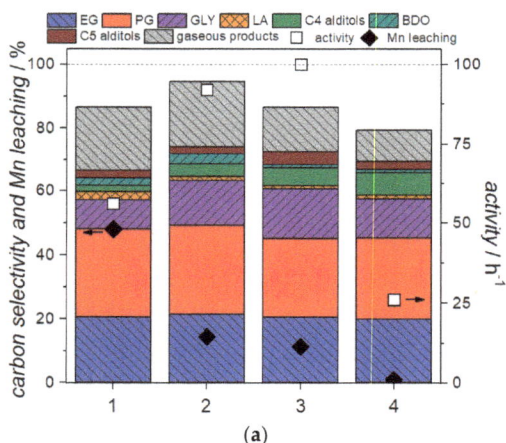

Reaction #	$m_{catalyst}$ [g]	$m_{xylitol}$ [g]	Molar Ratio xylitol/Ru
1	0.5	15	764
2	1.0	15	383
3	1.5	15	255
4	0.5	1.5	77

(b)

Figure 6. (a) Influence of catalyst mass on the activity, selectivity, and Mn leaching: (b) Table of the operating parameters. Conditions: xylitol 1–10 wt % (1.5–15 g), 135 mL H_2O, 0.5–1.5 g Ru-(Mn-Al)O$_X$ (molar ratio xylitol/Ru: 77–764), 60 bar H_2, 200 °C.

Normalized to the Ru moles, the catalytic activity increased from 57 h^{-1} to 101 h^{-1} as the catalyst mass increased (molar ratio from 764 to 255, Reaction #1–3), confirming that the Mn amount in the reaction system enhanced the activity, as observed in Table 2 and as reported previously [10]. When normalized to the total moles of Ru and Mn, the catalytic activity also varied by a factor of 2 between 1.1 and 2.0 mol$_{xyl}$ mol$_{Ru+Mn}^{-1}$ h^{-1}. Assuming that Mn is responsible for the retro-aldol reaction and Ru for the (de)hydrogenation reactions, the difference in the catalyst activity suggests a pseudo-order reaction higher than 1 for the retro-aldol reaction. The product distribution was notably similar whatever the mass of catalyst: the cumulated selectivity to GLY and glycols remained in the range 57–60%, while the selectivity to C$_5$ and C$_4$ products was 10%; the selectivity to carboxylic acids decreased from 5% to less than 1% as the xylitol/catalyst ratio decreased. Surprisingly, Mn leaching at the end of reaction dropped from 48% to 10% as the xylitol/Ru molar ratio decreased from 764 to 255. Meanwhile, Al leaching decreased from 15% to less than 1%, and the Ru leaching remained below the detection limit. Therefore, in order to emphasize this effect, the xylitol/catalyst molar ratio was further decreased to 77, using a 1.0 wt % xylitol solution (Figure 6, Reaction #4). Activity decreased to 25 h^{-1} suggesting mass transfer limitations. Interestingly, Mn leaching was below 1%, i.e., 45 mg L^{-1} Mn in the final reaction medium.

Furthermore, XRD analysis of the used Ru-(Mn-Al)O$_X$ catalyst (Figure 7) showed that the main crystalline phase was MnCO$_3$ after reactions using xylitol/Ru molar ratio between 764 and 255 (reactions #1–3). However, no peak attributed to MnCO$_3$ was detected in the used catalyst after reaction at molar ratio of 77 (reaction #4). Only peaks of Mn$_3$O$_4$ and large peaks of the hydrated MnO phase, Mn(OH)$_2$ were observed [42]. Therefore, the stability of Ru-(Mn-Al)O$_X$ was enhanced by diminishing the xylitol/catalyst mass ratio (molar ratio 77), which made the recycling of the catalyst interesting.

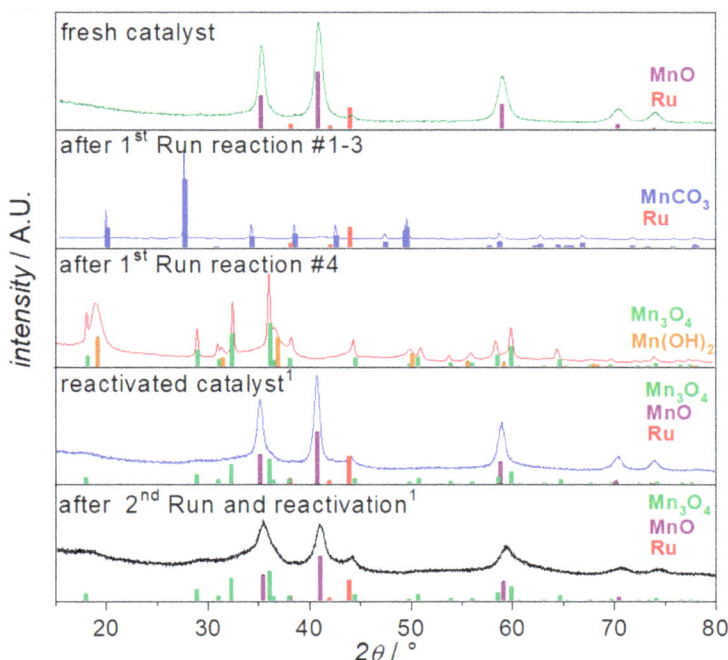

Figure 7. XRD patterns of catalyst Ru-(Mn-Al)O$_X$ after reactions #1–4 and reactivation. [1] Reduction at 450 °C under H$_2$ flow for 3 h.

After the first run, the used Ru-(Mn-Al)O$_X$ catalyst was washed with distillated water and reduced at 450 °C for 3 h in order to eliminate the surface hydroxyls and to reactivate Ru. The XRD analysis of the reactivated catalyst in Figure 7 shows that MnO was again the main crystalline phase. Small peaks associated with Mn$_3$O$_4$ can also be observed. It has been reported that formation of amorphous MnO$_X$ and Mn$_3$O$_4$ phases leads to an increase of specific surface area [43]; this was observed with the reactivated catalyst whose specific area was enhanced from 28 to 60 m^2 g^{-1}. These changes did not significantly modify the performance in hydrogenolysis of 1.0 wt % of xylitol in water with a molar ratio xylitol/Ru of 77, as shown in Table 3.

Table 3. Activity and product distribution in hydrogenolysis of xylitol over Ru-(Mn-Al)O$_X$ catalyst in two successive runs [1].

Run	Activity [h^{-1}]	Carbon Selectivity [2] [%]							CB$_T$ [4] [%]
		EG	PG	GLY	LA	C$_4$ Products [3]	C$_5$ Alditols	Gas Phase Products	
1	26	20	25	12	1	8	3	10	85
2	17	16	24	10	1	12	5	15	83

[1] Conditions: xylitol 1.0 wt % (1.5 g), 135 mL H$_2$O, 0.5 g catalyst (molar ratio xylitol/Ru: 77), 60 bar H$_2$, 200 °C; [2] determined at 80% conversion; [3] C$_4$ products: threitol, erythritol, butanediols; [4] CB$_T$: Total carbon balance.

Indeed, the activity only decreased from 26 to 17 h^{-1}, while the cumulated selectivity to GLY and glycols was 50% instead of 57%. The selectivity to C$_4$ and C$_5$ alditols increased slightly from 11% to 17% and the selectivity to gaseous products increased from 10% to 15%. Therefore, the recycling of catalyst favored the epimerization and cascade decarbonylation reactions and the Mn leaching after the second run was less than 2.0 wt % (55 mg L^{-1} in the final mixture). The XRD analysis of the reactivated catalyst (after 2nd run, Figure 7) revealed that the proportion of Mn$_3$O$_4$ in the crystalline bulk and amorphous phase had increased. These results showed that the structure of Ru-(Mn-Al)O$_X$ catalyst was slightly altered under these hydrothermal conditions. After a reactivation under H$_2$, the selectivity to GLY and glycols remained at a good level.

3. Experimental Section

3.1. Materials

D-Xylitol (99%) and titanium isopropoxide (>99% after distillation) were purchased from Sigma Aldrich (St. Quentin Fallavier, France); Ru(NO)(NO$_3$)$_3$·xH$_2$O (Ru > 31.3 wt %), Al(NO$_3$)$_3$,9H$_2$O (>99%), Mn(OAc)$_2$ and Na$_2$CO$_3$ (>99%) from Alfa Aesar (Karlsruhe, Germany); Mn(NO$_3$)$_2$·4H$_2$O (98%) from Merck (Fontenay sous Bois, France); Zn(NO$_3$)$_2$,6H$_2$O (99%) from AppliChem (Darmstadt, Germany); and ZrO(NO$_3$)$_2$,H$_2$O (99%) from Strem Chemical (Bischheim, France). Hydrogen (H$_2$ > 99.5%), argon (Ar > 99.5%), 1% v/v O$_2$/N$_2$, 1% v/v NH$_3$/He and 5% v/v CO$_2$/He gases were from Air Liquide (Paris, France).

3.2. Catalysts Preparation

The Ru-mixed oxide based catalysts were prepared by a standard co-precipitation method according to the literature [24,26]. In the synthesis of Ru-(Mn-Al)O$_X$ catalyst, an aqueous solution (330 mL) of Mn(NO$_3$)$_2$·4H$_2$O (9.1 g), Al(NO$_3$)$_3$·9H$_2$O (3.4 g) and Ru(NO)(NO$_3$)$_3$·xH$_2$O (0.4 g) and an aqueous solution (264 mL) of NaOH (10.5 g) and Na$_2$CO$_3$ (8.7 g) were added simultaneously dropwise under vigorous stirring at RT into a round-bottom 1 L flask and the pH was controlled at 12 ± 0.5. The obtained slurry was aged for 20 h at 80 °C, filtered, and washed a couple of times with hot deionized water (70 °C) until pH 7. Then, the cake was dried at 100 °C overnight, calcined under air flow (100 mL min^{-1}) at 460 °C for 2 h (3 °C min^{-1}) and reduced under H$_2$ flow (100 mL min^{-1}) for 3 h, and finally passivated under 1% O$_2$/N$_2$ gas mixture flow at RT for

30 min. The Ru-(Mn-Zr)O_X and Ru-(Mn-Zn)O_X catalysts were prepared analogously to this procedure. The Ru-(Mn-Ti)O_X was prepared as previously, however, Mn(OAc)$_2$ was used as the precursor due to the use of Ti(OiPr)$_4$ precursor. A support (Mn-Al)O_X was prepared by co-precipitation of Mn(NO$_3$)$_2$·4H$_2$O and Al(NO$_3$)$_3$·9H$_2$O, according to the literature [24,26]. The material was dried and calcined, as previously. Ru/(Mn-Al)O_X catalyst was then synthetized by wet impregnation of the aqueous solution of Ru precursor over (Mn-Al)O_X, followed by reduction and passivation, according to the previous procedure [25].

3.3. Catalysts Characterization

The metal loadings of the investigated catalysts were determined after mineralization of the solids in aqua regia at 150 °C for 12 h and in aqueous HF, and analysis by inductively coupled plasma-optical emission spectroscopy (ICP-OES, Activa Jobin-Yvon). Powder X-ray patterns were recorded using a Brucker D8A25 diffractometer with CuK$_\alpha$ radiation and equipped with a multi-channel fast detector LynxEye over the range $10° < 2\theta < 80°$ at $0.02°$ s^{-1}. The crystalline phases were identified using the JCPDS reference standards: MnCO$_3$ (44-1472), MnO (05-4310), Mn(OH)$_2$ (73-1604), Mn$_3$O$_4$ (06-6593), Mn$_5$O$_8$ (07-1171), Ru (06-0663). The crystallite sizes were calculated based on LVol-IB (d = 4/3 * LVol-IB) obtained by Rietveld refinement using Topas 5 software. The specific surface areas were determined by N$_2$ physisorption (ASAP 2020) at -196 °C after thermal outgassing (350 °C for 7 h under vacuum 10^{-4} mbar). The amount of basic and acidic sites on the catalyst surface were determined by temperature-programmed desorption (TPD, Belcat-M) of the probe CO$_2$ and NH$_3$, respectively, according to the procedure described previously [25]. The TEM images of replica were taken using a JEOL 2010 instrument operated at 200 keV.

3.4. Catalytic Testing

Hydrogenolysis of xylitol was carried out in a Hastelloy Parr autoclave (300 mL) at a stirring speed of 1000 rpm. In a typical run, 135 mL of 10 wt % xylitol aqueous solution and 0.5 g catalyst were introduced into the autoclave. After purging with Ar, the reactor was heated to the reaction temperature. It was then pressurized with H$_2$, which corresponded to the reaction time t = 0.

The products in the liquid samples were identified and quantified using two different HPLC instruments (Shimadzu, Marne la Vallée, France) connected to refractive index diffraction (RID10A) and UV detectors (SPD-M10A), as described previously [10]. The total organic carbon (TOC) in the liquid phase was measured using a TOC analyzer (Shimadzu TOC-VCHS equipped with ASI-automatic sampler) and compared to initial TOC to estimate the formation of products transferred to the gaseous phase. The carbon balance BC$_T$ was determined by comparison of TOC measured and TOC calculated from HPLC. Conversion, selectivity (on a carbon basis %), and initial reaction activity (mol$_{xylitol}$ mol$_{Ru}^{-1}$ h^{-1}) were calculated as previously [10]. The metal leaching of the catalyst was detected by ICP-OES analysis of the final reaction medium.

4. Conclusions

To conclude, the Ru-(Mn-M)O_X (M: Ti, Al, Zr) catalysts displayed high performances in the valorization of xylitol under base-free conditions. Interestingly, due to the high number of basic sites, the catalysts were able to selectively produce glycols and glycerol with a very low amount of carboxylic acids, such as lactic acid. The Ru-(Mn-Al)O_X catalyst presented the highest activity (57 h^{-1}) and high cumulated selectivity (58%) to glycols and glycerol. That makes Ru-(MnO-Al)O_X one of the most selective Ru-based catalysts under these conditions. The comparison of Ru/(Mn-Al)O_X and Ru-(Mn-Al)O_X catalysts revealed that Ru/(Mn-Al)O_X was more active (111 h^{-1}), however, it produced more C$_5$-C$_4$ alditols as well as gas phase products (i.e., CH$_4$, C$_2$H$_6$). The reason is the smaller Ru particles observed by XRD and TEM, which favor decarbonylation, epimerization as well as cascade decarbonylation reactions. However, Ru-(Mn-Al)O_X catalyst was not stable during the reaction since 42 wt % of Mn-leaching and MnCO$_3$ phase were detected after reaction. The amount of catalyst in the system has a strong effect on the catalyst performance. The Ru-(Mn-Al)O$_X$ activity increased up to 101 h^{-1} as the molar ratio xylitol/Ru was decreased from 764 to 255. Moreover, the decrease in Mn-leaching allowed us to perform a recycling test. After catalyst reactivation, low deactivation was observed in the second run. Therefore, under the appropriate reaction conditions, Ru-(Mn-Al)O$_X$ is a promising catalyst for the synthesis of glycols and glycerol from biomass-derived alditols.

Author Contributions: Funding acquisition, M.R.; Supervision, N.P., D.D., A.C., C.P. and M.B.; Writing—original draft, M.R. and M.B.; Writing—review & editing, N.P. and M.B.

Funding:: This work was funded by Ecole Doctorale de Chimie de Lyon and IFP Energies Nouvelles.

Acknowledgments: The scientific platform of IRCELYON (Pascale Mascunan, Yoann Aizac, Laurence Burel) are acknowledged for ICP, XRD, and TEM analysis of the catalysts.

Conflicts of Interest: The authors declare no conflict of interest.

Appendix A

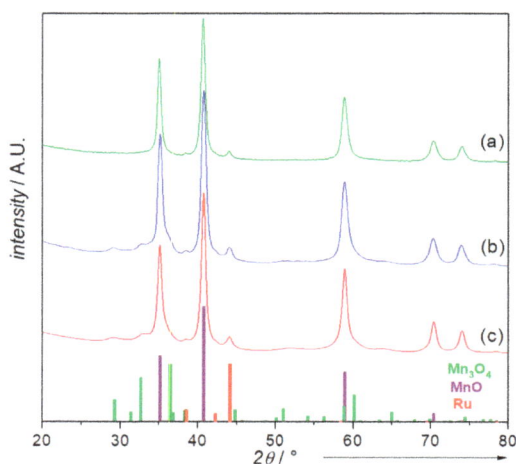

Figure A1. XRD patterns of Ru-(Mn-Al)O$_X$ catalyst: (**a**) 4.0 s/step of 0.02°. (**b**) 16.0 s/step of 0.04°. (**c**) Simulated pattern obtained after Rietveld refinement with TOPAS software.

References

1. Yue, H.; Zhao, Y.; Ma, X.; Gong, J. Ethylene glycol: Properties, synthesis, and applications. *Chem. Soc. Rev.* **2012**, *41*, 4218–4244. [CrossRef] [PubMed]
2. Dasari, M.A.; Kiatsimkul, P.-P.; Sutterlin, W.R.; Suppes, G.J. Low-pressure hydrogenolysis of glycerol to propylene glycol. *Appl. Catal. A Gen.* **2005**, *281*, 225–231. [CrossRef]
3. Zheng, M.; Pang, J.; Sun, R.; Wang, A.; Zhang, T. Selectivity control for cellulose to diols: Dancing on the eggs. *ACS Catal.* **2017**, *7*, 1939–1954. [CrossRef]
4. Climent, M.J.; Corma, A.; Iborra, S. Converting carbohydrates to bulk chemicals and fine chemicals over heterogeneous catalysts. *Green Chem.* **2011**, *13*, 520–540. [CrossRef]
5. Besson, M.; Gallezot, P.; Pinel, C. Conversion of biomass into chemicals over metal catalysts. *Chem. Rev.* **2014**, *114*, 1827–1870. [CrossRef] [PubMed]
6. Ruppert, A.M.; Weinberg, K.; Palkovits, R. Hydrogenolysis goes bio: From carbohydrates and sugar alcohols to platform chemicals. *Angew. Chem. Int. Ed. Engl.* **2012**, *51*, 2564–2601. [CrossRef] [PubMed]

7. Jin, X.; Thapa, P.S.; Subramaniam, B.; Chaudhari, R.V. Kinetic Modeling of Sorbitol Hydrogenolysis over Bimetallic RuRe/C Catalyst. *ACS Sustain. Chem. Eng.* **2016**, *4*, 6037–6047. [CrossRef]

8. Hausoul, P.J.C.; Beine, A.K.; Neghadar, L.; Palkovits, R. Kinetics study of the Ru/C-catalysed hydrogenolysis of polyols—Insight into the interactions with the metal surface. *Catal. Sci. Technol.* **2017**, *7*, 56–63. [CrossRef]

9. Sun, J.; Liu, H. Selective hydrogenolysis of biomass-derived xylitol to ethylene glycol and propylene glycol on supported Ru catalysts. *Green Chem.* **2011**, *13*, 135–142. [CrossRef]

10. Rivière, M.; Perret, N.; Cabiac, A.; Delcroix, D.; Pinel, C.; Besson, M. Xylitol Hydrogenolysis over Ruthenium-Based Catalysts: Effect of Alkaline Promoters and Basic Oxide-Modified Catalysts. *ChemCatChem* **2017**, *9*, 2145–2159. [CrossRef]

11. Deutsch, K.L.; Lahr, D.G.; Shanks, B.H. Probing the ruthenium-catalyzed higher polyol hydrogenolysis reaction through the use of stereoisomers. *Green Chem.* **2012**, *14*, 1635–1642. [CrossRef]

12. Hausoul, P.J.C.; Negahdar, L.; Schute, K.; Palkovits, R. Unravelling the Ru-Catalyzed Hydrogenolysis of Biomass-Based Polyols under Neutral and Acidic Conditions. *ChemSusChem* **2015**, *8*, 3323–3330. [CrossRef] [PubMed]

13. Huang, Z.; Chen, J.; Jia, Y.; Liu, H.; Xia, C.; Liu, H. Selective hydrogenolysis of xylitol to ethylene glycol and propylene glycol over copper catalysts. *Appl. Catal. B Environ.* **2014**, *147*, 377–386. [CrossRef]

14. Banu, M.; Sivasanker, S.; Sankaranarayanan, T.M.; Venuvanalingam, P. Hydrogenolysis of sorbitol over Ni and Pt loaded on NaY. *Catal. Commun.* **2011**, *12*, 673–677. [CrossRef]

15. Liu, H.H.; Huang, Z.; Kang, H.; Li, X.; Xia, C.; Chen, J.; Liu, H.H. Efficient bimetallic NiCu-SiO$_2$ catalysts for selective hydrogenolysis of xylitol to ethylene glycol and propylene glycol. *Appl. Catal. B Environ.* **2018**, *220*, 251–263. [CrossRef]

16. Zhao, L.; Zhou, J.H.; Sui, Z.J.; Zhou, X.G. Hydrogenolysis of sorbitol to glycols over carbon nanofiber supported ruthenium catalyst. *Chem. Eng. Sci.* **2010**, *65*, 30–35. [CrossRef]

17. Guo, X.; Dong, H.; Li, B.; Dong, L.; Mu, X.; Chen, X. Influence of the functional groups of multiwalled carbon nanotubes on performance of Ru catalysts in sorbitol hydrogenolysis to glycols. *J. Mol. Catal. A Chem.* **2016**. [CrossRef]

18. Beine, A.K.; Krüger, A.J.D.; Artz, J.; Weidenthaler, C.; Glotzbach, C.; Hausoul, P.J.C.; Palkovits, R. Selective production of glycols from xylitol over Ru on covalent triazine frameworks—Suppressing decarbonylation reactions. *Green Chem.* **2018**, *20*, 1316–1322. [CrossRef]

19. Chen, X.; Wang, X.; Yao, S.; Mu, X. Hydrogenolysis of biomass-derived sorbitol to glycols and glycerol over Ni-MgO catalysts. *Catal. Commun.* **2013**, *39*, 86–89. [CrossRef]

20. Vijaya Shanthi, R.; Sankaranarayanan, T.M.; Mahalakshmy, R.; Sivasanker, S. Fly ash based Ni catalyst for conversion of sorbitol into glycols. *J. Environ. Chem. Eng.* **2015**, *3*, 1752–1757. [CrossRef]

21. Jin, X.; Shen, J.; Yan, W.; Zhao, M.; Thapa, P.S.; Subramaniam, B.; Chaudhari, R.V. Sorbitol Hydrogenolysis over Hybrid Cu/CaO-Al$_2$O$_3$ Catalysts: Tunable Activity and Selectivity with Solid Base Incorporation. *ACS Catal.* **2015**, *5*, 6545–6558. [CrossRef]

22. Murillo Leo, I.; López Granados, M.; Fierro, J.L.G.; Mariscal, R. Selective conversion of sorbitol to glycols and stability of nickel–ruthenium supported on calcium hydroxide catalysts. *Appl. Catal. B Environ.* **2016**, *185*, 141–149. [CrossRef]

23. Vijaya Shanthi, R.; Mahalakshmy, R.; Thirunavukkarasu, K.; Sivasanker, S. Hydrogenolysis of sorbitol over Ni supported on Ca- and Ca(Sr)-hydroxyapatites. *Mol. Catal.* **2018**, *451*, 170–177. [CrossRef]

24. Du, W.C.; Zheng, L.P.; Shi, J.J.; Xia, S.X.; Hou, Z.Y. Production of C2 and C3 polyols from D-sorbitol over hydrotalcite-like compounds mediated bi-functional Ni-Mg-Al-Ox catalysts. *Fuel Process. Technol.* **2015**, *139*, 86–90. [CrossRef]

25. Rivière, M.; Perret, N.; Delcroix, D.; Cabiac, A.; Pinel, C.; Besson, M. Solvent Effect in Hydrogenolysis of Xylitol over Bifunctional Ru/MnO/C Catalysts under Alkaline-Free Conditions. *ACS Sustain. Chem. Eng.* **2018**, *6*, 4076–4085. [CrossRef]

26. Neațu, F.; Petrea, N.; Petre, R.; Somoghi, V.; Florea, M.; Parvulescu, V.I. Oxidation of 5-hydroxymethyl furfural to 2,5-diformylfuran in aqueous media over heterogeneous manganese based catalysts. *Catal. Today* **2016**, *278*, 66–73. [CrossRef]

27. Jeong, D.; Jin, K.; Jerng, S.E.; Seo, H.; Kim, D.; Nahm, S.H.; Kim, S.H.; Nam, K.T. Mn$_5$O$_8$ Nanoparticles as Efficient Water Oxidation Catalysts at Neutral pH. *ACS Catal.* **2015**, *5*, 4624–4628. [CrossRef]

28. Hem, J. *Chemical Equilibria and Rates of Manganese Oxidation*; Water-Supply Paper 1667-A; United States Government Publishing Office: Washington, DC, USA, 1963. [CrossRef]

29. Robinson, D.M.; Go, Y.B.; Mui, M.; Gardner, G.; Zhang, Z.; Mastrogiovanni, D.; Garfunkel, E.; Li, J.; Greenblatt, M.; Dismukes, G.C. Photochemical Water Oxidation by Crystalline Polymorphs of Manganese Oxides: Structural Requirements for Catalysis. *J. Am. Chem. Soc.* **2013**, *135*, 3494–3501. [CrossRef] [PubMed]

30. Gao, T.; Norby, P.; Krumeich, F.; Okamoto, H.; Nesper, R.; Fjellvåg, H. Synthesis and Properties of Layered-Structured Mn_5O_8 Nanorods. *J. Phys. Chem. C* **2010**, *114*, 922–928. [CrossRef]

31. Rekha, V.; Raju, N.; Sumana, C.; Paul Douglas, S.; Lingaiah, N. Selective Hydrogenolysis of Glycerol Over Cu-ZrO_2-MgO Catalysts. *Catal. Lett.* **2016**, *146*, 1487–1496. [CrossRef]

32. Mazzieri, V.; Coloma-Pascual, F.; Arcoya, A.; L'Argentière, P.C.; Figoli, N.S. XPS, FTIR and TPR characterization of Ru/Al_2O_3 catalysts. *Appl. Surf. Sci.* **2003**, *210*, 222–230. [CrossRef]

33. Ishikawa, M.; Tamura, M.; Nakagawa, Y.; Tomishige, K. Demethoxylation of guaiacol and methoxybenzenes over carbon-supported Ru–Mn catalyst. *Appl. Catal. B Environ.* **2016**, *182*, 193–203. [CrossRef]

34. Watanabe, M.; Aizawa, Y.; Iida, T.; Nishimura, R.; Inomata, H. Catalytic glucose and fructose conversions with TiO_2 and ZrO_2 in water at 473K: Relationship between reactivity and acid–base property determined by TPD measurement. *Appl. Catal. A Gen.* **2005**, *295*, 150–156. [CrossRef]

35. Bancquart, S.; Vanhove, C.; Pouilloux, Y.; Barrault, J. Glycerol transesterification with methyl stearate over solid basic catalysts: I. Relationship between activity and basicity. *Appl. Catal. A Gen.* **2001**, *218*, 1–11. [CrossRef]

36. Tajvidi, K.; Hausoul, P.J.C.; Palkovits, R. Hydrogenolysis of cellulose over Cu-based catalysts-analysis of the reaction network. *ChemSusChem* **2014**, *7*, 1311–1317. [CrossRef] [PubMed]

37. Zhou, Z.; Li, X.; Zeng, T.; Hong, W.; Cheng, Z.; Yuan, W. Kinetics of Hydrogenolysis of Glycerol to Propylene Glycol over Cu-ZnO-Al_2O_3 Catalysts. *Chin. J. Chem. Eng.* **2010**, *18*, 384–390. [CrossRef]

38. Hirano, Y.; Sagata, K.; Kita, Y. Selective transformation of glucose into propylene glycol on Ru/C catalysts combined with ZnO under low hydrogen pressures. *Appl. Catal. A Gen.* **2015**, *502*, 1–7. [CrossRef]

39. Sun, J.; Liu, H. Selective hydrogenolysis of biomass-derived xylitol to ethylene glycol and propylene glycol on Ni/C and basic oxide-promoted Ni/C catalysts. *Catal. Today* **2014**, *234*, 75–82. [CrossRef]

40. Ketchie, W.C.; Maris, E.P.; Davis, R.J. In-situ X-ray Absorption Spectroscopy of Supported Ru Catalysts in the Aqueous Phase. *Chem. Mater.* **2007**, *19*, 3406–3411. [CrossRef]

41. Trinh, T.-K.-H.; de Hemptinne, J.-C.; Lugo, R.; Ferrando, N.; Passarello, J.-P. Hydrogen Solubility in Hydrocarbon and Oxygenated Organic Compounds. *J. Chem. Eng. Data* **2016**, *61*, 19–34. [CrossRef]

42. Hu, C.-C.; Wu, Y.-T.; Chang, K.-H. Low-Temperature Hydrothermal Synthesis of Mn_3O_4 and MnOOH Single Crystals: Determinant Influence of Oxidants. *Chem. Mater.* **2008**, *20*, 2890–2894. [CrossRef]

43. Menezes, P.W.; Indra, A.; Littlewood, P.; Schwarze, M.; Göbel, C.; Schomäcker, R.; Driess, M. Nanostructured manganese oxides as highly active water oxidation catalysts: A boost from manganese precursor chemistry. *ChemSusChem* **2014**, *7*, 2202–2211. [CrossRef] [PubMed]

catalysts

MDPI

Article

Acid–Base Bifunctional Hf Nanohybrids Enable High Selectivity in the Catalytic Conversion of Ethyl Levulinate to γ-Valerolactone

Weibo Wu, Yan Li, Hu Li *, Wenfeng Zhao and Song Yang *

State Key Laboratory Breeding Base of Green Pesticide & Agricultural Bioengineering,
Key Laboratory of Green Pesticide & Agricultural Bioengineering, Ministry of Education,
State-Local Joint Laboratory for Comprehensive Utilization of Biomass, Center for Research & Development of
Fine Chemicals, Guizhou University, Guiyang, Guizhou 550025, China; weibo_wu666@126.com (W.W.);
ly15761698544@163.com (Y.L.); zwf2018gudx@126.com (W.Z.)
* Correspondence: hli13@gzu.edu.cn (H.L.); jhzx.msm@gmail.com (S.Y.);
  Tel.: +86-183-0261-4827 (H.L.); +86-851-8829-2171 (S.Y.)

Received: 26 May 2018; Accepted: 26 June 2018; Published: 29 June 2018

Abstract: The catalytic upgrading of bio-based platform molecules is a promising approach for biomass valorization. However, most solid catalysts are not thermally or chemically stable, and are difficult to prepare. In this study, a stable organic phosphonate–hafnium solid catalyst (PPOA–Hf) was synthesized, and acid–base bifunctional sites were found to play a cooperative role in the cascade transfer hydrogenation and cyclization of ethyl levulinate (EL) to γ-valerolactone (GVL). Under relatively mild reaction conditions of 160 °C for 6 h, EL was completely converted to GVL with a good yield of 85%. The apparent activation energy was calculated to be 53 kJ/mol, which was lower than other solid catalysts for the same reaction. In addition, the PPOA-Hf solid catalyst did not significantly decrease its activity after five recycles, and no evident leaching of Hf was observed, indicating its high stability and potential practical application.

Keywords: heterogeneous catalysis; transfer hydrogenation; biomass conversion; biofuels; catalytic materials

---

1. Introduction

Nowadays, with the continuous development of human society, the demand for energy and fine chemicals is increasing [1]. However, the use and gradual depletion of fossil resources need people to face and solve a series of problems, such as environmental pollution, greenhouse effects, and the energy crisis [2]. The development of renewable energy is, thus, highly demanded, where biomass sources are the only sustainable organic carbon feedstock, showing great potential for practical applications [3]. Lignocellulose is abundant in nature and was demonstrated as capable of being efficiently converted to a series of highly valuable chemicals and biofuels through chemo-catalytic reactions coupled with well-designed processes [4,5].

As a biomass-derived versatile platform molecule, γ-valerolactone (GVL) can be used as a green solvent for biomass conversion and organic transformations [6]. In addition, it can be employed for producing biofuels and fuel additives (e.g., 2-methyltetrahydrofuran) [7], and as a key intermediate in the synthesis of fine chemicals (e.g., pentenoic acid and α-methylene-γ-valerolactone (MeGVL)), as shown in Scheme 1 [8,9]. Typically, GVL can be prepared from lignocellulose via sequential catalytic pathways involving various reactions such as hydrolysis, isomerization, dehydration, etherification, esterification, hydrogenation, and lactonization [10–16]. Amongst these conversion processes, cascade hydrogenation and cyclization were deemed in recent years as the key step in catalytically upgrading levulinic acid and its esters to GVL [17,18].

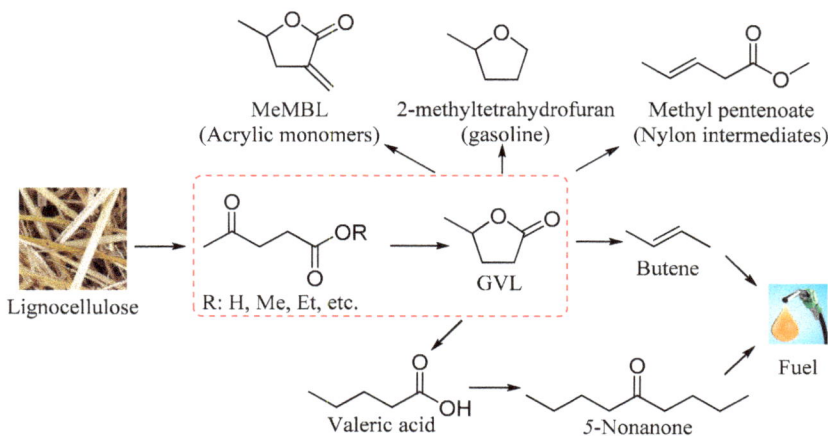

Scheme 1. Synthesis and application of γ-valerolactone (GVL).

Catalytic transfer hydrogenation (CTH) is widely used in the reduction of carbonyl compounds, where H_2 gas is replaced by other liquid molecules (e.g., formic acid and alcohols) as hydrogen donors [19–21]. It is worth noting that there is a certain requirement for the quality of the used reactor and the catalyst stability to tolerate acidic reaction conditions [22,23]. In this regard, alcohols seem to be a better candidate for CTH. A well-known example, Zr-incorporated zeolites with appropriate Lewis acidity, were reported as able to efficiently catalyze the CTH of biomass-derived carboxides, alike to the Meerwein–Ponndorf–Verley (MPV) reduction [24–26].

In the past decade, Zr-based catalysts were reported as efficient for the CTH of levulinic acid and its esters to GVL, and their reactivity is highly dependent on the catalyst composition. For instance, ZrO_2 [27,28], $Zr(OH)_4$ [29], $ZrFeO_x$ [30], Al–Zr [31], Zr–B [32], and $ZrOCl_2$ [33] need relatively harsh conditions (ca. 200 °C or >6 h) to achieve moderate yields of GVL (Table 1). The stability and catalytic performance can be improved via the incorporation of Zr species into solid supports like SBA-15 (Santa Barbara Amorphous-15); however, it generally involves complicated preparation procedures and relatively high production costs due to the use of expensive template agents [34]. Therefore, it is necessary to overcome these shortcomings by improving the stability of solid catalysts with a facile preparation method.

Table 1. Activity contrast of previous studies in the conversion of ethyl levulinate (EL) into γ-valerolactone (GVL).

Entry	Catalyst	Temp (°C)	Time (h)	GVL Yield (%)	EL Conversion (%)	Reference
1	ZrO_2	180	4	80	93	[27]
2	ZrO_2	250	1	63	82	[28]
3	$Zr(OH)_4$	240	3	80	99	[29]
4	ZrFeO(1:3)-300	230	3	87	93	[30]
5	Al_7Zr_3-300	220	4	83	96	[31]
6	Zr_1B_1	200	4	88	95	[32]
7	PPOA–Hf-1:1.5	160	6	85	100	This work

Organic phosphonates as ligands combined with metal ions can react quickly under liquid-phase conditions to afford corresponding inorganic–organic metal phosphonates with enhanced chemical and thermal stability [35–41]. Through the coordination of metal ions with organic phosphoric acid, additional micro- and mesopores are introduced into the resulting catalysts, thus greatly increasing specific surface areas [42,43], and the catalyst functionalities can be simply tuned by changing the organic ligand or preparation method [44]. In the presented study, a novel and stable inorganic–organic metal phosphonate catalyst (PPOA–Hf) was prepared from phenylphosphonic acid (PPOA) and hafnium (Hf, in the same group as Zr on the periodic table) chloride using a simple assembly method. This acid–base bifunctional solid catalyst was able to efficiently promote the conversion of EL into GVL under mild reaction conditions, and systematic studies were thereby conducted.

2. Results and Discussions

2.1. Catalyst Characterization

Prior to conducting catalytic reactions, HfO_2 and PPOA–Hf-1:1.5 were characterized using Brunauer–Emmett–Teller (BET), scanning transmission electron microscopy (STEM), transmission electron microscopy (TEM), thermogravimetry (TG), X-ray diffraction (XRD), and pyridine-absorbed infrared (PY-IR) spectroscopy. Compared to HfO_2 having a low surface area (23 m^2/g) with a large average pore diameter (28.9 nm), PPOA–Hf-1:1.5 was examined to be mesoporous (average pore diameter: 3.4 nm) with a surface area of 215 m^2/g and a pore volume of 0.16 cm^3/g using N_2 adsorption–desorption isotherms (Figures 1A and S1). From the PY-IR spectra in Figure 1B, a characteristic peak can be seen at 1450 cm^{-1}, which is indicative of Lewis acid sites in PPOA–Hf-1:1.5. Moreover, there were small peaks at 1490 cm^{-1} and 1520 cm^{-1}, showing the presence of weak Brønsted acid sites in PPOA–Hf-1:1.5. In contrast, HfO_2 lacks Brønsted acid sites. The hydrogen transfer reaction is typically catalyzed by Lewis acid sites, while Brønsted acid sites present in PPOA–Hf-1:1.5 may promote the adsorption of substrates, as well as the lactonization step. TG analysis was conducted to study the thermal stability of PPOA–Hf and HfO_2 (Figure 1C). It is interesting to note that less than 10% of the catalyst weight was lost until 400 °C, beyond which PPOA–Hf showed better stability than HfO_2. Increased weight loss was observed upon reaching 550 °C. This result indicates that the PPOA–Hf catalyst is thermally stable, and can be a good candidate for chemical reactions under thermal conditions. XRD analyses show that the commercial catalyst, HfO_2, has a good crystalline form (JPCDS # 06-0318), while PPOA–Hf does not have a highly crystalline nature (Figure 1D), with some broad bands belonging to tetragonal (t) and monoclinic (m) phases. Furthermore, the peak at $2\theta = 6°$ possibly resulted from the interlayer clearance of the phosphate [45].

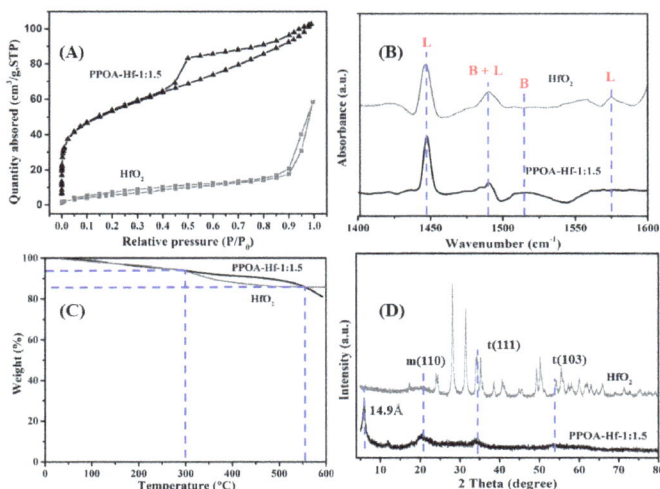

Figure 1. N_2 adsorption–desorption isotherms (**A**); pyridine-adsorbed infrared (IR) spectra (**B**); thermogravimetry (TG) curves (**C**); and X-ray diffraction (XRD) patterns (**D**) of HfO_2 and the phosphonate–hafnium solid catalyst (PPOA–Hf-1:1.5); L = Lewis acid; B = Brønsted acid.

TEM images further confirmed that PPOA–Hf-1:1.5 (Figure 2A) is non-crystalline when compared with HfO_2 (Figure 2B), and the detected lamellar structure is consistent with the result ($2\theta = 6°$) clarified using XRD (Figure 1A). Gratifyingly, the particle size of PPOA–Hf-1:1.5 seems to be smaller than that of HfO_2 (Figure 2), due to its amorphous structure. Furthermore, STEM elemental mappings indicate the even and well-connected combination of the organic ligand and Hf (Figure S2).

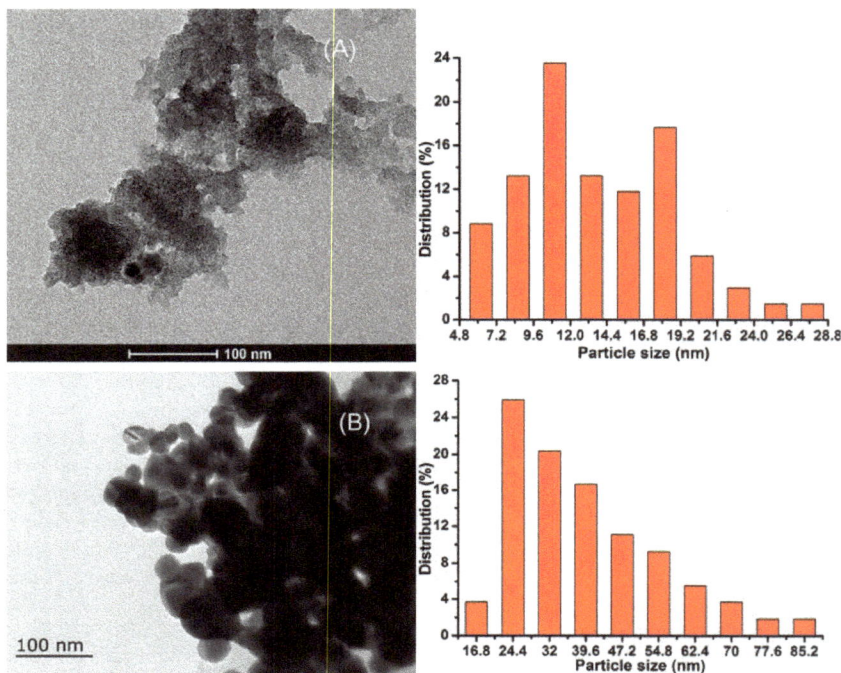

Figure 2. Transmission electron microscopy (TEM) images and particle size distribution of (**A**) PPOA–Hf-1:1.5, and (**B**) HfO_2.

The base and acid properties of the PPOA–Hf-*x* catalysts in various molar ratios of PPOA/Hf were determined using CO_2-temperature programmed desorption (TPD) and NH_3-TPD, respectively. As shown in Figure 3, it can be clearly observed that PPOA–Hf is an acid–base bifunctional catalyst, and the contents of the acid and base sites increase with the increase in Hf relative to PPOA. The Lewis acid and base sites were Hf^{4+} and O^{2-} species, respectively, and their contents increased with the augmentation of Hf, due to the appropriate formation of Hf–O–P. Notably, the acid–base site density remained nearly unchanged after recycling five times (Table S1), indicating that the PPOA–Hf catalyst was stable and reusable.

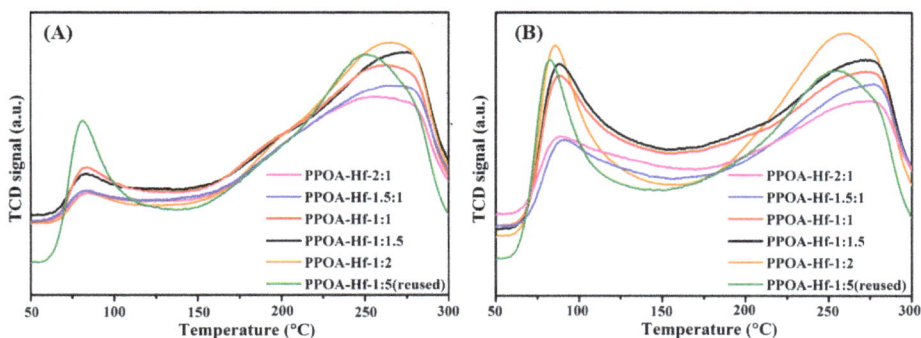

Figure 3. CO_2- temperature programmed desorption (TPD) (**A**), and NH_3-TPD (**B**) patterns of PPOA–Hf-*x* with various PPOA/Hf molar ratios.

The strengths of the acid and base sites were investigated using X-ray photoelectron spectroscopy (XPS) analysis, and the results are provided in Figures 4 and S3, and Table S2. Typically, low and high binding energy of O 1s and Hf 4f are indicative of the high strength of both base and acid sites [46]. The binding energy intensity of Hf 4f (Lewis acid) in PPOA–Hf shows the positive charge on the Hf atoms, which resulted in a stronger Lewis acidity with a higher Hf ratio [47]. Furthermore, the binding energy of the O element was also studied, where the lower binding energy of O 1s was correlated with a higher negative charge on the O atom, which resulted in stronger Lewis basicity of O [48]. It is not difficult to see that the strength of acidic and basic sites in the PPOA–Hf-*x* catalysts is negatively correlated with the increase in PPOA/Hf ratio (Figure 4). In this respect, both the content and strength of acid–base sites in PPOA–Hf-*x* can be controlled by adjusting the molar ratio of PPOA/Hf.

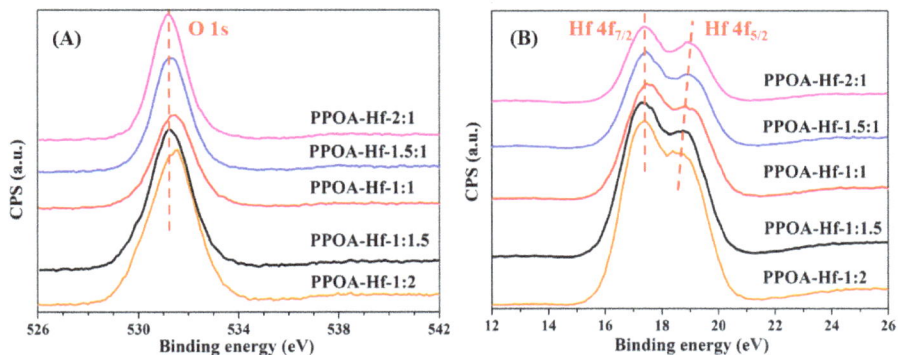

Figure 4. X-ray photoelectron spectroscopy (XPS) spectra of (**A**) O 1s, and (**B**) Hf 4f in PPOA–Hf-*x* with various PPOA/Hf molar ratios.

To clearly elucidate the functional structure of PPOA–Hf-*x*, Fourier-transform infrared (FT-IR) spectroscopy was conducted, and the spectra are shown in Figure 5. In area A, the peak of Hf–O $(500–800 \text{ cm}^{-1})$ is clearly visible in the spark line of HfO_2. On the other hand, the peaks at 600 cm^{-1} (corresponding to an aromatic ring) and 700 cm^{-1} (out-of-plane bending vibration of a C–H bond) decreased sharply with an increase in Hf content, implying the tight combination of hafnium and the organic ligand. Interestingly, with the combination of PPOA and Hf, the positions of the O–H $(2400–3200 \text{ cm}^{-1})$ bond and the Ph–H $(1900–2300 \text{ cm}^{-1})$ bond shifted to 3200–3700 cm^{-1} and

3000–3100 cm^{-1}, respectively. In addition, the characteristic peaks of the P–O (1000 cm^{-1}), C–P (1150–1200 cm^{-1}), and C=C (1400–1700 cm^{-1}) bonds were all present, indicating the strong connection between the organic ligand and Hf, and the possible origin of acid and base sites (e.g., Brønsted acid sites: −OH; Lewis acid-base sites: –Hf–O–P–).

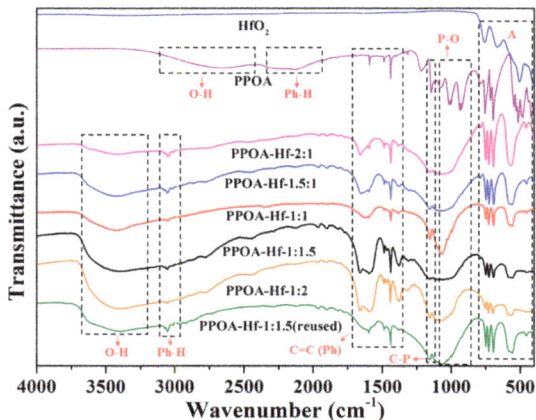

Figure 5. Fourier-transform infrared (FT-IR) spectra of PPOA–Hf–*x* with various PPOA/Hf molar ratios.

2.2. Catalytic Performance of PPOA–Hf-x

The catalytic activity of PPOA–Hf-x with various PPOA/Hf molar ratios in the CTH reaction of EL to GVL was investigated at first, and the results are shown in Figure 6. It is obvious that as the ratio of PPOA/Hf changed from 2:1 to 1:1.5, the conversion rate of EL and the yield of GVL increased from 72% and 55% to 100% and 85%, respectively. This tendency is approximately consistent with the content and strength of the acid–base sites (Figures 3 and 4B). However, a further increase in the PPOA/Hf ratio to 1:2 did not improve the yield of GVL, showing that the acidity and basicity of PPOA–Hf-1:1.5 is appropriate for GVL synthesis from EL. Therefore, PPOA–Hf-1:1.5 was considered as the optimal catalyst for subsequent studies.

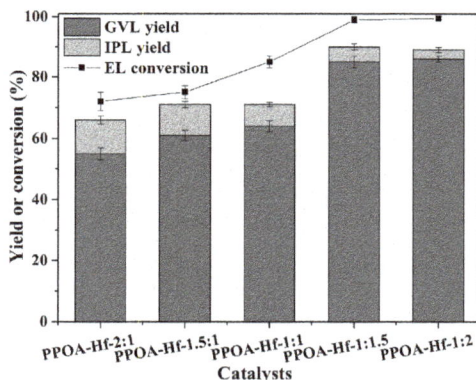

Figure 6. Catalytic results of PPOA–Hf–*x* with various PPOA/Hf molar ratios in the conversion of ethyl levulinate (EL) to GVL. Reaction conditions: EL, 1 mmol; catalyst, 72 mg; 2-propanol, 5 mL; T, 160 °C; and *t*, 6 h.

In the study of the mechanism, the acid–base sites in PPOA–Hf-1:1.5 played a cooperative role in the reaction of EL to GVL (Scheme 2). Firstly, 2-propanol was activated via combination with the Lewis acid sites (Hf^{4+}) and Lewis base sites (O^{2-}) of PPOA–Hf. Concurrently, the carbonyl group in EL was activated via the Lewis acid sites (Hf^{4+}). As a result, a hydrogen transfer reaction was carried out successfully with PPOA–Hf-1:1.5. Subsequently, lactonization occurred via a five-membered ring, which was facilitated by the Brønsted acid. Notably, more than 90% carbon balance was detected for the conversion of EL to GVL.

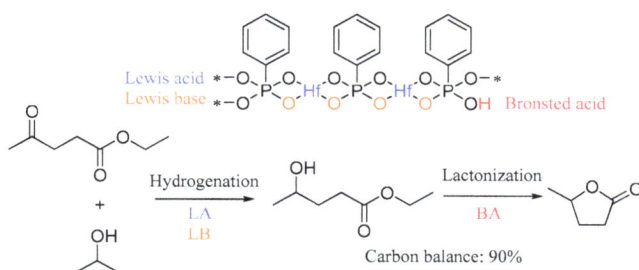

Scheme 2. Reaction pathways for the conversion of ethyl levulinate (EL) to GVL, catalyzed by the phosphonate–hafnium solid catalyst (PPOA–Hf-1:1.5); LA = Lewis acid; LB = Lewis base; BA = Brønsted acid.

2.3. Effect of Reaction Time and Temperature

Reaction temperature and time are important parameters for reactivity control in chemical reactions. The effect of reaction temperature (120–180 °C) on the conversion of EL to GVL catalyzed by PPOA–Hf-1:1.5 over 6 h was studied, and the results are shown in Figure 7. Relatively high yield of GVL (ca. 85%) was obtained in 6 h at 160 °C, and in 4 h at 180 °C. Isopropyl levulinate (IPL) was detected as the dominant byproduct, which reduced the selectivity of GVL. From an economic point of view, the optimal reaction conditions should be 160 °C and 6 h.

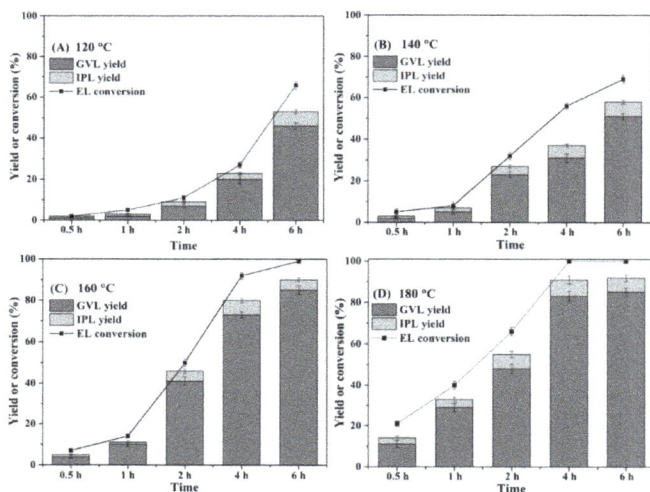

Figure 7. Effect of reaction temperature (120–180 °C) and time on the conversion of EL to GVL. Reaction conditions: EL, 1 mmol; PPOA–Hf-1:1.5, 72 mg; and 2-propanol, 5 mL.

The kinetics of the EL-to-GVL conversion were studied under the assumption of a pseudo first-order reaction, as illustrated in previous reports [49]. In this study, four reaction temperatures of 120 °C (393 K), 140 °C (413 K), 160 °C (433 K), and 180 °C (453 K) were selected. Well-fitted linear curves were obtained by plotting $-\ln(1-X)$ (X = EL conversion) with time for various reaction temperatures, as shown in Figure 8A. The apparent activation energy (E_a) was calculated to be 53 kJ/mol according to the Arrhenius equation, after the obtained reaction rate constant ($\ln k$) was plotted with temperature ($1/T$; Figure 8B). This value is much lower than other previously reported catalytic systems, such as Ti-Beta (69 kJ/mol) [49], Shvo-Ru (69 kJ/mol) [50], and Ru tris(m-sulfonatophenyl)phosphine (61 kJ/mol) [51], indicating that PPOA–Hf-1:1.5 is a more effective and promising solid catalyst for the catalytic conversion of EL to GVL.

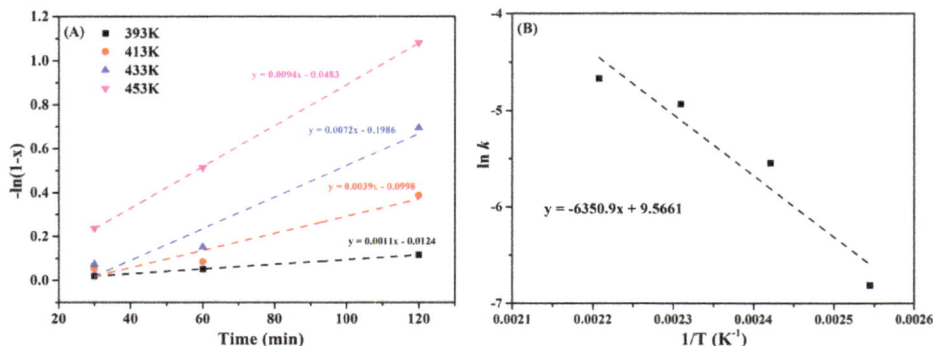

Figure 8. (**A**) Kinetic profiles, and (**B**) Arrhenius plot of the PPOA–Hf-1:1.5-catalyzed conversion of EL to GVL. Reaction conditions: EL, 1 mmol; PPOA–Hf-1:1.5, 72 mg; and 2-propanol, 5 mL.

2.4. Effect of Catalyst Dosage and Reactivity Comparison with Various Catalysts

In this section, the influence of catalyst dosage was examined (Figure 9). The absence of catalyst led to almost no GVL formed, indicating that the reaction required the participation of the catalyst. When the catalyst dosage increased, the yield of GVL rose accordingly. At a catalyst dosage of 72 mg, both the GVL yield and EL conversion reached a maximum. There was no further increase in GVL yield when more than 72 mg PPOA–Hf-1:1.5 was used, indicating that 72 mg (mass ratio 2:1) is the best catalyst dosage.

For comparison, several oxides and the precursor, PPOA, were used for the catalytic conversion of EL to GVL (Table 2). Weak acids or base oxides, such as SiO_2, TiO_2, and MgO, had nearly no catalytic activity for the reaction. In contrast, amphoteric oxides, including HfO_2, Al_2O_3, and ZrO_2, produced GVL in low yields of 6%, 1%, and 2%, respectively. PPOA, due to its high acidity, could catalyze the conversion of EL into IPL (13% yield). Notably, when a strong base (CaO) was used, a 17% yield of GVL and a 43% yield of IPL were obtained. Considering the above results, it can be deduced that base sites are helpful for GVL synthesis, while potentially resulting in the formation of IPL. On the other hand, strong acids in the absence of Lewis acid species are unable to catalyze the production of GVL, while potentially promoting the cyclization or lactonization steps. Due to the appropriate acid–base site content and strength, PPOA–Hf-1:1.5 can efficiently catalyze the synthesis of GVL (85% yield) from EL, with a yield much higher than other tested catalysts. Moreover, hafnium has stronger metal properties than zirconium. Accordingly, there is a weaker Lewis acid in the Zr-based organic phosphonic acid ligand catalyst (PPOA–Zr-1:1.5), which, when produced using the same method as PPOA–Hf-1:1.5, resulted in PPOA–Zr-1:1.5 showing a relatively poor catalytic performance (Table 2, entry 10).

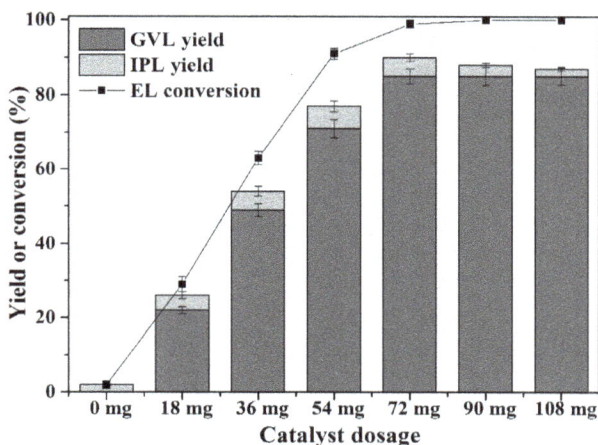

Figure 9. Effect of catalyst dosage on the conversion of EL to GVL. Reaction conditions: EL, 1 mmol; 2-propanol, 5 mL; T, 160 °C; and *t*, 6 h.

Table 2. Activity comparison of various solid catalysts in the conversion of EL to GVL [a].

Entry	Catalyst	GVL Yield (%)	IPL Yield (%)	EL Conversion (%)	Average Rate (μmol g^{-1}h^{-1}) [b]
1	SiO_2	<1	<1	1	<20
2	TiO_2	<1	<1	1	<20
3	MgO	<1	<1	1	<20
4	HfO_2	6	1	10	140
5	Al_2O_3	1	5	8	23
6	ZrO_2	2	-	5	46
7	PPOA	0	13	16	-
8	CaO	17	43	81	400
9	PPOA–Hf-1:1.5	85	5	100	1970
10	PPOA–Zr-1:1.5	73	6	90	1690

[a] Reaction conditions: EL, 1 mmol; catalyst, 72 mg; 2-propanol, 5 mL; T, 160 °C; and *t*, 6 h. [b] Average rate is defined as (mol of formed GVL)/(catalyst weight × time).

2.5. Catalyst Leaching Experiments and Recycling Study

Based on the results of the TG analysis, the catalyst has good thermal stability. To examine the chemical stability of the solid catalyst (PPOA–Hf-1:1.5), a hot filtration experiment was carried out. During the catalytic conversion of EL to GVL, PPOA–Hf-1:1.5 was filtered out after 3 h as soon as the reaction system cooled to ca. 80 °C. Then, the GVL yield was measured per hour for the next three hours. No significant increase in GVL yield was observed (Figure 10), and ICP analysis shows that almost no Hf and P were leached (<0.01 ppm). This result demonstrates that PPOA–Hf-1:1.5 is a heterogeneous catalyst, and is chemically stable in the reaction system.

Furthermore, the reusability of the solid catalyst was investigated (Figure 11), and the PPOA–Hf catalyst could be reused five consecutive times without a significant reduction in GVL yield and EL conversion. The IR spectra in Figure 7 show that the reused catalyst kept an intact structure after being recycled five times. In addition, the acid and base activity sites were retained in the reused catalyst, as illustrated by the NH_3-TPD and CO_2-TPD (Figure 4). These results prove the good reusability of the solid catalyst, PPOA–Hf.

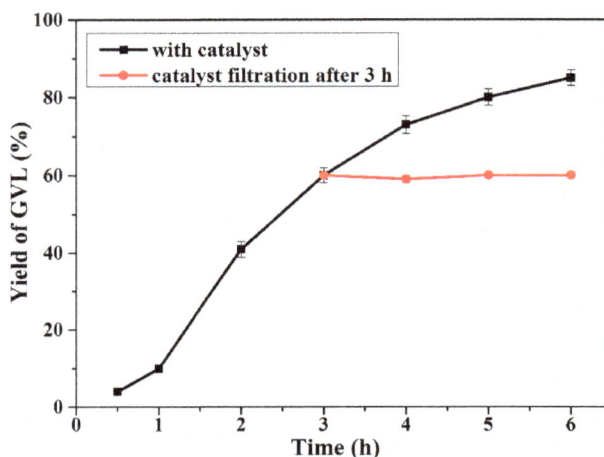

Figure 10. Hot filtration experimental results of the PPOA–Hf-1:1.5-catalyzed conversion of EL to GVL. Reaction conditions: EL, 1 mmol; PPOA–Hf-1:1.5, 72 mg; 2-propanol, 5 mL; and T, 160 °C.

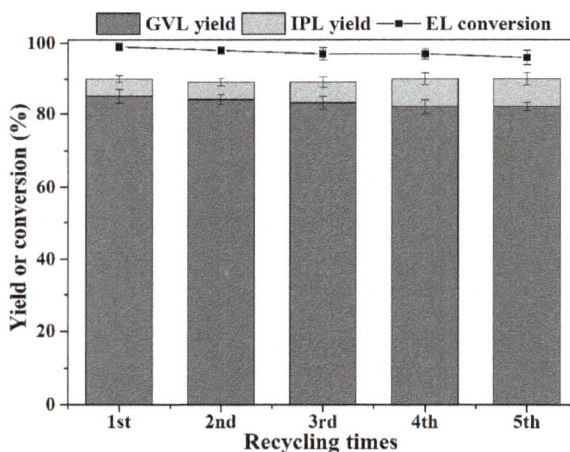

Figure 11. Reusability of PPOA–Hf-1:1.5 in the catalytic conversion of EL to GVL. Reaction conditions: EL, 1 mmol; PPOA–Hf-1:1.5, 72 mg; 2-propanol, 5 mL; T, 160 °C; and *t*, 6 h.

3. Materials and Experiments

3.1. Materials

Hafnium (IV) chloride (HfCl$_4$; 99%), hafnium (IV) oxide, and phenylphosphonic acid (PPOA, 98%) were purchased from Adamas Reagent Co., Ltd. (Shanghai, China) Ethyl levulinate (EL; 98%) was purchased from Alfa Aesar (Shanghai, China) Chemicals Co., Ltd. Naphthalene (99%), γ-valerolactone (GVL; 98%), 2-propanol (99.5%), and other reagents were purchased from Beijing Innochem Technology Co., Ltd. (Beijing, China), and were directly used for the study unless otherwise noted.

3.2. Catalyst Preparation

Hafnium phenylphosphonates (PPOA–Hf-*x*; *x* denotes the used molar ratio of PPOA to Hf) were prepared from PPOA and HfCl$_4$ with corresponding PPOA/Hf ratios. In a typical procedure for the synthesis of PPOA–Hf-1:1.5, 20 mmol PPOA (0.3162 g) was initially dissolved into 60 mL of *N*,*N*-dimethylformamide (DMF) in a Teflon-lined tube, followed by a dropwise addition of 30 mmol HfCl$_4$ (0.9609 g) under vigorous stirring conditions. After stirring for 20 min, the hydrothermal reactor was sealed and placed into an oven heated to 120 °C, for 24 h. Upon completion, the solid precipitates were separated from the liquid mixture via filtration, then successively washed with DMF (100 mL), ethanol (100 mL), and methanol (100 mL) 2–3 times, and finally, dried at 45 °C overnight to give the catalyst, PPOA–Hf-1:1.5.

3.3. Catalyst Characterization Techniques

Brunauer–Emmett–Teller (BET) surface areas of the porous materials were determined from nitrogen physisorption measurements at liquid-nitrogen temperature on a Micromeritics ASAP 2460 instrument (Micromeritics, Norcross, GA, USA). Fourier-transform infrared (FT-IR) spectra were measured by a Thermo Fisher Nicolet iS50 spectrometer (Thermo Fisher Scientific, Waltham, MA, USA) with a wavenumber range of 400–4000 cm^{-1}. The Lewis and Brønsted acid sites of PPOA–Hf were determined via a vacuum adsorption surface reaction infrared in situ characterization analysis system (Dalian Institute of Chemical Physics, Chinese Academy of Sciences, Dalian, China) with a Thermo Fisher Nicolet iS50 AEM with a wavenumber range of 1400–1600 cm^{-1}. Contents of Hf and P species in the reaction system were determined using an inductively coupled plasma-optical emission spectrometer (ICP-OES) on a PerkinElmer Optima 5300 DV (LabX, Midland, Canada). Scanning transmission electron microscopy (STEM) and transmission electron microscopy (TEM) tests were measured using a JEOL 2100 TEM/STEM (JEOL, Akishima, Japan). Particle size distribution was calculated via the Nano Measurer 1.2, Visual Basic 6.0 software. Thermogravimetry (TG) analysis was carried out using a NETZSCH STA 449 F3 Jupiter thermal gravimetric analyzer (NETZSCH, Selb, German). The acidity and basicity of the catalysts were measured by NH$_3$-temperature programmed desorption (TPD) and CO$_2$-TPD using a Micromeritics AutoChem 2920 chemisorption analyzer (Micromeritics, Norcross, GA, USA). The pyridine adsorption process was carried out at room temperature, followed by heating to 100 °C, and kept for 5 min. Then, the spectrum was obtained after cooling down to room temperature. X-ray photoelectron spectroscopy (XPS) measurements were recorded using a Physical Electronics Quantum 2000 Scanning ESCA Microprobe (Physical Electronics Inc., Chanhassen, MN, USA) equipped with a monochromatic Al Kα anode. The X-ray diffraction (XRD) (Rigaku, Akishima, Japan) data of the powder samples were obtained using a Rigaku International D/max-TTR III X-ray powder diffractometer with Cu Kα radiation, and 2θ scanned from 5° to 80°.

3.4. Catalytic Activity Measurements

The reactions for the conversion of EL to GVL were all carried out in a 10-mL Teflon-lined autoclave heated by an oil bath. In a general procedure, 1 mmol EL (144 mg), 72 mg of catalyst, and 5 mL of 2-propanol were added into the Teflon-lined reactor. The sealed reaction kettle was then put into the oil bath at a prefixed temperature, and the reaction time was accordingly recorded. After the reaction, the solid catalyst was removed by centrifugation, and the liquid mixture was passed through a filter membrane (0.22 μm) prior to GC and GC-MS analysis.

3.5. Product Analysis

Liquid products were identified using an Agilent 6890N GC-MS (Agilent, Santa Clara, CA, USA) with a 5973MS mass spectrometer. For the quantitative analysis of EL and GVL, an Agilent GC7890B equipped with a HP-5 19091J-413 column (30 m × 0.32 mm × 0.25 mm) and a flame ionization detector

(FID) was used. an internal standard method was adopted for the quantitative calculation based on standard curves ($R^2 > 0.99$) of EL and GVL, and naphthalene was used as the internal standard. Substrate conversion (X, %) and product yield (Y, %) were calculated using the following equations:

$$X\ (\%) = [1 - (\text{mole of substrate after reaction})/(\text{mole of initial substrate})] \times 100\%; \tag{1}$$

$$Y\ (\%) = (\text{mole of obtained product})/(\text{mole of initial substrate}) \times 100\%. \tag{2}$$

3.6. Catalyst Recycling

After each cycle of reactions, the catalyst was separated by centrifugation from the reaction mixture, successively washed with 10 mL of DMF, ethanol, and methanol, and then dried at 45 °C for 12 h. The resulting solid catalyst was directly used for the next run.

4. Conclusions

In summary, a stable Hf-containing acid–base bifunctional solid catalyst was prepared, and was determined to be highly efficient for the catalytic conversion of EL to GVL. A high GVL yield of 85% was obtained at 160 °C after 6 h, which was superior to the other tested catalysts. Acid and base sites were found to play a synergistic role in the hydrogenation step, while Brønsted acid species improved the adsorption of the substrate and the lactonization step, thus efficiently promoting the cascade reaction. In addition, the solid catalyst, PPOA–Hf-1.5, had good reusability, with five cycles of use without obvious activity decline.

Supplementary Materials: The following are available online at at http://www.mdpi.com/2073-4344/8/7/264/s1, Table S1. Quantitative analysis data of chemical adsorption. Table S2. Percentage of atomic concentration on the catalyst surface by XPS. Figure S1. Pore diameter of HfO2 and recovered PPOA-Hf-1:1.5. Figure S2. STEM image and elemental mappings of PPOA-Hf-1:1.5. Figure S3. XPS spectra of in PPOA-Hf-x with different PPOA/Hf molar ratios.

Author Contributions: W.W., H.L., and S.Y. conceived and designed the experiments; W.W., Y.L., and W.Z. performed the experiments; W.W. and H.L. analyzed the data; and W.W., H.L., and S.Y. co-wrote the paper.

Funding: This study was financially supported by the National Natural Science Foundation of China (21576059 & 21666008), Fok Ying-Tong Education Foundation (161030), Guizhou Science & Technology Foundation ([2018]1037 & [2017]5788), and Key Technologies R&D Program of China (2014BAD23B01).

Conflicts of Interest: There are no conflicts to declare.

References

1. Xu, C.; Arancon, R.A.D.; Labidi, J.; Luque, R. Lignin depolymerisation strategies: Towards valuable chemicals and fuels. *Chem. Soc. Rev.* **2014**, *43*, 7485–7500. [CrossRef] [PubMed]
2. Li, H.; Saravanamurugan, S.; Yang, S.; Riisager, A. Direct transformation of carbohydrates to the biofuel 5-ethoxymethylfurfural by solid acid catalysts. *Green Chem.* **2016**, *18*, 726–734. [CrossRef]
3. Li, H.; Bhadury, P.S.; Riisager, A.; Yang, S. One-pot transformation of polysaccharides via multi-catalytic processes. *Catal. Sci. Technol.* **2014**, *4*, 4138–4168. [CrossRef]
4. Holm, M.S.; Saravanamurugan, S.; Taarning, E. Conversion of sugars to lactic acid derivatives using heterogeneous zeotype catalysts. *Science* **2010**, *328*, 602–605. [CrossRef] [PubMed]
5. Li, H.; Zhao, W.; Fang, Z. Hydrophobic Pd nanocatalysts for one-pot and high-yield production of liquid furanic biofuels at low temperatures. *Appl. Catal. B Environ.* **2017**, *215*, 18–27. [CrossRef]
6. Alonso, D.M.; Wettstein, S.G.; Dumesic, J.A. Bimetallic catalysts for upgrading of biomass to fuels and chemicals. *Chem. Soc. Rev.* **2012**, *41*, 8075–8098. [CrossRef] [PubMed]
7. Corma, A.; Iborra, S.; Velty, A. Chemical routes for the transformation of biomass into chemicals. *Chem. Rev.* **2007**, *107*, 2411–2502. [CrossRef] [PubMed]
8. Liguori, F.; Moreno-Marrodan, C.; Barbaro, P. Environmentally friendly synthesis of γ-valerolactone by direct catalytic conversion of renewable rources. *ACS Catal.* **2015**, *5*, 1882–1894. [CrossRef]

9. Li, H.; Yang, S.; Riisager, A.; Pandey, A.; Sangwan, R.S.; Saravanamurugan, S.; Luque, R. Zeolite and zeotype-catalysed transformations of biofuranic compounds. *Green Chem.* **2016**, *18*, 5701–5735. [CrossRef]

10. Zhang, Z. Synthesis of γ-valerolactone from carbohydrates and its applications. *ChemSusChem* **2016**, *9*, 156–171. [CrossRef] [PubMed]

11. Qi, X.; Guo, H.; Li, L.Y.; Smith, R.L., Jr. Acid-catalyzed dehydration of fructose into 5-hydroxymethylfurfural by cellulose-derived amorphous carbon. *ChemSusChem* **2012**, *5*, 2411–2502. [CrossRef] [PubMed]

12. Zhou, P.; Zhang, Z. One-pot catalytic conversion of carbohydrates into furfural and 5-hydroxymethylfurfural. *Catal. Sci. Technol.* **2016**, *6*, 3694–3712. [CrossRef]

13. Guo, H.; Duereh, A.; Hiraga, Y.; Aida, T.M.; Qi, X.; Smith, R.L., Jr. Perfect recycle and mechanistic role of hydrogen sulfate ionic liquids as additive in ethanol for efficient conversion of carbohydrates into 5-ethoxymethylfurfural. *Chem. Eng. J.* **2017**, *323*, 287–294. [CrossRef]

14. Guo, H.; Qi, X.; Hiraga, Y.; Aida, T.M.; Smith, R.L., Jr. Efficient conversion of fructose into 5-ethoxymethylfurfural with hydrogen sulfate ionic liquids as co-solvent and catalyst. *Chem. Eng. J.* **2017**, *314*, 508–514. [CrossRef]

15. Saravanamurugan, S.; Van Buu, O.N.; Riisager, A. Conversion of mono- and disaccharides to ethyl levulinate and ethyl pyranoside with sulfonic acid-functionalized ionic liquids. *ChemSusChem* **2011**, *4*, 723–726. [CrossRef] [PubMed]

16. Saravanamurugan, S.; Riisager, A. Solid acid catalysed formation of ethyl levulinate and ethyl glucopyranoside from mono- and disaccharides. *Catal. Commun.* **2012**, *17*, 71–75. [CrossRef]

17. Isikgor, F.H.; Becer, C.R. Lignocellulosic biomass: A sustainable platform for the production of bio-based chemicals and polymers. *Polym. Chem.* **2015**, *6*, 4497–4559. [CrossRef]

18. Saravanamurugan, S.; Riisager, A. Zeolite catalyzed transformation of carbohydrates to alkyl levulinates. *ChemCatChem* **2013**, *5*, 1754–1757. [CrossRef]

19. Gilkey, M.J.; Xu, B. Heterogeneous catalytic transfer hydrogenation as an effective pathway in biomass upgrading. *ACS Catal.* **2016**, *6*, 1420–1436. [CrossRef]

20. Li, J.; Liu, J.; Zhou, H.; Fu, Y. Catalytic transfer hydrogenation of furfural to furfuryl alcohol over nitrogen-doped carbon-supported iron catalysts. *ChemSusChem* **2016**, *9*, 1339–1347. [CrossRef] [PubMed]

21. Wang, D.; Astruc, D. The golden age of transfer hydrogenation. *Chem. Rev.* **2015**, *115*, 6621–6686. [CrossRef] [PubMed]

22. Grasemann, M.; Laurenczy, G. Formic acid as a hydrogen source–recent developments and future trends. *Energy Environ. Sci.* **2012**, *5*, 8171–8181. [CrossRef]

23. Bigler, R.; Huber, R.; Stöckli, M.; Mezzetti, A. Iron(II)/(NH)$_2$P$_2$ macrocycles: Modular, highly enantioselective transfer hydrogenation catalysts. *ACS Catal.* **2016**, *6*, 6455–6464. [CrossRef]

24. Li, H.; He, J.; Riisager, A.; Saravanamurugan, S.; Song, B.; Yang, S. Acid-base bifunctional zirconium N-alkyltriphosphate nanohybrid for hydrogen transfer of biomass-derived carboxides. *ACS Catal.* **2016**, *6*, 7722–7727. [CrossRef]

25. Wang, J.; Okumura, K.; Jaenicke, S.; Chuah, G.K. Post-synthesized zirconium-containing beta zeolite in Meerwein-Ponndorf-Verley reduction: Pros and cons. *Appl. Catal. A* **2015**, *493*, 112–120. [CrossRef]

26. Assary, R.S.; Curtiss, L.A.; Dumesic, J.A. Exploring Meerwein-Ponndorf-Verley reduction chemistry for biomass catalysis using a first-principles approach. *ACS Catal.* **2013**, *3*, 2694–2704. [CrossRef]

27. Chia, M.; Dumesic, J.A. Liquid-phase catalytic transfer hydrogenation and cyclization of levulinic acid and its esters to γ-valerolactone over metal oxide catalysts. *Chem. Commun.* **2011**, *47*, 12233–12235. [CrossRef] [PubMed]

28. Tang, X.; Hu, L.; Sun, Y.; Zhao, G.; Hao, W.; Lin, L. Conversion of biomass-derived ethyl levulinate into γ-valerolactone via hydrogen transfer from supercritical ethanol over a ZrO$_2$ catalyst. *RSC Adv.* **2013**, *3*, 10277–10284. [CrossRef]

29. Tang, X.; Chen, H.; Hu, L.; Hao, W.; Sun, Y.; Zeng, X.; Lin, L.; Liu, S. Conversion of biomass to γ-valerolactone by catalytic transfer hydrogenation of ethyl levulinate over metal hydroxides. *Appl. Catal. B Environ.* **2014**, *147*, 827–834. [CrossRef]

30. Li, H.; Fang, Z.; Yang, S. Direct conversion of sugars and ethyl levulinate into γ-valerolactone with superparamagnetic acid-base bifunctional ZrFeO$_x$ nanocatalysts. *ACS Sustain. Chem. Eng.* **2016**, *4*, 236–246. [CrossRef]

31. He, J.; Li, H.; Lu, Y.; Liu, Y.; Wu, Z.; Hu, D.; Yang, S. Cascade catalytic transfer hydrogenation-cyclization of ethyl levulinate to γ-valerolactone with Al-Zr mixed oxides. *Appl. Catal. A Gen.* **2016**, *510*, 11–19. [CrossRef]

32. He, J.; Li, H.; Liu, Y.; Zhao, W.; Yang, T.; Xue, W.; Yang, S. Catalytic transfer hydrogenation of ethyl levulinate into γ-valerolactone over mesoporous Zr/B mixed oxides. *J. Ind. Eng. Chem.* **2016**, *43*, 133–141. [CrossRef]

33. Tang, X.; Zeng, X.; Li, Z.; Li, W.; Jiang, Y.; Hu, L.; Liu, S.; Sun, Y.; Lin, L. In situ generated catalyst system to convert biomass-derived levulinic acid to γ-valerolactone. *ChemCatChem* **2015**, *7*, 1372–1379. [CrossRef]

34. Kuwahara, Y.; Kaburagi, W.; Osada, Y.; Fujitani, T.; Yamashita, H. Catalytic transfer hydrogenation of biomass-derived levulinic acid and its esters to γ-valerolactone over ZrO$_2$ catalyst supported on SBA-15 silica. *Catal. Today* **2017**, *281*, 418–428. [CrossRef]

35. de los Reyes, M.; Majewski, P.J.; Scales, N.; Luca, V. Hydrolytic stability of mesoporous zirconium titanate frameworks containing coordinating organic functionalities. *ACS Appl. Mater. Interfaces* **2013**, *5*, 4120–4128. [CrossRef] [PubMed]

36. Li, H.; Yang, T.; Fang, Z. Biomass-derived mesoporous Hf-containing hybrid for efficient Meerwein-Ponndorf-Verley reduction at low temperatures. *Appl. Catal. B Environ.* **2018**, *227*, 79–89. [CrossRef]

37. Gelman, F.; Blum, J.; Avnir, D. Acids and bases in one pot while avoiding their mutual destruction. *Angew. Chem. Int. Ed.* **2001**, *40*, 3647–3649. [CrossRef]

38. Yang, Y.; Liu, X.; Li, X.; Zhao, J.; Bai, S.; Liu, J.; Yang, Q. A yolk-shell nanoreactor with a basic core and an acidic shell for cascade reactions. *Angew. Chem.* **2012**, *124*, 9298–9302. [CrossRef]

39. Margelefsky, E.L.; Zeidan, R.K.; Davis, M.E. Cooperative catalysis by silica-supported organic functional groups. *Chem. Soc. Rev.* **2008**, *37*, 1118–1126. [CrossRef] [PubMed]

40. Li, H.; Fang, Z.; Smith, R.L., Jr.; Yang, S. Efficient valorization of biomass to biofuels with bifunctional solid catalytic materials. *Prog. Energy Combust. Sci.* **2016**, *55*, 98–194. [CrossRef]

41. Veliscek-Carolan, J.; Hanley, T.L.; Luca, V. Zirconium organophosphonates as high capacity, selective lanthanide sorbents. *Sep. Purif. Technol.* **2014**, *129*, 150–158. [CrossRef]

42. Zhu, Y.; Ma, T.; Liu, Y.; Ren, T.; Yuan, Z. Metal phosphonate hybrid materials: From densely layered to hierarchically nanoporous structures. *Inorg. Chem. Front.* **2014**, *1*, 360–383. [CrossRef]

43. Ma, T.; Yuan, Z. Metal phosphonate hybrid mesostructures: Environmentally friendly multifunctional materials for clean energy and other applications. *ChemSusChem* **2011**, *4*, 1407–1419. [CrossRef] [PubMed]

44. Bhanja, P.; Bhaumik, A. Organic-inorganic hybrid metal phosphonates as recyclable heterogeneous catalysts. *ChemCatChem* **2016**, *8*, 1607–1616. [CrossRef]

45. Silbernagel, R.; Martin, C.H.; Clearfield, A. Zirconium (IV) phosphonate-phosphates as efficient ion-exchange materials. *Inorg. Chem.* **2016**, *55*, 1651–1656. [CrossRef] [PubMed]

46. Li, H.; Liu, X.; Yang, T.; Zhao, W.; Saravanamurugan, S.; Yang, S. Porous zirconium-furandicarboxylate microspheres for efficient redox conversion of biofuranics. *ChemSusChem* **2017**, *10*, 1761–1770. [CrossRef] [PubMed]

47. Tang, B.; Dai, W.; Sun, X.; Wu, G.; Guan, N.; Hunger, M.; Li, L. Mesoporous Zr-Beta zeolites prepared by a post-synthetic strategy as a robust Lewis acid catalyst for the ring-opening aminolysis of epoxides. *Green Chem.* **2015**, *17*, 1744–1755. [CrossRef]

48. Song, J.; Zhou, B.; Zhou, H.; Wu, L.; Meng, Q.; Liu, Z.; Han, B. Porous zirconium–phytic acid hybrid: A highly efficient catalyst for Meerwein–Ponndorf–Verley reductions. *Angew. Chem. Int. Ed.* **2015**, *54*, 9399–9403. [CrossRef] [PubMed]

49. Luo, H.; Consoli, D.F.; Gunther, W.R.; Román-Leshkov, Y. Investigation of the reaction kinetics of isolated Lewis acid sites in Beta zeolites for the Meerwein-Ponndorf-Verley reduction of methyl levulinate to γ-valerolactone. *J. Catal.* **2014**, *320*, 198–207. [CrossRef]

50. Assary, R.J.; Curtiss, L.A. Theoretical studies for the formation of γ-valero-lactone from levulinic acid and formic acid by homogeneous catalysis. *Chem. Phys. Lett.* **2012**, *541*, 21–26. [CrossRef]

51. Chalid, M.; Broekhuis, A.A.; Heeres, H.J. Experimental and kinetic modeling studies on the biphasic hydrogenation of levulinic acid to γ-valerolactone using a homogeneous water-soluble Ru-(TPPTS) catalyst. *J. Mol. Catal. A Chem.* **2011**, *341*, 14–21. [CrossRef]

Review

Metal–Organic Frameworks-Based Catalysts for Biomass Processing

Vera I. Isaeva [1,2,*], **Oleg M. Nefedov** [1] and **Leonid M. Kustov** [1,2]

[1] N.D. Zelinsky Institute of Organic Chemistry, Russian Academy of Sciences, Leninsky Prospect 47, 119991 Moscow, Russia; svitanko@mail.ru (O.M.N.); LMK@ioc.ac.ru (L.M.K.)
[2] National University of Science and Technology "MISiS", Leninsky Prospect, 4, 119991 Moscow, Russia
[*] Correspondence: sharf@ioc.ac.ru; Tel./Fax: +7-499-1372-935

Received: 5 August 2018; Accepted: 21 August 2018; Published: 31 August 2018

Abstract: Currently, metal–organic frame works (MOFs) as novel hybrid nanoporous materials are a top research interest, including endeavors in heterogeneous catalysis. MOF materials are promising heterogeneous catalytic systems due to their unique characteristics, such as a highly ordered structure, a record high surface area and a compositional diversity, which can be precisely tailored. Very recently, these metal-organic matrices have been proven as promising catalysts for biomass conversion into value-added products. The relevant publications show that the structure of MOFs can contribute essentially to the advanced catalytic performance in processes of biomass refining. This review aims at the consideration of the different ways for the rational design of MOF catalysts for biomass processing. The particular characteristics and peculiarities of the behavior of different MOF based catalytic systems including hybrid nanomaterials and composites will be also discussed by illustrating their outstanding performance with appropriate examples relevant to biomass catalytic processing.

Keywords: biomass valorization; value-added products; heterogeneous catalysis; hybrid materials; metal–organic frameworks (MOFs)

1. Introduction

In modern industrial society, non-renewable fossil fuels such as petroleum, coal and natural gas is a worldwide source of the energy and with the rapid technology development their reserves decrease with an exponentially growing rate [1]. In addition, the non-renewable fossil fuels supply the industrial processes relevant to the production of a large diversity of chemicals [2]. This concern is a stimulus for the search for new sustainable and efficient alternate sustainable feedstocks instead of non-renewable sources. In recent decades, the biomass has become a candidate No. 1 in this respect that can be converted through bio-refinery and catalytic valorization into liquid fuels, value-added chemicals and new bio-based materials such as bioplastics [3].

The transformation of biomass to chemicals and fuels can be generally realized by three different techniques: thermal, biochemical and chemical routes. Thermal techniques, like pyrolysis and gasification, can take full advantage of the entire organic substance of this resource. However, this pathway features a low selectivity and high energy input. On the contrary, bioconversion or fermentative conversion of biomass is characterized by good selectivity. An example is a fermentation of carbohydrates affording non-conventional energy sources like bio-ethanol and bio-gas as well as butanol and CO_2 [1,4]. However, bioconversion processes feature often a low efficiency. The most efficient route is the catalytic transformation of biomass into non-conventional fuels like bioethanol and biogas and highly valuable chemicals. The necessary requirements for this conversion are a moderate temperature regime, liquid phase combined with a high processes selectivity. A variety of processes for the production of green chemicals derived from biomass have been developed in the last few years [3,5].

In this context, catalysis provides efficient tools for converting and upgrading biomass into highly demanding products. It is important to note in this respect that the heterogeneous catalytic processes obey the principles of green chemistry with the result of a significant reduction of wastes [6].

The catalytic conversion of biomass and derivatives to chemicals has been the subject of intensive research efforts during the past decade. A relatively large number of reviews are focused on specific biomass feedstocks such as carbohydrates [5,7], triglycerides [8], glycerol [9], 5-hydroxymethylfurfural, cellulose, hemicelluloses and pentoses, polyols like mannitol and sorbitols, lignin and lignocellulose. Some reviews were dedicated to specific reaction types such as hydrogenation, hydrogenolysis/dehydroxylation, telomerization, metathesis and oxidation [10]. These review articles consider the important issues relating to the choice of starting feedstocks along with peculiarities in their biorefinery processing and the best way to obtain the target chemicals [11]. The available literature data show that the development of new synthetic routes for catalytic processes for transformation of biomass into known bio-based products or new building blocks and materials has a great impact.

The third category of reviews (relatively few) related to this topic (to which belongs this review) deals with particular catalytic system types, including metal catalysts [12] by highlighting their specific characteristics and advantages/disadvantages or comparing different catalysts with each other (MOFs and zeolites) in order to evaluate and develop cost-effective catalytic processes adapted to the molecular structure of highly functionalized biomass molecules [13].

Numerous solid catalytic systems for the effective conversion of biomass feedstocks into value-added chemicals and fuels have been developed. Solid catalysts exploited for biomass upgrading can be classified into four main groups: (a) micro- and mesoporous materials, (b) metal oxides, (c) supported metal catalysts and (d) sulfonated polymers [12,14]. As to metal-based catalysts, many reviews describe conversion of biomass to biofuels, comparatively less attention has been paid to catalysts adapted to the biomass-to-chemical-value chain [12].

High surface acidity and porous nanostructures (large surface area) of the nanomaterials play crucial role in these heterogeneous catalytic processes. Accordingly, several nanoporous solid acid catalysts such as porous resins, micro/mesoporous carbons, micro-/mesoporous zeolites, mesoporous metal oxides, functionalized mesoporous silicas and porous organic polymers are successfully employed in selective biomass conversion reactions [1].

In addition to nanoporosity, bifunctionality is another important feature of the efficient catalysts designed for biomass processing. Commonly, the biomass conversion process comprises a series of cascade reactions [15]. Therefore, the efficient catalytic system has to possess bifunctional properties like Brønsted-Lewis acidic, acidic–basic and metal nanoparticles-acidic or basic ones to accomplish these transformations. Accordingly, bifunctional catalytic materials are used now in one-pot multiple transformations of biomass into biofuels and related chemicals [16]. An example is the use of appropriate bifunctional catalysts that simplify processing of lignocellulosic biomass for producing biofuels and chemicals. The advanced performance of the bifunctional catalysts like metal phosphonates and relatively novel catalytic nanomaterials—metal–organic frameworks (MOFs) is compared to those of the conventional solid acid catalysts in [1].

Thus, the literature reports that the ideal catalyst for biomass should be highly porous and bifunctional. The modern class of the hybrid nanoporous inorganic–organic materials—MOFs are now among the currently most studied solids in heterogeneous catalysis [17–22]. They almost fully fit these requirements thanks to their intrinsic compositional, structural and physicochemical properties, which can be rationally designed. Essential in the context of design of the advanced catalytic system for biomass upgrading, MOFs are an ideal platform for preparation of functional materials on their basis, including MOFs-based and MOFs-derived composites or hybrid materials, which can also show the enhanced intrinsic catalytic behavior in the considered processes.

The scope of the review is to justify the opinion (*pro* and *contra*) that namely MOFs are the most promising catalytic materials for biomass valorization and upgrading for the preparation of valuable chemicals and biofuel. We intend to consider the specific features of catalytic MOFs-materials and MOFs-derived hybrids and illustrate their performance by appropriate (best and most illustrative) examples relevant to biomass catalytic processing, including synthesis and conversion of glycerol and glycerides, 5-hydroxymethylfurfural, polyols, carbohydrates as well as lignocellulosics. The specific characteristics of the MOF based materials will be also discussed.

2. MOFs-Based Catalysts and Their Main Characteristics

In this section, we will analyze briefly the characteristic features of MOF materials relevant to the sustainable catalysis of the biomass conversion.

Metal–organic frameworks (MOFs) are an important class of relatively new hybrid materials in the vast field of metal organic materials (MOMs) [6,23,24]. These inorganic–organic materials MOFs are porous coordination polymers constructed from inorganic nodes (metal ions or clusters) coordinated with multi-topic organic bridging linkers through strong bonds to form three-dimensional (3D) coordination networks [25,26]. Over past two decades, they capture an increased attention of the worldwide researcher community due to their outstanding properties [27]. The crystalline nature with a low crystal density (up to 0.13 g cm^{-3}), ultrahigh surface area (up to 10 400 m^2 g^{-1}), a large pore aperture (up to 98 Å), structural diversity as well as tailored pore functionality make MOFs the promising functional materials with potential applications in diverse fields, such as gas sorption and separation, chemical sensing, proton conductivity, biomedicine and so forth [28]. In particular, heterogeneous catalysis is one of the most important applications for MOFs [29]. The open porous system and high porosity and large inner surface area of MOFs allow fast mass transport and/or interactions with substrates [6,30]. The efficiency of MOF catalytic materials in shape- and size-selective and regio- and enantioselective reactions can be realized due to unique active site structures and adjusted micro/meso porosity of a number of the metal–organic frameworks [31,32]. MOF-based catalysts can have active sites both in inorganic nodes (metal centers) and organic linkers [33]. Especially, the organic bridging linkers may be used as scaffolds to which distinct catalytic complexes, bio-molecules and homogeneous catalysts can be immobilized or encapsulated.

MOF catalysts exhibit the following distinct traits: (i) uniformly dispersed catalytic sites on the pore surface, which contributes to selectivity; (ii) appropriate hydrophilic and hydrophobic pore nature to facilitate the recognition and transportation of reactant and product molecules; (iii) multifunctional microenvironment to realize synergistic catalysis; and (iv) simple separation and recovery for long-term usage [34]. The presence of metal coordination centers in MOF frameworks can promote a wide range of organic reactions [35], in particular, relevant to biomass conversion. Additionally, thanks to the versatility of synthesis and their structural and compositional diversity MOFs have a reputation of eco-friendly alternatives for catalysis [36].

However, the full realization of the potential of these catalytic materials is limited often by a deficiency of functional sites in the MOF frameworks [37]. The adequate solution of this problem is fine tuning of the MOF structures and porosity for a specific catalytic process. Among the possible routes to introduce the supplementary active sizes in MOF host matrices, one can use the direct synthesis by careful choice of organic and inorganic building blocks or their post-synthetic functionalization (PSM) [38], grafting of active groups on the open metal sites [39] and encapsulation of active species like metal and metal oxide nanoparticles as well as small molecules (active homogeneous catalysts) in the pore voids [26,40–42]. The diversity of the modification routes distinguishes MOFs from other nanoporous materials such as zeolites and activated carbons.

MOFs are compared very often with zeolites that is, another type of highly ordered nanoporous materials. To date, zeolites have been used as heterogeneous catalysts in many industrial processes [43,44]. However, these materials exhibit limitations for the manufacturing of bulky and high value added organic molecules due to the small pore size of zeolite channels and cavities. Multi-functionality, high porosity, tenability and original flexibility are placing MOF materials at the frontier between zeolites and enzymes [45,46]. However, in most cases, MOFs show a good thermal stability (400–500 °C) but not a bit approaching the stability of zeolites (as high as ~700 °C.

As it was mentioned above, in terms of catalytic biomass valorization, a very large number of MOF materials have a great potential due to their high porosity and compositional versatility. The particular property of MOFs is that their bifunctionality, or rather multifunctionality comprises the possession of active sizes of different sorts on and in their porous matrices. Bifunctional MOFs containing unsaturated metal centers and metal NPs or organic linkers with Lewis/Brønsted acidic/basic sites are used as solid catalysts in organic transformations [6,35]. Some multi-functional MOFs have been also developed and demonstrate an enhanced performance in different organic reactions including transformations relevant to the biomass processing [13].

Thus, the structure of MOFs is highly attributed to the catalytic activity and selectivity in processes of biomass refining. Accordingly, we intend to highlight structure–activity/selectivity correlation in this review.

3. Rational Design of MOF Catalysts for Biomass Valorization

Mainly, the available excellent reviews related to the MOF application in biomass valorization highlight the processes and reactions [13,21,36]. On the contrary, we are attempting to focus on the MOF materials essential properties and approaches to their design, which involve the versatile hybrid nature of highly porous metal–organic frameworks: a careful choice of organic and inorganic building blocks, control of the size of MOF catalyst materials in the nanoscale range, post-synthesis modification of inorganic nodes and organic linkers, encapsulation of functionalities (catalytically active molecules and species) in MOF host matrices, preparation of MOFs-based composites and MOFs-derived materials.

In the next section, we present the ways for the rational design of MOF structures with illustration by the relevant catalytic processes. The most representative process from this point of view is carbohydrate depolymerization, resulting in the formation of platform compounds. This reaction is an efficient strategy in order to obtain the value-added chemicals and bio-fuels, such as methyl lactate, levulinic acid, hydroxymethylfurfural (HMF), furfural and so forth. As a rule, these transformations demand for the presence of acidic active sites, because the reaction pool comprises dehydration, isomerization, esterification as well as cascade reactions involving these reactions. In this process, all features or possibilities of MOF structures can be involved.

Practical routes to introduce the appropriate catalytic functionalities needed for efficient biomass valorization include direct synthesis or post-synthetic modification [47], grafting of active groups on the open metal sites of certain structures [48] and encapsulation of active species [49,50]. However, a simpler strategy involving a judicious selection of MOF structures with desired active sites in their building blocks, inorganic nodes and organic linkers allows one to obtain sometimes good catalytic results.

3.1. Evaluation of Appropriate MOF Structures Relevant to Biomass Valorization

An appropriate choice of the parent (precursor) MOF structure is a first step in this engineering of the MOFs-based catalysts. Generally speaking, there are some popular MOF materials that are most frequently used as heterogeneous catalysts [34]. Moreover, the overwhelming number of publications relevant to biomass valorization discussed a catalytic performance of these MOF structures, which are, first of all, HKUST-1, MIL-101(Cr), UiO-66 and UiO-67 (amino-modified analog of the UiO-66 analog) frameworks [13]. Their structures have coordinatively-unsaturated (open) sites with Lewis acidity in inorganic nodes (metal ions) in the networks. These Lewis acid sites as structure constituent elements are of paramount importance for cascade processes of catalytic biomass valorization (involving depolymerization, dehydration and isomerization).

The first MOF structure, which is preferred as a biomass upgrading catalyst is the MIL-101(Cr) $(Cr_3(F)O(BDC)_3$, BDC = 1,4-benzenedicarboxylate) framework [51]. It has two types of zeotypic mesoporous pores (Figure 1) with free diameters of ~2.9 and 3.4 nm, respectively, accessible through two microporous windows of ~1.2 and 1.6 nm. The MIL-101(Cr) framework has a very high specific surface area of 4100 m^2 g^{-1} (BET). On the other hand, this framework contains unsaturated (open metal) sites (Cr^{3+}, up to 3 mmol g^{-1}) in which solvent or substrate molecules can be adsorbed. These unsaturated coordination positions can be used as Lewis acid sites in many organic transformations. The MIL-101(Cr) material is thermally stable up to 275 °C (air), which allows its application for a wide range of catalytic reactions. For this reason, the MIL-101(Cr) structure is one of the most studied MOF catalysts for biomass upgrading [52].

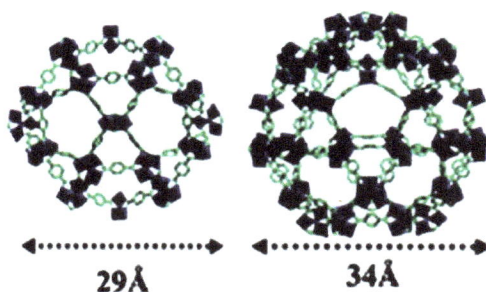

Figure 1. Two types of cages present in the structure of MIL-101(Cr). Reprinted with permission from reference [34]. Copyright 2017 Royal Society of Chemistry.

HKUST-1 (Cu$_3$(BTC)$_2$, BTC = benzene-1,3,5-tricarboxylate) is another metal-organic framework extensively studied as a catalyst for biomass valorization (Figure 2). It is one of the most known MOF compounds with a long history due to its interesting structure and adsorption characteristics and open metal sites (Cu^{2+}) with Lewis acidity, which are potential active centers [53,54]. Noteworthy, the concentration of these open Lewis centers in the HKUST-1 framework is high thereby they can contribute in particular improvement of the catalytic activity [55]. To date, a number of studies reported an important catalytic behavior of [Cu$_3$(BTC)$_2$], highlighting its potential as a promising acidic heterogeneous catalyst for the production of fine chemicals, including value-added products of biomass-upgrading [56] due to Lewis acid sites located at the inorganic nodes and Brønsted acid sites situated in 1,3,5-benzenetricarboxylate linkers.

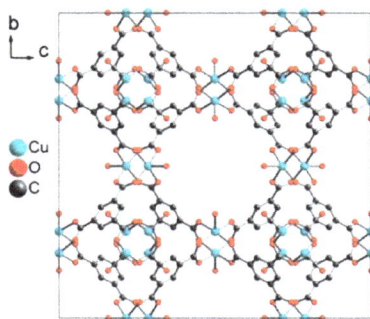

Figure 2. View of the packing of the cubic unit cell of HKUST-1. Reprinted with permission from reference [34]. Copyright 2017 Royal Society of Chemistry.

The third MOF structure evaluated for heterogeneous catalysis is the UiO-66(Zr) framework based on BDC linkers and its isostructural analogue, UiO-67(Zr) based on the 4,4'-biphenyldicarboxylate (BPDC) linker (Figure 3) [57]. The UiO-66(Zr) and UiO-67(Zr) metal–organic frameworks are studied for different application fields due to their exceptional thermal and chemical stability [58].

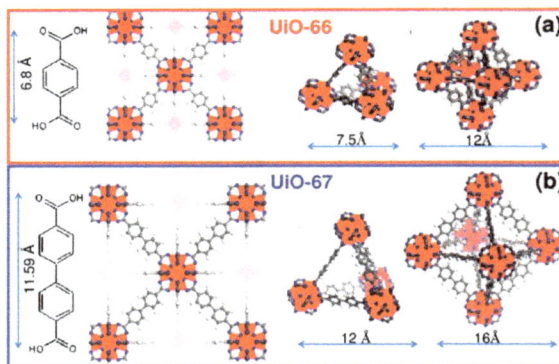

Figure 3. Structures of UiO-66 (**a**) and UiO-67 (**b**). Reprinted with permission from reference [34]. Copyright 2017 Royal Society of Chemistry.

The importance of open metal sites (coordinatively unsaturated sites) for catalytic activity of MOF materials should be especially emphasized. In many cases, the Lewis acid character of the metal–organic frameworks is derived from the creation of a coordination vacancy upon thermal removal of a solvent molecule (usually, H_2O) initially bound to the metallic centers (in inorganic nodes of the framework). In this context, the HKUST-1 framework is a typical example.

As it was mentioned above, the most representative examples of MOF-based catalysts for biomass transformation are the MIL-101(Cr) and UiO-66(Zr) (as well as UiO-67(Zr)) frameworks and their derivatives. The reasons for this are their Lewis acidity, open metal sites (Cr^{3+}, Zr^{2+}) and thermal stability of these metal-organic matrices. For instance, zirconium terephthalate UiO-66(Zr) is a highly active, stable and reusable heterogeneous system for acid catalysis [59]. Moreover, when defects are present in the structure of UiO-66(Zr), coordinatively unsaturated Zr ions with Lewis acid properties appear, which assists to the catalytic activity of the resulting defective framework. Therefore, the catalytic activity increases along with the number of missing linkers [13]. However, UiO-66(Zr) type frameworks with an ideal crystalline structure without coordination vacancies also display catalytic activity as Lewis acid solid in some organic transformations like Fisher fatty acid esterification related to biomass upgrading [60]. More detailed characteristics of the UiO-66(Zr) matrix as a catalyst relevant to biomass valorization will be given below.

One more type of metal–organic frameworks is very prospective for biomass valorization. This is a series of zeolite imidazolate frameworks (ZIFs), representing a subfamily of MOF materials that display exceptional chemical and thermal stability, as high as 550 °C. ZIF frameworks are composed of divalent metal nodes linked by imidazolate bridging ligands [61]. Their 145° metal-imidazolate-metal angles in ZIF structures (Figure 4) are very close to the zeolite Si-O-Si angle, this zeolite-like topology provides the unique stability and robustness of the framework [62]. Due to their unique stability [63] and N-heterocyclic moiety providing specific catalytic performance and hydrophobic properties [64], ZIF matrices are very attractive as heterogeneous catalysts for biomass valorization.

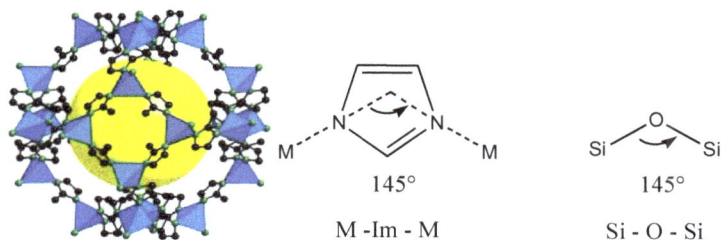

Figure 4. The structure of the ZIF-8 zeolite imidazolate framework. The M–Im–M (2-methylimidazolate–metal angle in ZIF structures is similar to the angle between Si–O–Si bonds in zeolites (145°). Reprinted with permission from reference [61]. Copyright 2010 American Chemical Society.

Another important issue, which can be taken into account is the hydrothermal stability of MOF based catalysts. Essentially, biomass valorization reactions, like dehydration or esterification processes involve an aqueous reaction medium or water generated along with products. Therefore, hydrothermally stable MOF materials, such as MIL-53(Al), MIL-101(Cr), UIO-66(Zr), UIO-67(Zr) and ZIF structures are highly demanded for discussed catalytic processes.

These MOF and ZIF materials can be used as catalysts in a pristine form, as well as post-synthesis and in situ modified for tuning their acid strength regarding the biomass valorization process. For instance, the BDC linker in the UiO-66 and MIL-101(Cr) frameworks can be functionalized by partially or integrally replacing it with analogous organic linkers via direct or post-synthesis modification due to the robustness of the framework [65]. The porous systems in the HKUST-1, UiO-66 and MIL-101(Cr) (2.9 and 34 nm) structures contribute to the encapsulation of the catalytic species, for example, polyoxometallates, metal nanoparticles or poly(N-bromomaleimide) in these host matrices [13]. Particularly, the confinement effect of these frameworks prevents the embedded metal nanoparticles from agglomeration.

In particular, to enhance the functional properties of ZIF materials, a number of efficient methods were developed. There are (1) linker functionalization by post-synthesis modification (PSM) or solvent-assisted ligand exchange (SALE) (or post-synthetic exchange (PSE)); (2) encapsulation of metal and metal oxide nanoparticles (NPs) inside ZIF porous matrices and (3) adsorption or encapsulation of biomolecules [66].

All these strategies regarding the design of the MOF based catalytic materials for biomass valorization will be discussed in the next sections.

3.2. Choice of MOF Building Blocks

Since most biomass transformation processes including carbohydrates depolymerization, involve often reactions demanding acidic sites, one of the conditions of MOF-catalyst efficiency is the presence of Lewis and/or Brønsted functionality at metallic centers (inorganic nodes) in the framework. Therefore, a careful selection of MOF building blocks for catalytic biomass valorization is of paramount importance.

The study of four isostructural MOF-74 catalytic materials differed in metallic centers (Co, Ni, Mg, Zn) shows clearly a strong impact of the metal ions in the sugar conversion into methyl lactate (Figure 5) [67]. The yields of methyl lactate were 16%, 19% and 20% while using MOF-74(Ni), MOF-74(Zn) and MOF-74(Co), respectively. The MOF-74(Mg) material exhibited the best catalytic activity and a methyl lactate (as the main product) yield of 47% was obtained from sucrose, while the highest product yield from glucose of 35% was achieved at optimal reaction conditions (220 °C, 6 h). MOF-74(Mg) was also most active for the retro-aldol reaction of fructose to methyl lactate as a main product with a yield of 37%. Simultaneously, the conversion rate of glucose was higher and a slightly higher yield of side products like methyl glycolate and glycol aldehyde dimethylacetal was obtained from glucose. As for disaccharides, except sucrose, lower yields of methyl lactate (19% for both lactose and maltose) were achieved. This may be explained by the resistance of these disaccharides to methanolysis under reaction conditions [68].

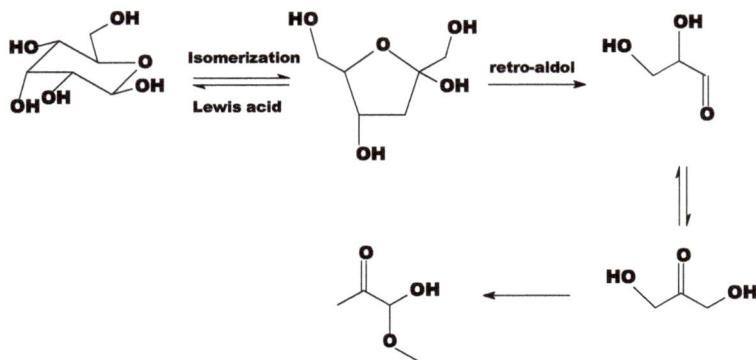

Figure 5. The conversion of glucose into methyl lactate in near-critical methanol solution over MOF-74 catalytic materials.

The reason for the enhanced catalytic performance of MOF-74(Mg) is the presence of stronger Lewis acid sites in this structure [69], which are more efficient for the conversion of glucose [70]. A smaller crystal size and a larger surface area of the MOF-74(Mg) catalyst contribute probably also to its activity. These factors may favor the accessibility of reactants to the active sites. This Mg-based catalyst can be reusable for at least three cycles without deactivation.

The same transformation was studied using other types of MOF materials as catalysts. The ZIF-8 and ZIF-67 zeolitic imidazolate frameworks were tested as catalysts in the conversion of sugars (sucrose, glucose and fructose) into methyl lactate [71]. These matrices have the same sodalite type structure and similar textural properties, namely cavities with sizes of 11.6 Å accessible by small windows of 3.4 Å for the ZIF-8 framework and cavities of 11.4 Å and small windows of 3.3 Å for the ZIF-67 framework [63]. Due to the framework flexibility, the cavities of the ZIF-8 and ZIF-67 catalysts are large enough to accommodate the molecular diameters of glucose and fructose (~8.5 Å) [72].

However, these ZIF catalytic materials display a different catalytic behavior in the sugar conversion into methyl lactate. The reason for this difference is the presence of different metallic centers (Zn^{2+} in ZIF-8 and Co^{2+} in ZIF-67) in these ZIF frameworks. The ZIF-8 catalytic material is the most active (methyl lactate yield of 42%, 160 °C, 24 h) and can be reused in four catalytic cycles. On the contrary, while using ZIF-67 as a catalyst, the final product yield was only 19.1%. Two factors contributing to this different catalytic behavior were suggested: the crystal size and the Lewis acidity of the metal sites. The crystal size of ZIF-67 is larger (about 1 μm) than that of ZIF-8 (100–150 nm). Therefore, the smaller crystal size improves the accessibility of reactants to the active sites. Another reason is related to the difference in the acid strengths of both catalysts: the ZIF-8 matrix has stronger acid sites, while the acidity of the ZIF-67 material is moderate.

The systematic study of the impact of the metallic centers was carried out with isostructural metal–organic frameworks belonging to the MIL family based on M^{3+} ions (Al^{3+}, Fe^{3+}, Cr^{3+}, V^{3+}) [73]. MIL materials have a great potential from the practical point of view due to their high thermal and chemical stability (see previous section). The synthesis of solketal from acetone and glycerol (I) was investigated over catalytic MOF materials, namely, MIL-100(M) and MIL-53(M) (M = V, Al, Fe and Cr), as well as mixed MIL-53(Al,V) (Al/V–100/0, 75/25, 50/50, 25/75 and 0/100 atom/atom). The main products were a five-membered solketal (2,2-dimethyl-1,3-dioxane-4-methanol, (II)) and a six-membered acetal (2,2-dimethyl-dioxane-5-ol, (III)) (Figure 6). It was demonstrated that the reaction rate and isomer selectivity depend on different parameters such as the type of the metal ion, the length of the M-O bond, the rate constant for the exchange of the water molecules from the first coordination sphere of the metal ion and the value of the zero point of the surface charge (pH_{PZC}). In particular, the glycerol conversion decreases in the following order: $V^{3+} > Al^{3+} > Fe^{3+} > Cr^{3+}$. This order correlates with the value of pH_{PZC}. The reaction rate and selectivity towards II increase with increasing the V^{3+} content in the mixed MIL-53(Al,V) framework. V-containing MOFs possess a high activity and selectivity at 25 °C. The MIL-100(V) and MIL-47(V) metal-organic materials featured the higher efficiencies as compared to strong acids, such as H_2SO_4, $SnCl_2$ and p-toluenesulfonic acid (25 °C). The MIL-100(V) catalyst is reusable for 4 cycles.

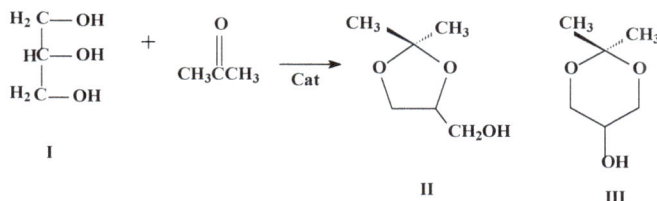

Figure 6. Glycerol acetalization over MIL-100(M) and MIL-53(M) catalysts.

Another reason for the differences in the catalytic activity is related to the exchange of water molecules in the first coordination sphere of the metal ion. The isomer selectivity possibly depends on the length of the M-O bond in the MIL framework. The decrease in the length of the M-O bond contributes to the increased II yield. Thus, both high activity and enhanced selectivity towards II are achieved over the V-containing frameworks, MIL-100(V) and MIL-47(V).

Zr-containing hybrid materials have been proven as efficient catalysts for biomass conversion, due to rather strong Lewis acidity [74]. An example of the design of the catalytic MOF materials using Zr^{2+} ions as inorganic building blocks with strong Lewis acidity are Zr-based metal–organic frameworks with terephthalate (UiO-66(Zr)) or 2-aminoterephthalate ligands (NH$_2$-UiO-66(Zr)), which are preferential structures for biomass valorization catalysis (see above). Both matrices show activity in the esterification of levulinic acid with EtOH, n-BuOH and long-chain fatty alcohols (Figure 7). This process leads to biomass-derived alkyl levulinates of industrial importance [75]. The catalytic efficiency of the UiO-66(Zr) and NH$_2$-UiO-66(Zr) materials is comparable (and in some cases superior) to the heterogeneous acid catalysts previously reported in the literature, such as supported heteropolyacids and zeolites. In this respect, they are somewhat inferior to highly acidic sulfated mixed Zr and Ti oxides [76]. The reason for such a superior activity is a synergy of acidity of the metal centers and organic linker functionality in the NH$_2$-UiO-66(Zr) catalyst. This synergy is realized as a dual acid–base activation mechanism. In this case, levulinic acid is activated on Zr^{2+} sites, while alcohol is activated at the amino groups of the linkers, which are Brønsted basic sites.

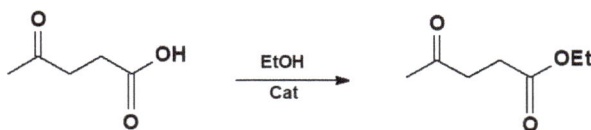

Figure 7. Esterification of levulinic acid with EtOH over UiO-66(Zr) and NH$_2$-UiO-66(Zr) catalytic materials.

Another activity source in the discussed UiO-66(Zr) and NH$_2$-UiO-66 catalysts are open metal centers as Zr^{4+} ions. They represent defect sites formed due to linker deficiency in the framework (see above) and their concentration is dependent critically on the synthesis conditions [13,36,75].

The particle size impacts also on the catalytic activity of the UiO-66(Zr) material (see below the relevant section).

In addition to organic linkers, metal clusters composed MOF frameworks may be another source of intrinsic Brønsted acidity (and Lewis acidity as well). Z. Hu reported [77] the preparation of two highly stable sulfonated and hierarchically porous MOF structures containing Zr and Hf clusters, namely, NUS-6(Zr) ([$Zr_6O_4(OH)_8L$]$_{3.5}$·xH_2O, L = SO_3-BDC) and NUS-6(Hf) {[$Hf_6O_4(OH)_8L$]$_{3.5}$·xH_2O), respectively, by a modulated hydrothermal approach. Both frameworks can be regarded as UiO-66 framework derivatives and topologically are similar to it. On the contrary, the NUS-6(Zr) and NUS-6(Hf) frameworks feature partially missing linkers and Hf_6 (Zr_6) clusters. The hierarchically porous structure of NUS-6(Zr) and NUS-6(Hf) is composed of squashed mesopores (~2.6 nm × 3.6 nm, measured from Hf to Hf vertexes) that are interconnected with microporous tetrahedral and octahedral cavities as in the UiO-66 framework. Such a hierarchically porous structure contributes to the catalyst performance thanks to the dense catalytic sites in micropores and easy mass transfer through mesopores [78].

In the NUS-6(Hf) framework, the μ_3-OH groups in Hf clusters (which are the inorganic building blocks) could act as Brønsted acid sites. It was shown by acid-base titration that the Brønsted acidity of the NUS-6(Hf) framework is stronger than that of NUS-6(Zr). These MOF materials were used as solid acid catalysts for dehydration of fructose to 5-hydroxymethylfurfural (HMF) (Figure 8). Due to a synergy of strong Brønsted acidity of the Hf-based nodes and sulfonated linkers, the NUS-6(Hf) material exhibited better reaction kinetics and chemoselectivity than that of their Zr-based counterparts (NUS-6(Zr)) in the dehydration of fructose. By using the NUS-6(Hf) catalyst, the quantitative fructose conversion and a HMF yield of 98% were achieved, which makes this framework one of the best heterogeneous catalysts for such conversion. The superior catalytic activity of NUS-6(Hf) was attributed to its stronger Brønsted acidity as well as more suitable pore size that can inhibit side reactions. Therefore, sulfonated Hf-containing MOF frameworks containing both organic and inorganic Brønsted acid sites should be excellent solid acid catalysts.

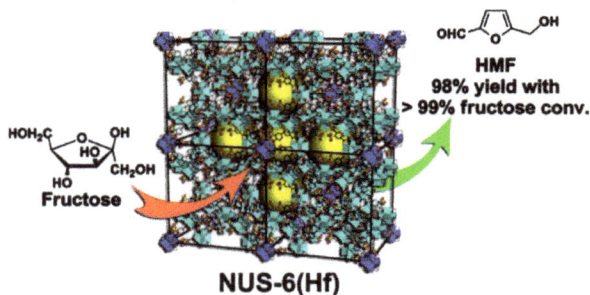

Figure 8. Synthesis of HMF from fructose over hierarchically porous NUS-6(Hf) catalyst. Reprinted with permission from reference [77]. Copyright 2016 American Chemical Society.

Another biomass valorization process governed very much by Lewis acidity is the selective synthesis of monoglycerides by the esterification of glycerol with fatty acids. This is a difficult reaction because of immiscibility of reagents and the formation of di- and tri-glycerides as by-products [79]. A rather distinct amorphous tin-based organic framework with a pore size of ca. 1 nm demonstrates superiority as the catalyst in the reaction of oleic acid with glycerol compared to zeolites, which were rather inactive in this process. On the contrary, a tin–organic framework (Sn-EOF) with Lewis acidity showed an excellent performance in esterification of oleic acid with glycerol (150 °C). Sn–EOF is an elemental-organic framework (EOF) material constructed from Sn^{4+} ions connected to the 4,4′-dibromobiphenyl organic linker via element-carbon bonds.

Sn-EOF displayed a higher activity as compared to other studied MOF-based catalysts with both Lewis and Brønsted acid sites, including their composites with highly acidic heteropolyacids (Keggin anions), such as HPW/HKUST-1, HSiW/HKUST-1, NH$_2$-MIL-53(Al), NH$_2$-MIL-68(In), UiO-66, NH$_2$-UiO-66 and ZnF(NH$_2$-TAZ) materials. Using the Sn-EOF catalyst resulted in achieving an excellent selectivity (≥98% monoglyceride) at a 40% conversion. Ii is important that the Sn-beta zeolite was inactive with an oleic acid conversion of 4% (20 h), even below the conversion observed in absence of the catalyst (20%).

Leaching of tin from the Sn–EOF catalyst was suppressed by limiting the amount of oleic acid in the starting mixture, because normally oleic acid may cause decomposition of metal–organic frameworks by coordination with metallic centers. This study demonstrates the potential of the unique MOF feature–fine tuning of the catalytic activity by judicious choice of inorganic building blocks in the metal-organic framework for the synthesis of monoglycerides.

As it can be seen from the observed literature examples, the MOF catalysts are mostly bifunctional due to the presence of both Brønsted and Lewis sites in inorganic nodes. However, in most publications relevant to catalytic biomass valorization, the hybrid (or dual) nature of metal-organic framework is involved. Therefore, the catalytic active sites in inorganic nodes of the framework (as a rule with Lewis acidity) are supplemented very often by Brønsted centers in organic linkers. An impact of the organic moiety on the MOF catalyst performance will be discussed below.

3.3. Organic Linker Choice for MOF Catalyst in Biomass Valorization

One more efficient strategy for designing MOF catalytic materials for biomass valorization is based on tailoring the functional groups in organic linkers. In these processes, acid functionality is of prime interest. An appropriate organic linker choice allows tuning the catalytic activity by supplementing the Lewis and Brønsted acidity at inorganic ions by Brønsted acidic functionality at organic linkers. Compared to the relatively comprehensive work that has been conducted on the Lewis acidity of MOFs, exploring Brønsted acidity of MOF materials is more challenging and remains less explored [77], mainly because of the weakened framework stability caused by the introduction of Brønsted acidity typically through sulfonic acid groups. There are only very few successful examples of such strategy usage for heterogeneous catalyst design for biomass valorization. One of them is the preparation of the mesoporous SO$_3$H-MIL-101(Cr) framework containing both Lewis acid (as Cr^{3+} ions) and Brönsted acid sites starting from sulfonated organic building blocks (monosodium 2-sulfoterephthalic acid (BDC-SO$_3$Na)) [80,81]. The sulfonic groups are accessible—Brønsted acid sites distributed with a high density on the SO$_3$H-MIL-101(Cr) pore surface. The SO$_3$H-MIL-101(Cr) matrix has a high surface area and well dispersed nanocrystals. Thanks to these strong Brønsted acid sites, SO$_3$H-MIL-101(Cr) catalytic material shows a high activity in cellulose hydrolysis into mono- and disaccharides, such as glucose, xylose and cellobiose and demonstrates high durability in the catalytic reaction (Figure 9).

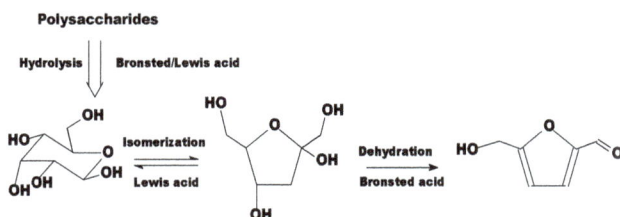

Figure 9. Hydrolysis of polysaccharides and cascade reactions of glucose as a platform chemical to HMF transformation over SO$_3$H-MIL-101(Cr) material.

The bifunctional SO$_3$H-MIL-101(Cr) material mentioned above was extensively studied in challenging and practically demanding processes—cascade reactions of glucose as a platform chemical to HMF transformation [80,82]. This process involves isomerization over Lewis acid sites followed by the dehydration to HMF over Brönsted acid sites (Figure 9). It could be complicated by the formation of undesirable side products through an extra-isomerization step leading to a decrease of the HMF yield. A number of solid acid catalysts were examined in this process and relevant experimental results showed that these heterogeneous systems afforded lower HMF yields than those obtained using homogeneous catalysts. Some MOF materials were also studied in the glucose into HMF conversion. However, using them gave only low yields of the target products–below 16%. The modification of organic benzene-1,4-dicarboxylate linkers of the MIL-101(Cr) material with sulfonic acids has been proven as a promising strategy for an enhancement of its catalytic activity. The bifunctional MIL-101-SO₃H catalyst can promote an isomerization of glucose to fructose, while the –SO₃H groups are active Brönsted acid sites for the fructose dehydration (Figure 10).

This strategy resulted in achieving a 29% conversion of glucose (24 h, THF:H₂O (*v*:*v* = 39:1) mixture) over a functionalized SO₃H-MIL-101(Cr)) catalyst. This conversion level is lower as compared to traditional acid catalysts, such as mesoporous tantalum phosphate and Sn montmorillonite [13,80]. As a rematch, this MOF catalyst features a high selectivity of 80% to HMF over levulinic acid. This selectivity level is much higher than that achieved using other inorganic systems, such as Amberlyst-15 and sulfuric acid, which gave mostly levulinic acid. It was found that the SO₃H-MIL-101(Cr)) material is inactive in the isomerization of glucose to fructose.

The SO₃H-MIL-101(Cr) matrix retains its porous and crystalline structure during catalytic cycles. However, this catalyst has a sufficient drawback, such as pore blocking in the catalysis course by the formation of bulky humins. Therefore, a reactivation step is needed during exploitation of the SO₃H-MIL-101(Cr) material.

The same sulfonic acid functionalized MIL-101(Cr)-SO₃H material was examined as the catalyst for HMF synthesis in an integrating process involving biomass-derived γ-valerolactone-mediated dehydration of glucose [80]. Under the optimal reaction conditions (150 °C, 120 min) in a fixed-bed reactor, an HMF yield was 44.9% and selectivity of 45.8%. Noteworthy, using this MOF catalyst resulted in a steady HMF yield.

The next example of an enhanced catalytic performance of a sulfonated SO₃H-MIL-101(Cr) material is the conversion of furfuryl alcohol (FA) into ethyl levulinate (EL) [83]. A target product yield of 79.2% at a 100% FA conversion in ethanol was achieved over this bifunctional MOF catalyst. The SO₃H-MIL-101(Cr) material shows a superior performance to other sulfonic acid-functionalized solid catalysts in this transformation (Figure 10). In particular, a density and strength of -SO₃H Brönsted acid sites placed in the organic linker sites is a key factor impacted on this catalyst performance. Actually, the SO₃H-MIL-101(Cr) material displayed a higher catalytic activity as compared to the performance of other sulfonated MOF matrices, such as the SO₃H-UiO-66(Zr) framework (23.4% FA conversion, 3.1% EL yield, 20.8% 2-ethoxymethylfuran yield), which is characterized by a lower density of Brönsted acid sites.

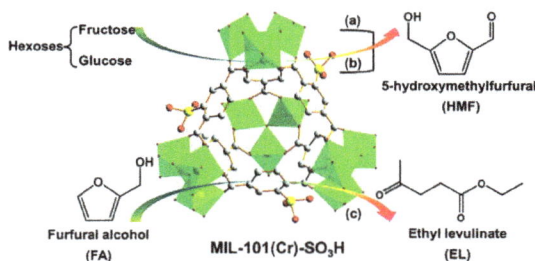

Figure 10. Transformations of (**a**) fructose and (**b**) glucose into HMF and (**c**) furfural alcohol into ethyl levulinate over SO₃H-MIL-101(Cr). Reprinted with permission from reference [36]. Copyright 2016 Viley-VCH.

It can be seen from the observed literature, that the choice of the functionalized organic linkers as well as archetypical metal–organic frameworks in whole for engineering an acid MOF catalyst is rather limited. Other approaches, that is, the post-synthesis modification and in-situ synthesis of the functionalized MOF matrices, allow one to expand a range of possible catalytic MOF materials for biomass processing.

3.4. Post-Synthesis Modification of MOF Structures

As it was mentioned above, one of the most distinct MOFs characteristics is a possibility of their rational design by post-synthetic modification of their organic or inorganic building blocks with appropriate functional groups or introducing the functional species inside pore voids via template synthesis in-situ. To date, numerous works reported the efficiency of the preparation of MOFs-based catalytic systems using these ways [84–86].

Here we present some important results concerning tuning the MOF catalyst activity in biomass upgrading by post-synthesis modification through grafting organic molecules containing appropriate functional groups to coordinatively unsaturated sites and introducing organic molecules with a desirable function in organic linkers. Encapsulation of functional species and molecules in MOF host matrices will be discussed in the "MOF-based composite" section. We consider below the illustration of these approaches taking catalytic depolymerization of carbohydrates as a most representative example.

As it was mentioned above, despite of their rather strong Lewis acidity, most MOF materials are lacking Brønsted acidity for efficient catalytic processes of biomass valorization, first of all, for sugar dehydration reactions. Besides the one-pot synthesis using organic building blocks with appropriate functionalities, the most common method of introducing Brønsted acidity to MOFs catalysts is by post-synthetic modifications. In most cases, sulfonic acid groups are introduced using this strategy. However, such supplementary acidification of MOF materials can result in some stability and crystallinity loss. Therefore, only robust structures can be modified by this post-synthetic approach for introducing functional organic sites [87] that could be used directly or after post-synthesis modifications. An example is a post-synthetically modified UiO-66(Zr) framework in which Brønsted acid sites are introduced [88]. A benefit of such post-synthesis modification is the creation of the bifunctionality in the framework, that is, generation of both Lewis and Brönsted acid sites. Such dualism of active sites is achieved with simultaneous appropriate selection of inorganic nodes (Zr^{4+}) with Lewis acidity and post-synthesis modification of organic linkers by sulfo functionalities.

3.5. Post-Synthesis Modification of Inorganic Nodes

The grafting of ligands with additional functional moieties to inorganic nodes is not trivial, since such groups may directly coordinate to the metal ions [50]. An elegant way of tuning the selectivity of active acid sites by post-synthesis modification of inorganic nodes in the NU-1000 metal–organic framework was demonstrated in [89]. The NU-1000 structure is composed of the $Zr_6(\mu_3–OH)_4(\mu_3–O)_4(OH)_4(OH_2)_4$ nodes (Zr_6 nodes) and tetratopic 1,3,6,8-tetrakis(*p*-benzoate)pyrene (TBAPy) linkers. While the UiO-66 and UiO-67 structures discussed above feature 12-connected Zr6 nodes, NU-1000 is composed of 8-connected Zr6 nodes and contains terminal water and hydroxyl ligands in addition to μ3-bridging hydroxyl and oxido groups [90]. The NU-1000 framework has wide mesoporous channels with a diameter of 31 Å and smaller pores connected perpendicularly to these large pores that allow the diffusion of a substrate throughout the material. This hierarchical nanoporous structure contributes to exploring this matrix as a catalyst or a catalyst support [91,92].

The NU-1000 metal-organic framework was studied in glucose transformation to HMF. Noteworthy, its strong Lewis acid sites in Zr_6 nodes are too reactive and catalyze undesired side reactions. To enhance the selectivity of this MOF catalyst, a partial phosphate modification of zirconia-cluster nodes in the NU-1000 framework was carried out in order to poison unselective active sites with phosphoric acid by forming the oxo substituents of grafted phosphate species. The NU-1000 catalyst modified by this method demonstrates both a high yield of HMF and selectivity (~64%), at an initial glucose concentration of 1 mM in water/2-propanol. In contrast, complete replacement of Lewis sites with Brønsted acid sites using the same strategy reduces the catalytic performance of the NU-1000 material. The authors conclude that the ideal catalyst for the studied process should be bifunctional one. In other words, both moderate Lewis acidity as well as Brønsted acid sites in the NU-1000 matrix are demanded. On the contrary, similar partial phosphate modification of bulk zirconia results in a considerable reduction of the HMF selectivity. Ii is important to note that this selectivity drop is observed despite the fact that this inorganic material has almost identical Brønsted acidity to the selective NU-1000 system.

An important issue or critical point for practical exploitation of MOF catalysts is their insufficient hydrothermal stability and poor resistance to acidic medium, which hinder their usage as catalysts in dehydration reactions to produce biofuels or fine chemicals. As it was mentioned above, the mesoporous MIL-101(Cr) framework is a bifunctional material featuring the dual character of Cr^{3+} ions in inorganic nodes as open sites assisting to substrate coordination and simultaneously serving as Lewis acidic sites catalyzing cascade reactions relevant to the biomass transformation. Thanks to this bifunctionality, the MIL-101(Cr) material displays an enhanced catalytic performance in xylose dehydration yielding 49% of target furfural (170 °C). However, this material can be utilized only in 4 catalytic cycles. Post-synthesis modification of coordinatively-unsaturated Cr^{3+} centers in the MIL-101(Cr) framework with hydrophobic octadecyltrichlorosilane (OTS) allows one to improve the hydrothermal stability of this matrix. The protection of the Cr^{3+} active sites by silane groups contributes to the retention of the crystalline structure of the OTS-MIL-101(Cr) framework with a different extent of OTS modification as compared to the pristine MIL-101(Cr) material under severe conditions of xylose dehydration (an elevated temperature in combination with eliminated water). Therefore, the modified framework 0.5-OTS-MIL-101(Cr) (with OTS content of 50%) allows one to increase the furfural yield (56%). Simultaneously, a dramatic improvement of its catalytic durability is another benefit of this functionalization. As a result, the 0.5-OTS-MIL-101(Cr) material can be exploited without any activity loss during up to 8 catalytic cycles [93].

3.6. Post-Synthesis Modification of Organic Linkers

A series of sulfonic acid-functionalized metal–organic frameworks (MOF-SO$_3$H), namely, MIL-101(Cr) [SO$_3$H-MIL-101(Cr)], UiO-66(Zr) [SO$_3$H-UiO-66(Zr)] and MIL-53(Al) [SO$_3$H-MIL-53(Al)], were prepared by post-synthetic modification of the organic linkers with chlorosulfonic acid [94]. The effect of this functionalization with Brønsted acid sites was studied in fructose transformation to HMF over MOF-SO$_3$H catalysts. Important to note, this is the main process, in which MOF catalysts with linkers functionalized with Brønsted acid (mostly, sulfo-modified linkers) are extensively studied. The SO$_3$H-MIL-101(Cr) framework was selected as the most representative catalyst among modified MOF-SO$_3$H matrices. Using this catalyst resulted in a high HMF yield of 90% with a complete fructose conversion (120 °C, 60 min, dimethylsulfoxide (DMSO)). The concentration (or density) of sulfonic groups in MOF-SO$_3$H materials corresponded to the –SO$_3$H grafting degree. At a lower –SO$_3$H grafting extent, a good linear correlation between the catalytic activity (turnover frequency) and sulfonic acid site density of MOF-SO$_3$H materials was found. Moreover, sulfonic acid groups as Brønsted acid sites are equivalent as catalytically active centers in all modified MOF-SO$_3$H matrices, regardless of the pristine MOF frameworks. Both the conversion of fructose and selectivity towards HMF increase with the sulfonic acid site density in MOF-SO$_3$H at the initial stage of fructose transformation. It was found that only fructose in sucrose and inulin was converted into HMF. The obtained modified MOF materials, in particular, the SO$_3$H-MIL-101(Cr) catalyst, are reusable and can be utilized several times. Noteworthy, (15%) SO$_3$H-MIL-101(Cr) and NUS-6(Hf) are among the best heterogeneous catalysts reported so far for the conversion of fructose into HMF [13,77].

Glycerol carbonate (GC), derivative of glycerol, is a valuable intermediate for the production of polycarbonates and polyurethanes [95]. To date, it is produced by glycerolysis of urea over inorganic salts such as $ZnSO_4$, $MgSO_4$ and ZnO and some heterogeneous systems, such as hydrotalcites based on these oxides [96]. It was found that the catalyst bifunctionality, that is, the simultaneous presence of the acid and basic sites, contributed significantly to the catalyst performance in the synthesis of cyclic carbonates from urea and diols. A bifunctional MOF catalyst having an appropriate combination of Lewis acid and Lewis basic sites was developed specifically for this process [97]. This material, that is, the IRMOF-3 framework ($Zn_4O(BDC)_3$) with amino functions in organic linkers was modified by a post-synthesis strategy by alkyl halides resulted in the conversion of –NH_2 to a quaternary ammonium salt species. Functionalized F-IRMOF-3 materials have quaternary ammonium groups (1.0–4.3 mg of N^+/g-cat). Therefore, these systems possess electrophilic sites (Zn_4O, inorganic building blocks) and nucleophilic centers (X^-). Thanks to this bifunctionality, F-IRMOF-3 materials display an enhanced activity in the solvent-free glycerolysis of urea into glycerol carbonate as compared to other heterogeneous systems. The F-IRMOF-3 catalyst with a larger alkyl chain in the organic linker and a more nucleophilic counter anion is more active in the synthesis of GC. The conversion of glycerol over the F-IRMOF-3 material with I^- species was higher as compared to analogous modified catalysts with Cl^- and Br^- counter anions. This effect could be attributed to the nucleophilicity of the counter anion, so the I^- counter ion is much more nucleophilic as compared to other studied halide anions.

The preparation procedure impacts significantly on the catalytic properties of the F-IRMOF-3 matrices, because defects in the framework, such as ZnO and Zn-OH species are formed mainly during the fast precipitation procedure as compared to the convenient solvothermal method, thanks to the introduction of acid-base bifunctional active sites. The combination of the ZnO and Zn-OH species serving as supplementary acid sites and free amino groups in the linkers that are basic sites enhances also the F-IRMOF-3 catalyst performance.

Thus, the reported results show an efficiency of post-synthesis modification strategy for the design of MOF catalysts exploited in diverse biomass valorization processes.

4. MOF Based Hybrid Nanomaterials and MOF-Derived Nanostructures in Biomass Valorization

4.1. MOF-Based Hybrid Nanomaterials

The fascinating properties of MOFs pave the way for their numerous applications. The obvious obstacles on this route are the insufficient thermal, mechanical and chemical stability as well as the lack of appropriate catalytic sites. These MOF materials weak points do not allow one to exploit fully their potential in practice. Therefore, there is a need to further enhance and tune the functional properties of MOF materials for a specific process using appropriate procedures. Integration of MOF matrices with a variety of functional materials and species is an efficient strategy to introduce new functionality leading to further improvement of their performance for practical use. To date, a number of comprehensive reports deal with a combination of MOFs with various materials in order to produce new functional hybrids or composites [66,98]. These hybrid nanomaterials were prepared by assembling metal–organic frameworks and functional species, organic molecules, metal nanoparticles, metal oxides, polyoxometallates (POMs) and metal complexes as well as other types of matrices including graphene, carbon nanotubes (CNTs), MMS silicas and nanorods [99,100]. Some of MOF based hybrid nanomaterials were tested in heterogeneous catalysis including biomass conversion. According to imparted extra properties, the MOFs based hybrids can be divided into two main groups: (1) imparting the additional catalytic properties and (2) acquiring the stability (hydrothermal, mechanical strength and so forth.) [101–103].

Despite the advantages of MOFs for catalysis indicated above, the type of active sites on pristine metal–organic frameworks is mainly limited to the following two sources: unsaturated metal centers and organic ligands, which catalyze a limited pool of reactions (see previous sections). In addition to approaches aimed at introducing active sites in MOF matrices mentioned in the previous sections (a careful evaluation of network building blocks and their post-synthesis modification), MOFs as nanoporous materials are able to accommodate guest species (e.g., metal complexes, enzymes and metal NPs (MNPs)) into their pore space by a noncovalent interaction. The guest@MOF prepared in this way can display a synergistic effect in catalytic reactions. The main benefit of this synergy is extending the potential applications of MOF materials in catalysis.

In terms of the catalytic application, MOF hybrids show some superior characteristics because of the unique features of MOFs mentioned above. For example, the high porosity with ordered crystalline pores and high surface areas contribute to the uniform dispersion and the high density of catalytic sites, which can improve the catalytic efficiency [104]. The confined pore sizes can limit active species (such as metal nanoparticles) growth and agglomeration and selectively transport different substrate molecules for size-selective (or shape-selective) catalysis [105]. These superior characteristics provide a potential of these hybrid materials in heterogeneous catalysis, in which MOF hybrids could display an enhanced activity, stability and reusability.

4.2. Embedding Catalytic Active Sites in MOF Host Matrices by Template Synthesis

The extra-large cavities and high surface area of MOF matrices make them ideal hosts for embedding large molecules, such as dyes, metalloporphyrins, metal nanoparticles, or nanosized POM clusters [83,103]. In view of functionalizing the MOF pores or cavities, catalytically active species or molecules could be encapsulated in the pores by two ways: (1) introducing the functional molecules by a template or in-situ approach ("a ship around the bottle," template synthesis) or (2) encapsulation of nanoparticles by post-synthesis modification ("ship in a bottle") [103,106,107]. Template in-situ synthesis involves the incorporation of functional molecules or particles during the MOF material synthesis. Currently, the potential of this pathway for the preparation of MOF-based hybrids as catalytic materials for biomass upgrading was justified by a number of examples [108,109]. The template in-situ synthesis for preparation of MOFs-based hybrid nanomaterials, first at all, for polyoxometallates (POMs) embedding will be reviewed below.

POMs along with MOFs are another type of highly ordered catalytic materials. They are soluble metal oxygen anion clusters of d-block transition metals in high oxidation states: W(VI), Mo(V), Mo(VI), V(IV), V(V) [110]. Two most important features of POMs contributing to catalysis applications are their strong acidity and redox properties, which can be controlled by changing the chemical composition [111]. These materials demonstrate thermal and chemical stability, particularly in multi-electron redox reactions. It is possible to tailor the catalytically active sites of POMs at the molecular or atomic level. All these characteristics make POMs promising oxidation and acid catalysts.

Particularly, heteropolyacids as a most important type of POMs are widely used for the conversion of biomass due to their Brønsted acidity and high proton mobility [112,113]. Unfortunately, POMs have some drawbacks that limit their applications such as small surface areas (<10 m^2 g^{-1}), low stability under catalytic conditions and high solubility in aqueous solutions [111].

The optimal way to exploit the catalytic properties of POMs is to deposit them on a porous support with a high surface area and to combine the activity of POMs with the reusability and recyclability of heterogeneous catalysts [113]. MOF materials are promising matrices for the encapsulation of POMs due to their textural and structural characteristics. Additionally, metal–organic frameworks have often own active sites, so the catalysis synergy between them and POMs reactivity can be obtained [114].

In the previous section, imparting of the Brønsted acidity to the mesoporous MIL-101(Cr) framework by using sulfonated organic linkers was discussed. Embedding POMs clusters is another strategy to create Brønsted acid sites in this matrix. The porous MIL-101(Cr) host is considered as a material of choice for POM encapsulation due to large cavities allowing the diffusion of reactants after introducing these guest species. The hybrid material obtained in this way was studied in the tandem reaction of fructose into 2,5-diformylfuran (DFF) conversion involving the dehydration of fructose to HMF catalyzed by an acid catalyst and subsequent oxidation of HMF to DFF by a redox catalyst [115]. To carry out this transformation, the synergy of bifunctional acid sites in the MIL-101(Cr)-based catalyst was imparted by encapsulation of phosphomolybdic acid (PMA) in this porous host. The PMA@MIL-101(Cr) host-guest material features a strong Lewis acidity (Cr^{3+}

ions) and Brønsted acidity (PMA clusters), as well as a moderate redox potential. The generated multifunctionality of active sites provides a high activity and selectivity of the PMA@MIL-101(Cr) hybrid in one-pot synthesis of 2,5-diformylfuran (DFF) directly from fructose, yielding 75.1 % of DFF as a target product. This bifunctional catalyst also demonstrated the recyclability in the studied process.

A series of hybrid materials was prepared by encapsulation of another type of POM clusters—phosphotungstic acid, $H_3PW_{12}O_{40}$ (PTA), in the MIL-101(Cr) host matrix by one-step in-situ synthesis. PTAx@MIL-101 hybrids modified with Brønsted acid functionalities from PTA (with different PTA weight loading) were exploited in selective dehydration of fructose and glucose [116]. Noteworthy, PTA is needed as a source of the Brønsted acid functionality, because Lewis sites (Cr^{3+} ions) in the MIL-101(Cr) framework are not active in the dehydration of carbohydrates (see previous sections). The catalytic activities of the resulting PTAx@MIL-101 hybrids depended on the PTA loading and the optimal one was 0.3 wt %. By using the PTA (3.0)@MIL-101(Cr) hybrid, an HMF yield of 63% from fructose was obtained (30 min, 130 °C, DMSO). This catalyst is quite stable and almost no PTA leaching (~2%) is observed during the reaction course.

Fine tuning of the catalytic performance of the MIL-101(Cr) framework as a solid acid used for the challenging transformation of glucose to HMF in has been reported [117]. For this purpose, the combined way to simultaneous modification both with sulfonic acid in organic linkers and with PTA in the cavities was studied. A host-guest material PTA@MIL-101(Cr)-SO_3H hybrid was prepared by self-assembly in-situ (or template synthesis) under hydrothermal conditions. It is regarded as the optimal catalyst (as compared to its counterpart MIL-101(Cr)-SO_3H without PTA guest species), which induced the isomerization and cyclodehydration of glucose in one pot. The PTA@MIL-101(Cr)-SO_3H multifunctional catalyst provided over 99.9% glucose conversion and a HMF yield of 80.7% (80 °C, 4 h, in water). Importantly, the PTA@MIL-101(Cr)-SO_3H catalyst allows the authors to achieve the HMF yield in the cheap glucose conversion, which is comparable to the HMF yield (85.3%) afforded in the expensive fructose conversion. The catalyst is quite reusable and recyclable without any noticeable activity loss.

Along with the MIL-101(Cr) matrix, the microporous HKUST-1 framework based on Cu^{2+} ions is among the MOF matrices most studied for POMs embedding [118]. The HKUST-1 material having a large surface area (1500 m^2/g) and suitable cavities can be used as a host matrix for the encapsulation of the POM nanoclusters to form an insoluble and recyclable catalyst for biomass conversion. It should be noted that due to the relatively small size of the HKUST-1 cavities (7 and 9.6 Å), only monomeric Keggin-type POMs can be incorporated. Noteworthy, these guest species exhibit a template effect during the formation of the MOF material [119].

Some of POMs@HKUST-1 hybrids were studied as heterogeneous catalysts in biomass valorization. An example is encapsulation of Keggin-type POMs (namely HPM (HPM = phosphomolybdic acid hydrate) in the HKUST-1 porous host [120]. The resulted HPM@HKUST-1 hybrid or NENU-5 was prepared by a one-step co-precipitation method. According to physico-chemical characterization, the as–synthesized material features the localization of HPM molecules in the largest pores of the Cu-BTC host (9.6 Å nm). The homogeneous distribution of HPM in the pores of the HKUST-1 matrix provides the combination of the catalytic activity of HPM and the insolubility, large surface area and hierarchical pore structure of the metal-organic framework. The catalytic performance of the HPM@HKUST-1 hybrid was tested in the etherification of HMF with ethanol, reaching a 55% yield of EMF (5-ethoxymethylfurfural) and an 11% yield of EL (100 °C, 12 h). This catalyst showed a good reusability after five cycles.

A series of nanosized hybrid materials was synthesized using the HKUST-1 host matrix for encapsulation of different POMs (Keggin ions) including $H_3PW_4O_{12}$, $H_5PMo_{12}O_{40}$, $H_5PVMo_{10}O_{40}$, $H_5PV_2Mo_{10}O_{40}$, $H_5PV_3Mo_{10}O_{40}$ [121]. The structural examinations showed that the Keggin ions acted as template species during the synthesis of the metal-organic framework. POM@HKUST-1 host-quest materials were exploited in selective oxidation of glycerol with H_2O_2 (40 °C, water) into valuable esters. POM served as the active sites for this oxidative esterification process demanding weak acidity provided by H_2O_2 as the oxygen source. The selectivity to esters was 34.5% in this one-pot oxidative esterification process. Such selectivity is caused by the porous HKUST-1 structure that provides the specific environment for POM active sites. Thanks to the pore limitation effect, diffusion was restricted assisting to ester formation over encapsulated POM (Figure 11). These POM@HKUST-1 catalysts could be recovered after the reaction, thereby exhibiting good stability and reusability without POM leaching.

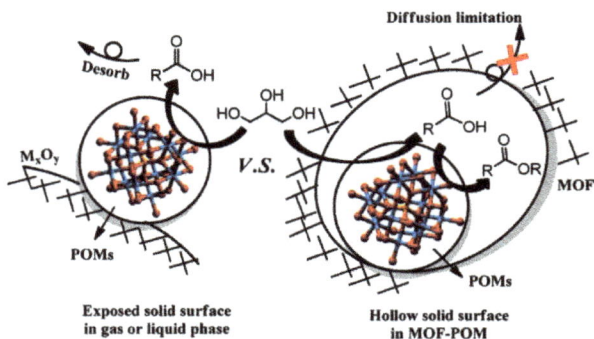

Figure 11. Diffusion limited glycerol transformation over a POMs@HKUST-1 hybrid catalyst. Reprinted with permission from reference [121]. Copyright 2015 Royal Society of Chemistry.

To date, examples of in-situ modification of MOF matrices with guest particles other than POMs are rather scarce. An efficient route for tuning the MIL-101(Cr) catalytic activity by deposition of co-precipitated $Cr(OH)_3$ nanoparticles was demonstrated in [122]. A $Cr(OH)_3$/MIL-101(Cr) hybrid with bifunctional Lewis (Cr^{3+} ions) and Brønsted acid sites ($Cr(OH)_3$) was prepared via a one-pot synthesis method. The combination of base-like chromium hydroxide particles and Lewis acidic sites of MIL-101(Cr) results in a highly selective conversion of glucose to fructose (100 °C, 24 h, ethanol) with a product distribution of 23.5% glucose, 59.3% fructose and 2.9% mannose. The fructose yield obtained over the $Cr(OH)_3$/MIL-101(Cr) material is comparable with the performance of optimized Sn-containing zeolites demonstrating strong Lewis acidity, such as Sn-Beta [123].

4.3. Hybrid Materials in the Form of MOF Host Matrices Containing Embedded Metal Nanoparticles

To date, there are limited reports on the application of MOFs as catalysts for biomass valorization, which mainly focus on functionalized MOFs as solid acid catalysts for dehydration and hydrolysis of biomass. There are very few examples of application of NPs@MOFs nanocatalysts for these processes [36]. As far as MOFs are successfully employed in catalytic applications, they must be stable and resistant to degradation under the established conditions, which can include changes in chemical and thermal environments in challenging transformations of biomass. As to the supported active phase, the structure of metal–organic frameworks can provide different anchoring sites for the reactants. In this case, the MOF-based heterogeneous system acts as a bifunctional catalyst thanks to both interesting adsorption and catalytic properties [124].

As a rule, most M NPs@MOF hybrids are used in hydrogenation as a key step in many processes of biomass transformations. The relevant investigations show that using these MOF based catalysts allows significant selectivity improvement in these reactions [125]. For instance, the MIL-101(Cr) matrix containing embedded Ru nanoparticles demonstrated a high efficiency in the selective hydrogenation of furfural to cyclopentenone, which is a useful way to the sustainable synthesis of important chemicals from biomass [126]. Particularly, cyclopentanone is a versatile reagent in the synthesis of insecticides, medicines and perfume and can serve as a solvent in the electronics industry [127]. Using a Ru@MIL-101(Cr) nanocatalyst resulted in a complete conversion of furfural with a selectivity higher than 96% in an aqueous media (2.5 h, 160°C, 4.0 MPa H_2 pressure). The Ru@MIL-101(Cr) catalyst features the homogeneous dispersion of very small Ru NPs (4–5 nm) and density of Lewis acid sites due to specific acidic functionality and structural properties of the MIL-101(Cr) framework. This bifunctionality assists to the sequential dehydrogenation step of 4-hydroxy-2-cyclopentenone conversion into cyclopentanone in a one-pot process.

The transformation of a carbonyl group of vanillin (a large component of pyrolysis oil derived from the lignin fraction) into a methyl group could proceed via two paths: (1) direct hydrogenolysis and (2) hydrogenation/hydrogenolysis. For Pd@MIL-101(Cr) nanohybrids, it was found that both the realization of a specific pathway of vanillin hydrodeoxygenation and the selectivity level depended critically on the dispersion of Pd NPs (0.5–5.0 nm) [109]. The catalyst with a higher dispersion of Pd NPs demonstrates an enhanced performance in vanillin conversion (49%, 1 h) with a 100% selectivity into 2-methoxy-4-methylphenol (2 bar, H_2, 348 K). However, Pd@MIL-101 with larger Pd NPs produced a significant amount of vanillin alcohol associated with a low yield of 2-methoxy-4-methylphenol under the same conditions according to pathway (2). The higher activity of the first catalyst Pd@MIL-101(Cr) is attributed to the smaller size of Pd NPs. Its outstanding selectivity to 2-methoxy-4-methylphenol can be explained by the steric hindrance and strong interaction of vanillin alcohol caused by the encapsulation of ultrafine Pd NPs within the MIL-101(Cr) host matrix. This results in the elimination of the reaction pathway (2) [104].

An efficient Pd-containing heterogeneous catalyst based on amino-modified NH_2MIL-101(Cr) with aminoterephthalate linkers (Pd@NH_2MIL-101(Al)) was prepared by using a direct anionic exchange approach and subsequent gentle reduction. The uniform dispersion of Pd nanoparticles in the NH_2MIL-101(Cr) matrix composite is attributed to the presence of amino groups in the parent NH_2MIL-101(Cr) framework. This multifunctional catalyst Pd@NH_2MIL-101(Cr) promotes the selective hydrogenation of biomass-based furfural to tetrahydrofurfuryl alcohol under green conditions (water media, 40 °C). A complete hydrogenation of furfural is achieved at a low temperature with the selectivity to tetrahydrofurfuryl alcohol close to 100%. Such an almost absolute selectivity is caused by the improvement of hydrogen bonding interactions between the intermediate furfuryl alcohol and the amine-functionalized host matrix. These interactions assist to the further hydrogenation of furfuryl alcohol to tetrahydrofurfuryl alcohol coordinated with Pd sites [128].

The amino-modified mesoporous NH_2-MIL-101(Al) metal organic framework based on Al^{3+} ions with the same texture and structural characteristics as MIL-101(Cr) and NH_2-MIL-101(Cr) was also used for Pd nanoparticles embedding [129]. As in the case of NH_2-MIL-101(Cr), the presence of free amine moieties in the frameworks of NH_2-MIL-101(Al) plays a key role in the formation of homogeneous and well-dispersed palladium nanoparticles in the resulted Pd@NH_2-MIL-101(Al) nanocomposite. This catalyst was applied in selective hydrogenation of HMF. It was found that the amine-functionalized NH_2-MIL-101(Al) material shows a preferential adsorption to hydrogenation of intermediate 2,5-dihydroxymethylfuran (DHMF) than in the case of HMF as a reactant. It is explained by a more hydrophilic nature of DHMF as well as by improved hydrogen bonding interactions between DHMF and the NH_2-MIL-101(Al) matrix. This bonding promotes a further hydrogenation of DHMF to 2,5-dihydroxymethyltetrahydrofuran (DHMTHF) upon the in-situ formation of DHMF over 3%Pd@NH_2-MIL-101(Al). The study results reveal that the high selectivity toward DHMTHF (~96%) is related to the synergy between the Pd sites and the amine moiety of the NH_2-MIL-101(Al) support.

Pt nanoparticles were encapsulated in a UiO-67(Zr) metal–organic framework via a linker design method [130]. It was found that embedding Pt NPs (4–6 nm) in the UiO-67 matrix can enhance the hydrogen adsorption capacity of UiO-67 (50 °C, PH_2 = 1 atm). This characteristic assists the implementation of the hydrogenation processes over Pt NPs@UiO-67 nanohybrids, which were studied in the oxidation and hydrogenation of HMF in aqueous solutions (90 °C). The Pt NPs@UiO-67 catalyst displays a higher activity and selectivity in hydrogenation than in oxidation under the same conditions (except the gas atmosphere), that is, conversion of HMF: 31.2% (hydrogenation) versus 13.6% (oxidation). Additionally, the Pt NPs@UiO-67 nanohybrid retains its crystalline structure in the catalytic cycles.

A bifunctional catalyst Ru/@SO$_3$H-UiO-66(Zr) was prepared by depositing Ru nanoparticles on a sulfonic acid-functionalized Zr-based metal–organic framework (SO$_3$H-UiO-66) [131]. Its efficiency has been proven in a high-yield one-pot upgrade strategy for converting biomass-derived methyl levulinate (ML) into γ-valerolactone (GVL) by obtaining a quantitative (100%) yield of GVL (80 °C, 0.5 MPa H$_2$, 4 h, aqueous media). In contrast, a very poor yield of GVL was achieved using a reference metal catalyst without any acidity (e.g., Ru/C). The catalytic activity and selectivity of the Ru@SO$_3$H-UiO-66(Zr) nanohybrid were significantly suppressed upon neutralization of its acidic sites. This fact confirms the essential role of the sulfonic groups of organic linkers in the SO$_3$H-UiO-66 framework. This acid functionality promotes the intramolecular dealcoholation of the 4-hydroxypentanoic methyl ester (4-HPME) intermediate (Figure 12). The Ru@SO3H-UiO-66 catalyst was recyclable over five cycles without any activity loss.

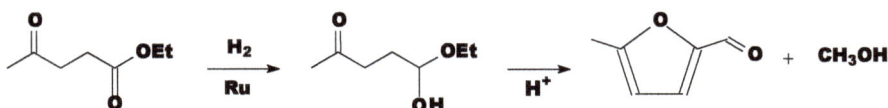

Figure 12. Cascade conversion of ML to GVL via hydrogenation followed by intramolecular dealcoholization. Modified with permission from reference [33]. Copyright 2016 Royal Society of Chemistry.

Two MIL-53(Al) (MIL-53(Al)-BDC and MIL-53(Al)-ADP) materials with the same topology but different linkers (aromatic and aliphatic ones) were prepared using 1,4-benzenedicarboxylic acid (BDC) and adipic acid (ADP) as organic building blocks, respectively [132]. The characterization results indicated that the MIL-53(Al)-BDC material showed a much higher surface area (761 m^2 g^{-1}) than its MIL-53(Al)-ADP analogue (4 m^2 g^{-1}). Therefore MIL-53(Al)-BDC showed a larger adsorption capacity towards furfural. Both materials were utilized for embedding Ru nanoparticles with a particle size in the range of 1–4 nm via a simple deposition–reduction method. Ru nanoparticles immobilized on MIL-53(Al) supports catalyze the catalytic hydrogenation of furfural under mild conditions (20 °C and 0.5 MPa H$_2$). Using the Ru@MIL-53(Al)-BDC catalyst, a full conversion of furfural and 100% selectivity to furfuryl alcohol was achieved.

As it was pointed out in the previous section, post-synthesis modification can assist to the MOF catalyst stability in harsh conditions of biomass processing. Liquid furanic hydrocarbons typically achieve a moderate product selectivity and yield at a high temperature and H$_2$ pressure in multi-step processes over different catalysts. Using MOF based catalyst, a single-step catalytic process was developed for direct conversion of various saccharides into furanic biofuels such as 2,5-dimethylfuran and 2-methylfuran at a decreased temperature [133]. This process involved the transfer-hydrogenation over a Pd NPs-containing MIL-53(Al) matrix with high yields (>95%) at 110–130 °C. Hydride (H$^-$) in polymethylhydrosiloxane (PMHS) serving as a green H-donor did not interfere with upstream reactions (e.g., hydrolysis, isomerization and dehydration) for the in-situ formation of furanic aldehydes/alcohols from sugars. This H-donor facilitates selectively the subsequent hydrodeoxygenation of carbonyl and hydroxyl groups other than the furanic ring in one pot according to deuterium-labeling study. Coating with polydimethylsiloxane resulted in stability improvement of the Pd@MIL-53(Al) catalyst (Pd/MIL-53(Al)-P) during at least five consecutive cycles.

The critical impact of the support on the catalytic properties of the deposited metal nanoparticles is well established. The role of the compositional and structural properties of MOF host matrices in catalytic transformations of biomass was studied in [134]. Prototypical MOF-5 ((Zn$_4$O(BDC)$_3$) with Zn^{2+} ions and Zr-containing UiO-66(Zr) and NH$_2$-UiO-66(Zr) metal–organic frameworks were selected as host matrices for Pt nanoparticles embedding. The high BET surface area, the abilities for reversible adsorption and desorption and appropriate pore sizes of these frameworks assist to the efficient control of the Pt nanoparticles dispersion and prevent them from aggregating. The resulted Pt@MOFs

(Pt@MOF-5, Pt@UiO-66 and Pt@UiO-66-NH$_2$) nanohybrids were studied extensively in selective conversion of bio-furfural with ethanol into furan-2-acrolein. The studied Pt@MOFs catalysts retain the framework crystallinity in bio-furfural conversion. Using a Pt@MOF-5 catalyst, the maximum conversion (up to 84.1%) of furan and the selectivity to furan-2-acrolein about 90.1% were achieved (150 °C, 4 h, O$_2$). Particularly, this selectivity is higher than the selectivity demonstrated by other catalytic systems, such as Au/Al$_2$O$_3$. A pronounced cooperative effect between Pt nanoparticles and the framework structure is responsible for the enhanced performance of the Pt@MOF-5 catalyst. On the contrary, the Pt@UiO-66(Zr) and Pt@NH$_2$-UiO-66(Zr) catalytic systems show a lower conversion of furfural and poor selectivity.

The role of the organic linker nature and network topology of the MOF support on the catalytic performance of immobilized Ru NPs in the selective hydrogenation of furfural to furfuryl alcohol reaction was evaluated also by Q. Yuan et al. [135]. This is a challenging process, because the product can be further hydrogenated into tetrahydrofurfuryl alcohol and, also, because furfuryl alcohol can be polymerized. For this purpose, two series of Zr-based metal–organic frameworks, UiO-66(Zr), UiO-67(Zr), Zr6-NDC, MIL-140A, MIL-140B and MIL-140C with different organic linkers such as BDC, 2,6-naphthalenedicarboxylic acid (NDC) and BPDC were prepared. These MOF materials were used as host matrices for Ru nanoparticles. The catalytic behavior of the obtained Ru nanocomposites was studied in hydrogenation of furfural to furfuryl alcohol under mild conditions (20°C, 5 bar H$_2$, 4 h). In all cases, furfuryl alcohol was a single product. The activity order was (TOF in parentheses): Ru@UiO-66(Zr) (11) ≈ Ru@UiO-67(Zr) (11) > Ru@Zr-NDC (8.0) ≈ Ru@MIL-140B (5.1), Ru@MIL-140A (4.8) > Ru@MIL-140C (2.1). So, the highest catalytic activity was exhibited by the Ru@UiO-66 nanohybrid. By using this catalyst, a 94.9% yield of furfuryl alcohol can be achieved. This catalytic performance is close to that of Ru NPs on nanosized mesoporous Zr-promoted silica. Importantly, the Ru@UiO-66(Zr) catalyst could be reused in five consecutive reaction cycles without any activity loss. It is interesting to note that the activities of the Ru@UiO-66(Zr) and Ru@UiO-67(Zr) nanocomposites were almost similar, despite the much lower surface area of the UiO-67(Zr) material as compared to UiO-66(Zr). The Ru NPs embedded in MIL-140 type matrices featured a lower activity than Ru-containing UiO materials. An explanation of the catalytic performance presented in this work involves the metal-support interactions. The organic moiety and the framework structure of the Zr-MOF matrices have a profound effect on the reducibility of embedded Ru NPs. Too strong metal-support interactions result in hard reducibility of Ru nanoparticles.

A more sophisticated way is the simultaneous use of the different routes—post-synthesis and in-situ with additional functional molecules and species for the modification of one metal-organic framework. An example is post-synthesis modification of one more member of the mesoporous MIL-101(M) framework family—metal-organic framework NH$_2$-MIL-101(Fe) with a neocuproine ligand through an amide (CONH) bond. Thus, modified neocuproine-NH$_2$-MIL-101(Fe) was used as a host matrix for the immobilization of palladium and cerium nanoparticles [136]. The hybrid catalyst Pd-Ce@neocuproine-NH$_2$-MIL-101(Fe) prepared by this way demonstrated a high activity in selective glycerol oxidation towards 1,3-dihydroxyacetone (DHA). The efficiency of Pd-Ce@neocuproine-NH$_2$-MIL-101(Fe) in this conversion surpasses the efficiency of commercial Pt-Bi/C catalysts. The bimetallic catalyst is recyclable with keeping a sustainable activity.

The next illustration of the potential of modifying the MOF pore voids both by metal nanoparticles (post-synthesis modification) and functional molecules (encapsulation by in-situ strategy) is introducing phosphotungstic acid (PTA) clusters by an in-situ approach and Ru NPs encapsulation in a mesoporous MIL-100(Cr) host matrix (Fe$_3$O(BTC)) [137]. The amount and strength of acid sites in the PTA@MIL-100(Cr) composite was achieved through the effective control of encapsulated PTA loading in MIL-100(Cr). This design and preparation method led to an appropriately functionalized Ru-PTA@MIL-100(Cr) material in terms of the Ru dispersion and their hydrogenation potential, on the one hand and the acid site density of PTA/MIL-100(Cr) (responsible for acid-catalyzed hydrolysis), on the other hand. The remarkable feature of this host-guest nanocomposite is its water-tolerance.

While using this engineered Ru-PTA@MIL-100(Cr) catalyst, cellulose and cellobiose were selectively converted into sorbitol by hydrogenation in aqueous media. The correlation between the acid/metal balance of bifunctional catalysts Ru-PTA@MIL-100(Cr) and their performance in the conversion of cellulose and cellobiose into sugar alcohols was found. The ratio of acid site density to the number of Ru surface atoms (nA/nRu) of Ru-PTA@MIL-100(Cr) was used to evaluate the balance between hydrogenation and hydrolysis functions. The optimum balance between these two catalytic functions is 8.84 < nA/nRu < 12.90. Its implementation allows the authors to achieve the maximum conversion of cellulose and cellobiose into hexitols. Under the optimal reaction conditions, a 63.2% yield of hexitols with a selectivity for sorbitol of 57.9% at the complete conversion of cellulose and a 97.1% yield of hexitols with a selectivity for sorbitol of 95.1% at the complete conversion of cellobiose were achieved with loadings of 3.2 wt % for Ru and 16.7 wt % for PTA. This research demonstrated efficiency of the rational design of acid/metal bifunctional catalysts for biomass conversion.

4.4. MOF-Based Composites with Other Matrices

As it was mentioned above, MOF materials in the form of bulk solids do not allow exploitation of their porosity, because they are generally prepared as insoluble microcrystalline powders featuring rather low mechanical and chemical stabilities, particularly in the acidic environment, which creates some obstacles for subsequent technological application. In particular, a low thermal stability and insufficient Brønsted acidity are the deficiencies of MOF materials [138]. These drawbacks would greatly limit their applications, particularly for reactions occurring at high temperatures and pressures, such as dehydration of carbohydrates. A prospective approach to improve some characteristics of MOF materials, including their hydrothermal stability, mechanical strength and shaping possibility and thereby realization of the potential of MOFs as porous host matrices is based on combining them with suitable solid supports affording functional composite materials.

The composites based on MOFs or derived from MOFs can be used successfully in biomass valorization that is carried out often in rather harsh conditions. To date, an extremely limited number of works reported the catalytic application of the composites based on MOFs and other type matrices in biomass upgrading processes. In particular, a supplement of MOF matrices by carbon-based materials makes active sites more hydrophobic, thus resulting in the protection of the coordination bonds from water molecules as well as increasing the affinity of active sites towards the organic reactants [139].

As it was pointed out in the previous sections, modifying metal–organic frameworks with sulfonic groups in order to use the obtained Brønsted acid-functionalized SO$_3$H-MOFs catalysts for cellulose hydrolysis [137] can result in some stability decrease, particularly, in hydrothermal conditions [140]. Supporting the acid functionalized MOF material on an appropriate robust support is an efficient strategy for framework stability retaining. In Reference [141], hollow nanotubes (HNTs) [Al$_2$Si$_2$O$_5$(OH)$_4$·2H$_2$O] modified with amphiphilic non-ionic polymer polyvinylpyrrolidone (PVP) were used for the deposition of a SO$_3$H-UiO-66(Zr) material with sulfonated linkers in order to obtain the stable composite catalyst with a potential application in HMF synthesis from sugars. These inorganic matrices with hollow tubular structure are composed of a 1:1 layered aluminosilicate and feature an enhanced mechanical strength [142].

The composite MOF-based catalysts PVP-HNTs@X-SO$_3$H−UiO-66(Zr) (X = 0.5, 1, 2, 3, where X stands for the amount of immobilized SO$_3$H−UiO-66(Zr) material) show both enhanced hydrothermal stability and catalytic performance in the transformation of fructose into HMF as compared to the SO$_3$H−UiO-66(Zr) matrix in the powder form. By varying the ratio of the inorganic HNT support and SO$_3$H−UiO-66(Zr), the Lewis and Brønsted acidity of the PVP-HNTs@X-SO$_3$H−UiO-66(Zr) composites can be precisely tuned. As a result, the 92.4% HMF yield was achieved over these bifunctional composite catalysts, which demonstrate also a high stability and can be reused at least five times without a significant activity loss.

Another example of the composite MOF-based catalyst for the sugar conversion is a combination of mesoporous MIL-101(Cr) matrix and a carbon material in the form of activated fly ash [138]. As it was mentioned above, the MIL-101(Cr) framework is one of the most widely studied MOF catalytic materials for biomass valorization. This matrix has been considered mainly as a promising catalyst for dehydration of sugars to furan and furfural derivatives. However, the absence of Brønsted acid sites in the MIL-101(Cr) framework has a negative impact on the furfural selectivity. Therefore, there is an urgent need to enhance the Brønsted acidity of this MOF catalyst.

The obtained composite material based on MIL-101(Cr) structure and fly-ash shows a high stability under severe hydrothermal conditions even in the acidic medium. Its catalytic performance was examined in the xylose dehydration into furfural. The furfural yield and selectivity of 71% and 80%, respectively, were afforded over the MIL-101(Cr)–based composite. Its catalytic properties remain intact during ten consecutive reaction cycles. This catalytic performance is much better than the pristine MIL-101(Cr) matrix shows. It is proposed that an enhanced catalytic activity is mainly attributed to the synergy between the unsaturated Lewis acid sites (Cr^{3+} ions) in the MIL-101(Cr) matrix and hydroxyl groups (Brønsted acid sites) of activated fly ash acting as Lewis acid centers.

Another way to introduce Brønsted acidity and additional Lewis acid sites to the MIL-101(Cr) structure and simultaneously improve its hydrothermal stability is the preparation of a composite material using mesoporous tin phosphate nanoparticles [87]. Tin phosphate has been selected as a composite component, because, it is hydrothermally stable and the P-OH species act as strong Brønsted acid sites contributed to an enhanced selectivity towards furfural in the resulted composite. Simultaneously, the Cr^{3+} ions in the MIL-101(Cr) structure act in combination with Sn^{4+} ions imparting the Lewis acidity sites. As a result, this bifunctional composite catalyst is highly relevant to the furfural synthesis from sugars. It was justified by examination of its catalytic performance in the dehydration of xylose into furfural. The MIL-101(Cr)-based composite featured an improved hydrothermal stability and resistance to destruction in an acidic medium under dehydration conditions. On the contrary, the crystalline structure of the pure MIL-101(Cr) matrix degraded after 4 catalytic cycles in this process. The composite catalyst allows one to achieve an 86.7% furfural yield and 92.3% selectivity to the target product at a lower temperature (150 °C, 180 min). This furfural yield is higher than the MIL-101(Cr) matrix or tin phosphate can provide individually.

As it was mentioned above, the catalytic biomass upgrading comprises numerous competitive reactions such as hydrodeoxygenation (HDO), decarbonylation, decarboxylation, hydrogenation, hydrocracking, polymerization and reforming [143]. For these challenging processes, a non-trivial strategy is preferred in terms of the design of advanced catalytic systems. An example is rather sophisticated multifunctional catalysts elaborated for one-pot conversion of monosaccharides (fructose and glucose) into 2,5-DMF [144]. These heterogeneous systems, that is, 2.4%Pd/UiO-66@SGO and 4.8%Pd/UiO-66@SGO, were obtained by embedding Pd NPs in a UiO-66(Zr) matrix deposited on sulfonated graphene oxide (SGO). The Brønsted acid sites of sulfonated SGO carrier provides the fructose dehydration into HMF, while the Pd nanoparticles catalyze the hydrogenolysis and hydrogenation of HMF to 2,5-DMF in the next steps. The composite materials catalyze efficiently sugars conversion affording a high 2,5-DMF yield of 70.5 mol % with fructose and 45.3 mol% yield of the target product using glucose as a reagent (160 °C, 1 MPa H_2, 3 h). These catalytic results surpass the previously reported ones. It is suggested that the enhanced fructose adsorption onto the composite catalyst surface can assist seriously to its conversion. This adsorption is improved by a synergistic effect of SGO matrix (Brønsted acid in a combination with hydrophilic surface) and deposited UiO-66(Zr) material (high porosity and surface area associated with strong acidity of Lewis sites (Zr^{4+} ions) and Brønsted sites (hydroxyl Zr-OH species). Furthermore, the 4.8% Pd/UiO-66@SGO catalyst is reusable and can be exploited during five catalytic cycles.

4.5. MOF-Derived Hybrid Materials

As a new class of porous materials, MOFs have recently been considered as ideal sacrificial templates for the preparation of various nanoporous carbons and metal or metal oxides nanoparticles because of their excellent properties such as ordered structures, high surface area and tunable functionality. The principal weaknesses of MOF materials (low thermal, hydrothermal and chemical stability) can be turned into an advantage by using their controlled thermal decomposition for in-situ nanoparticle generation embedded in carbon matrices [145]. While the confinement effect of MOF matrices provides a high dispersion of the metal/metal oxides nanoparticles throughout porous carbon, which is derived from the organic ligands of MOFs and the carbon matrix can stabilize the metal/metal oxide particles and prevent them from aggregation [146].

4.6. MOF-Derived Carbon Materials

When compared to porous materials synthesized by conventional methods, MOF-derived carbon nanostructures generally have improved catalytic activities and selectivities due to their larger surface areas (with well-defined and interconnected pore system) and easily tailored functionality [144]. A sustainable and highly efficient catalytic system in the form of Fe–Co alloy nanoparticles as hollow microspheres embedded in a carbon matrix was elaborated for direct transformation of HMF to demanded 2,5-diformylfuran (DFF) [147]. This magnetically separable FeCo/C nanohybrid was prepared by thermolysis at selected temperatures, that is, 500, 600, 700, 800 °C, using a bimetallic MIL-45b ($[M_3(BTC)_3]\cdot 5H_2O$, M = Co^{2+}, Fe^{2+}) material as a sacrificial template. The hollow Fe-Co nanoparticles with an average exterior diameter of ~60 nm highly dispersed in the carbon matrix were obtained at the pyrolysis temperature of ~500 °C. Such specific nanoarchitecture has a pronounced impact on the FeCo/C catalytic properties. In particular, the hollow structure of the nanoparticles provides the adsorption of HMF and quick desorption of the formed DFF from the catalyst surface. Thanks to specific textural and structural characteristics, the novel FeCo/C catalyst with a non-noble active phase demonstrated a high activity and selectivity in the aerobic oxidation of HMF, affording DFF as the target product in a >99% yield (100 °C, 1 MPa O_2, toluene, 6 h). This excellent performance is comparable to that of noble metal catalysts studied in this process under similar conditions. The FeCo/C nanohybrid showed a good reusability during six runs without any activity loss.

As it was mentioned above, as a rule, biomass transformations into value-added products involve cascade reactions, which need very often alternating conditions. Catalysis by acidic sites in an inert atmosphere is used for sugars dehydration, while HMF oxidation proceeds with participation of molecular oxygen [77]. An example is the one-pot conversion of fructose to (DFF) via HMF as an intermediate. Up to date, the development of efficient one-pot strategies for fructose selective conversion into DFF is still challenging. An example to resolve this problem is a fabrication of a MOF-derived nanohybrid for one-pot conversion of fructose into DFF [148]. The hybrid catalyst was synthesized by pyrolysis using a Fe-containing metal-organic framework (MIL-88B, ($Fe_3O(BDC)_3$)) as a sacrificial template, a carbon precursor and S powder as a dopant. The obtained Fe-based catalyst had small quantities of acidic sulfur-derived functional groups with highly uniform octahedral Fe_3O_4 nanoparticles with exposed (111) crystal faces distributed homogeneously in the sulfur doped carbon matrix. The prepared Fe_3O_4/C-S nanocomposite showed extremely high DFF yield (>99%) from fructose. Its superior selectivity to DFF is related to the interactions of the low adsorption energy of DFF on Fe_3O_4 (111) crystal faces and the presence of a non-oxidized S dopant on the support, which makes the MIL-88B-derived catalyst less oxidative. The catalyst is magnetically separable after the reaction and can be reused over 6 runs without any activity loss.

A non-noble MOF-derived nanohybrid CuNi0.5@C was studied in furfural transformation to cyclopentanone [149]. This material was prepared by embedding CuNi bimetallic nanoparticles in the carbon matrix. The synthesis procedure involves a direct thermolysis of the HKUST-1 metal–organic framework with Cu^{2+} ions impregnated with nickel nitrate. At the molar ratio of nickel to copper around 0.5, the bimetallic CuNi0.5@C nanohybrid consists of very small copper and nickel nanoparticles (about 15 nm) embedded in the carbon matrix. A pronounced confinement effect of the HKUST-1–derived porous carbon matrix preventing the aggregation of Cu and Ni NPs was observed in this study. Thanks to texture properties and highly dispersed metal nanoparticles, the CuNi0.5@C nanohybrid demonstrates an excellent catalytic performance, that is, 99.3% conversion of furfural and 96.9% yield of cyclopentanone (130 °C, 5 h, 5 MPa). This HKUST-1 derived hybrid catalytic material is quite reusable and shows no activity loss during four catalytic cycles.

The influence of the calcination temperature on the performance of other HKUST-1-derived catalytic carbon material was studied in [146]. Bimetallic non-noble CuCo@C nanohybrids with various molar ratios of cobalt to copper were prepared using the HKUST-1 metal-organic framework ions doped by Co via one-step thermal decomposition in a nitrogen atmosphere at temperatures ranging from 773 K to 1073 K. The ordered porous structure and long distance in the Cu-Cu dimer (as an inorganic building block) in the HKUST-1 framework assist to the uniform dispersion of cobalt ions in the metal-organic matrix [150]. Therefore, the obtained nanohybrids have highly dispersed mixed CuCo metal/metal oxide embedded species in a porous carbon matrix. CuCo@C catalytic materials were tested in selective hydrogenation of furfural to furfuryl alcohol. The optimal conditions for the catalyst preparation were evaluated in respect to the activity and selectivity. These conditions are characterized by the cobalt to copper molar ratio 0.4 and the calcination temperature 873 K (CuCo0.4@C-873 catalyst). In this case, a 98.7% furfural conversion and 97.7% furfuryl alcohol selectivity are achieved (ethanol, 413 K, PH_2 3 MPa). These characteristics are among the best reported ones. This excellent catalytic performance is related to the small particle size (about 9 nm) and the synergistic effect of copper and cobalt. Other important factors contributing to both the high conversion and selectivity are a relatively high surface area and porosity. It was found also that the HKUST-1-derived catalyst features a good stability and reusability in this process.

A magnetic Co-containing carbon nanocomposite (MCo@C) was prepared by carbonization of a CoPDA metal organic matrix with 2,6-pyridinedicarboxylate linkers and Co^{2+} ions [150]. This material is a porous carbon matrix with embedded highly dispersed Co NPs and it exhibits magnetic properties. The MCo@C nanocomposite with a high density of cobalt active sites demonstrates a sufficient catalytic activity in conversion of the lignin model compound, vanillyl alcohol (VAL) to vanillin (VN). The selectivities for VN were achieved as high as 99% and 100% depending on the oxidant type - H_2O_2 and air, respectively.

Another type of catalytic metal-free material with acid active sites for biomass processing was reported in [151]. A MOF-derived carbon (MDC) was prepared by direct pyrolysis of a -5 material as a carbonaceous precursor. Then $-SO_3H$ groups with Brønsted acidity were grafted onto the carbonaceous support in order to tailor its catalytic properties for HMF synthesis from sugars. The sulfonated carbon matrix with a hierarchical pore structure, MDC-SO_3H, was examined as an acidic heterogeneous catalyst in the dehydration of fructose into HMF in an isopropanol-DMSO mixed solvent. It was found that the highest yield of HMF (89.57%) was obtained (120 °C, 2 h) using 90 vol % isopropanol as the cosolvent in reaction media. The MOF-5 derived carbonaceous catalyst operated during 5 reaction cycles without a loss of the catalytic activity.

4.7. MOF-Derived Metal Nanostructures

The upgrade method for biomass-derived oil is hydrotreatment over a solid catalyst at 250–450 °C and a high H_2 pressure. Bifunctional transition metal catalysts supported on γ-Al_2O_3 (such as NiMo, CoMo and CoW) in their oxidic, reduced, or sulfidic form are traditionally used in this process [152]. For this purpose, another type of MOF-derived nanostructures can be utilized. They are obtained by calcination of metal–organic frameworks in air thus affording metal oxides with a uniform dispersion.

An example of the relevance of MOFs-derived metal oxide in the catalytic biomass upgrading is glycerol conversion. Glycerol is one of the top 12 platform chemicals, it can be converted into value-added products including acrolein [153]. Commonly, the glycerol dehydration into acrolein is carried out over solid acid catalysts, usually γ-Al_2O_3 having acidic OH functionalities. The main challenge in the design of alumina catalysts for glycerol dehydration reaction lies in overcoming the low acrolein selectivity and obtaining a higher stability for industrial application [154]. With this goal, a MIL-96(Al) ($Al_{12}O(OH)_{18}(H_2O)_3(Al_2(OH)_4)[BTC]_6 \cdot 24H_2O$) metal-organic framework with a honeycomb topology was selected as a precursor (template) to prepare nanoclusters (50–100 nm) of alumina γ-Al$_2$O$_3$ (M-Al$_2$O$_3$) [155]. This M-Al$_2$O$_3$ material was prepared by one-step calcination at 650 °C for the removal of the organic linker and transformation of MIL-96(Al) to γ-Al$_2$O$_3$. The catalytic performance of this M-Al$_2$O$_3$ nanocomposite was tested in glycerol dehydration to acrolein. Compared with conventional bulk alumina [156] and nanorod alumina [157] prepared by traditional precipitation and dry gel methods, the M-Al$_2$O$_3$ catalyst with a smaller particle size featured a higher activity with a conversion of glycerol over ~80% and high acrolein selectivity (74%) associated with stability during 200 h. This efficient catalyst performance is explained by the specific architecture of MIL-96(Al)-derived acidic M-Al$_2$O$_3$ alumina oxide characterized by the abundance of accessible acid sites due to the open porous system and ordered crystalline 3D structure of the MOF precursor.

An elegant way to overcome the most important drawback of the MOF structures, that is, the low thermal stability, was demonstrated in [143]. A Ni-containing MIL-77(Ni) metal-organic framework with glutarate linkers thermally decomposes under reaction conditions required for the hydrotreatment of biomass-derived oil. Accordingly, Ni nanoparticles (~100 nm) were generated in-situ during the hydrotreatment of glycerol-solvolyzed lignocellulosic (LC) biomass over MIL-77(Ni) (300 °C, PH$_2$-8 MPa) in a slurry reactor. MIL-77(Ni)- derived Ni nanoparticles showed a high hydrogenolysis activity in conversion of the biomass-derived oil into a polar and a non-polar phase with a significantly lower viscosity (60 min, MIL-77(Ni) loading above 3 wt %). Kinetic modeling showed that the MIL-77(Ni)–derived active phase had a more than ten times higher mass activity for hydrogenolysis in comparison to commercially available Ni/SiO$_2$–Al$_2$O$_3$ with Ni NPs with an average size ~10 nm. It was suggested that the ordered crystalline 3D structure of the metal organic precursor contributes significantly to the superiority of the catalytic performance of MIL-77(Ni)-derived Ni NPs by governing their size, shape and dispersion in the calcination process.

An efficient method for the controlled synthesis of Cu/ZnO hybrid in the form of nanosized ZnO particles dotted on Cu was reported in Reference [158]. This strategy involves nanohybrid preparation by subsequent steps of the calcination and reduction using Cu(Zn)-HKUST-1 isostructural metal-organic matrices with different Cu:Zn ratios as precursor. The catalytic performance of the bimetallic Cu/ZnO material was studied in a continuous process of hydrogenolysis of glycerol in a fixed-bed reactor. It was found that the interface between Cu and ZnO in the novel nanohybrid is of paramount importance regarding its catalytic properties, which depend also on the Cu:Zn ratio. The Cu1.1/ZnO catalyst derived from the Cu$_{1.1}$Zn$_{1.9}$(BTC)$_2$·9.4(H$_2$O) sacrificial matrix was more active and stable than Cu/ZnO in the form of ZnO plates with supported Cu particles prepared by other procedures, that is, solid grinding and co-precipitation.

The reviewed literature data point out that the use of the different ways for the preparation of MOF–derived and MOF- based hybrid nanomaterials is an efficient strategy for the engineering of advanced catalysts for biomass processing. This approach allows one to involve much more prototypical MOF structures for the relevant catalyst design, than other methods, including the buil ding block choice and their post-synthesis modification. Such MOF structure versatility contributes more effectively to tailoring the activity and selectivity of the designed catalysts.

5. Play with Physical Properties: Texture, Geometry and Morphology

Besides versatile framework functionalities, MOF matrices offer outstanding physical properties such as a record large surface area and exceptionally high and simultaneously tunable porosity. In addition to rational design of the framework by imparting novel properties via post-synthesis tailoring and in-situ modification, another possibility is a simple choice of the appropriate structural and textural characteristics of MOF catalysts favorable for effective biomass valorization among the variety of the available and described structures.

The evaluation of the appropriate porous structure and favorable network topology can be illustrated by using the metal–organic framework MIL-53(Al) (Al(OH)bdc, bdc = benzene-1,4-dicarboxylate) for the hydrolysis of carboxymethyl cellulose (CMC) to HMF in an aqueous medium. Due to its specific porous structure, the MIL-53(Al) material was used without any functionalization. This framework is built from infinite chains of corner-sharing $AlO_4(\mu_2\text{-}OH)_2$ octahedra making 1D lozenge-shaped channels accessible for substrate molecules [159]. The MIL-53(Al) material features isolated weak Brønsted and Lewis acid sites, lattice dynamic flexibility and a porous structure arranged as a system of 1D channels with the diameter about ~8.5 Å for the fully desolvated MIL-53(Al)$_{ht}$ modification, which allows fast diffusion of the reactants [160]. The MIL-53(Al) framework can also retain its crystallinity and permanent porosity without structural collapse in neutral and acidic aqueous solutions until 100 °C [161]. The hydrothermal stability of the MIL-53(Al) material contributes to achieving an HMF molar yield of 40.3% and total reducing sugar (TRS) molar yield of 54.2% (200 °C, 4 h, water) over this catalyst. Moreover, thanks to this tolerance, it is quite reusable and could be exploited three catalytic cycles with retaining almost the same activity level intact [162].

The introduction of a secondary pore system in order to create a hierarchical porous structure has been proven as an advantageous strategy in zeolite catalysis [163]. Recently, this approach was developed for modifying nanoporous MOF matrices [164–168]. This way of the rational design of MOF porous structures extends essentially application fields in catalysis by MOF materials.

A combined approach to catalyst design for biomass upgrading, which involves both tuning the porous structure of the pristine metal-organic framework to hierarchical one and scaling down this material to nanoparticles is reported in [169] taking esterification of glycerol with oleic acid into monoglycerides as an example. This process is a very challenging, because currently industrial monoglyceride production proceeds at high temperatures (220–250 °C) with the use of a basic homogeneous catalyst. Therefore, some substantial drawbacks are associated with esterification of glycerol under these conditions, that is, limited monoglyceride selectivity along with the formation of di- and triglyceride side products and soap. Moreover, the high reaction temperature has a negative impact on the taste, aroma and color of the final product. Developing a heterogeneous catalytic process at a lower temperature for selective monoglyceride production is a major scientific challenge.

To resolve this problem, a catalyst presenting a hierarchical zeolitic imidazolate framework (ZIF-8) in nanocrystalline form, and featuring a capability to operate efficiently at a moderate temperature (150 °C), was developed. Ii is important to note that the ZIF-8 nanomaterial becomes hierarchically structured in the esterification course, that is, a local transformation of the zeolite imidazolate framework takes place under reaction conditions. Although the structure integrity of the recovered ZIF-8 catalyst was confirmed by XRD, significant alteration of the pore structure was observed with the formation of internal cavities. The presence of mesopores was confirmed with N_2 adsorption revealing a Type IV isotherm with a hysteresis loop superimposed on the original Type 1 isotherm. The morphological and textural changes are more significant than it would be expected based on the relatively small quantities of Zn leaching. It was suggested that a simple pretreatment of the ZIF-8 material with oleic acid before the catalytic process could result in hierarchization, supplementing the other known strategies such as surfactant-templated synthesis [170,171], the use of mixed linkers [172], or vapor assisted crystallization [173].

The advantages of the hierarchical porous system were demonstrated also in [87]. This work has been discussed in the previous section and deals with two hierarchically porous NUS-6(Zr) and NUS-6(Hf) structures possessing a unique porous structure composed of coexisting micropores (0.5, 0.7 and 1.4 nm) and mesopores (4 nm) framed with pending sulfonic groups. The hierarchical porosity contributes to the high selectivity of the NUS-6 catalytic materials in dehydration of fructose to HMF thanks to appropriate pore sizes, which can suppress the undesirable side reactions due to the geometry factor (or molecular sieving properties).

6. An Outlook and Perspectives

Obviously, the main goal of the design of the heterogeneous catalyst is their application in industrial processes. To date numerous studies have aimed to demonstrate that MOF materials can exhibit activity and stability that are high enough from a practical point of view [34]. This applies in full to the MOF based catalysts designed for the biomass conversion. In this context, the following critical points should be taken into account.

1. Comparison with other "traditional" catalytic systems;
2. Energy efficiency;
3. Resource and cost economy and optimization;
4. Relevance to Green chemistry and sustainable development.

As highly ordered nanoporous materials, MOFs are compared often with zeolites and silicoaluminophosphates in respect to adsorption and catalysis. The contemporary literature shows that MOFs catalysts are not only competitive to zeolites as the most perspective heterogeneous systems in chemical processes of industrial importance but in some cases even surpass them in the catalytic performance. It is highlighted that, in general, liquid-phase processes (and biomass valorization comprises mostly liquid-phase reactions) are most suitable for MOF catalysts as compared to gas phase reactions that are mostly catalyzed by zeolites.

Perhaps, it is plausible to say that the selectivity plays a more important role than the activity in catalysis. MOF materials and MOF-based hybrids offer the most precise and efficient selectivity control than other solid acid catalysts. An illustrative example is the phosphate modified Zr-based NU-1000 matrix (discussed above), which shows an enhanced selectivity in glucose conversion into HMF [89]. Simultaneously, the suppression of the side intermolecular condensation of glucose/fructose/HMF takes place. Noteworthy, such a selectivity increase is an intrinsic NU-1000 metal-organic framework property, because its inorganic counterpart, similarly modified bulk zirconia with acid sites of the same strength of the acid sites does not display a selectivity increase.

An example of the superiority of the MOF catalysts in the selectivity terms is the application of other Zr-based frameworks, namely, UiO-66(Zr) and NH$_2$-UiO-66(Zr) materials (see above) in levulinic acid esterification into ethyl levulinate [76]. These catalysts show a selectivity, which surpasses the selectivity of zeolites and supported heteropolyacids. Ii is important to note that Zr-based MOF catalytic materials, such as the NU-1000 and UiO-66(Zr) frameworks in addition to the high selectivity, demonstrate a high productivity and low cost reusability and regenerability [141]. Importantly, they are synthesized from non-expensive and commercially available reagents in a green solvent, that is, water, under rather mild conditions.

As to MOF-based hybrid catalytic materials, we can consider an example concerning M NPs embedded in metal-organic matrices. Biomass valorization involves often a hydrogenation step. It was demonstrated that MOF carriers for metal NPs can also serve as effective selectivity regulators for supported catalysts in biomass upgrading via both the confinement effect of the framework and a number of specific metal-support interactions provided by intrinsic host matrix properties.

As it was pointed out in the previous sections, the use of MOF matrices as catalytic materials allows one to develop new routes of synthesis of a large variety of valuable chemicals and products from biomass sources by the full exploitation of MOF characteristic features, that is, a high surface area, tunable pore sizes and tailored chemical properties by functionalization of the organic linkers, inorganic nodes and pore voids as well as bifunctional or even multifunctional nature. These properties contribute to the achievement (in some extent) by MOFs of the efficiency of the bio-catalysts–enzymes. An example is an interesting and efficient strategy involving an association of the catalytic properties of MOF matrices and their molecular recognition capability. Noteworthy, molecular recognition processes involving sugars may be crucial for the selective synthesis of glucose from biomass-based feedstock via cellulose depolymerization, as currently accomplished by enzymes (i.e., cellulases). In this way, MOF catalysts can show an enhanced performance due to a cooperative effect. The NU-1000 catalytic material mentioned above demonstrates an excellent adsorption of sugar dimers, such as cellobiose and lactose from aqueous solutions, while completely excluding the adsorption of the monomer glucose [174]. These properties allow one the application of MOF catalysts in the selective biomass catalytic upgrading through multistep processes starting from crude feedstock like cellulose and lignocellulose.

Another abundant biomass feedstock is lignin, which is composed of aliphatic and phenolic units connected through aryl ether and carbon–carbon bonds. Catalytic hydrogenolysis is an efficient way for lignin valorization into value added chemicals but very challenging as well, therefore solving this problem requires new approaches. The development of effective catalysts for controlled lignin depolymerization on a technical scale still remains very relevant. Recently, MOF catalysts have been proven as a promising alternative to traditional catalysts for this process. In particular, in addition to exceptional features of MOF materials, they have the pores larger than other solid acid catalysts, which can accommodate relatively large reagents, intermediates and products of the valorization of the biomass, as well as lignin fragments, in particular. The isostructural IRMOF-74 frameworks were selected as catalysts for lignin valorization thanks to particular properties, such as framework robustness and thermal and chemical stability [175]. In addition, the tunable pore size of the Mg-IRMOF-74(I) and Mg-IRMOF-74(II) frameworks contributes to accommodation of a range of lignin-derived fragments and molecules.

Noteworthy, the utilization of MOF-based catalytic materials for biomass valorization allows one to reduce the process temperature and accomplish it in a greener or energy-saving manner. Along this trend, Mg-IRMOF-74(I) and Mg-IRMOF-74(II) systems show a high activity in catalytic C-O bond cleavage to the corresponding hydrocarbons and phenols under mild conditions, such as 1 MPa hydrogen pressure and at unprecedentedly low temperature (for lignin depolymerization) −120 °C. These MOF-based catalysts show also a high selectivity, which is associated with their recycling possibility and pushes them into the leader among prospective lignin valorization catalysts.

It was pointed out that the synthesis of a number of target MOF structures used most often for biomass valorization is not expensive. Moreover, the cost of these syntheses is paid back with interest due to the possibility of the implementation of one-pot cascade, tandem and domino processes relevant to biomass valorization over appropriate multifunctional MOF-based catalysts. Such implementation of the cascade reactions provides an efficient strategy for process intensification and material economy, allowing one to reduce consumption of energy and solvents as well as to minimize the wastes [176]. Thus, the possibility to have or create active sites of different nature, sometimes controversial (acid and basic sites in the same catalyst) can make the biomass valorization processes more energy-saving and greener.

Thus, the next steps in order to adapt MOF-based catalysts to industrial requirements relevant to biomass valorization processes are the reduction of the synthesis expenses and the further improvement of the catalyst stability and reusability/recyclability. A sustainability aspect related to the most cited MOF catalytic materials, such as MIL-101(Cr), HKUST-1 and UIO-66(Zr) widely studied in biomass valorization is an elucidation of the possibility of their densification for shaping and pelletization under various pressures without porosity and activity loss [177]. So, even there is still a long way to the eventual commercialization, especially in terms of the regeneration, reproducibility and production scale-up, the perspectives look really promising.

7. Conclusions

The literature data show that the exploitation of MOF structures for catalytic biomass processing into value-added products has started very recently, in 2010s. However, the obtained experimental results revealed a high potential in these cascade processes. The main benefits of the usage of MOF based catalytic materials is related to significant activity and selectivity improvement in some cases as well as reaction temperature decrease or even an implementation of one-pot processes instead of multi-step ones. In this relation, some MOF-based catalytic materials demonstrate a superiority over other solid acid catalysts, first at all, zeolites and the number of such examples is still growing.

Noteworthy, catalytic biomass valorization is one of the most fascinating potential applications of MOF materials. This area provides a huge field of the research activity relevant to various tuned MOF structures for cascade processes of biomass processes. It is of paramount importance from academic and practical point of view. In this way, a researcher can meet a number of challenging and non-trivial tasks to be resolved. For instance, it is controllable creation of different active sites with various content and appropriate accessibility, density and dispersion in one catalytic material that contributes to precise tailoring of the catalytic performance. With this goal, all methods that introduce the activity can be combined. For this purpose, an involvement of all efficient synthesis tools for the design of catalytic materials based on the comprehensive analysis of the MOF structure may be efficient.

The particular prospects are related to the development of the MOF based hybrid materials. MOF structures provide a possibility to introduce Lewis and Brønsted acid sites, metal nanoparticles and polyoxametallates in MOF pores. Another possibility is a combination of MOF structures with other materials, such as aluminosilicates and carbons as well as preparation of the MOF-derived nanostructures, such as carbonaceous matrices and metal oxide clusters. A vector is the implementation of an integrated approach to the design of new catalysts based on MOFs. The modification of MOF materials is directed towards both their catalytic performance enhancement and improvement of their hydrothermal stability and reusability.

Tailoring of the MOF porous structure and crystal dispersion deserves a particular attention in context of catalytic properties control. For instance, using MOF materials with an hierarchical pore structure allows one to manage the mass-transfer processes and contribute to extension of reagent variety as well as achieving a high process selectivity thanks to specific molecular-sieving properties. In case of exploitation of nanoscale MOF catalytic materials, minimizing the limitations of internal diffusion as well as creation of the specific surrounding of the active sites are possible. In this way, the synergy or cooperative effects between outer and inner surface characteristics and interesting interfacial properties can arise. Therefore, an associate task is a development of efficient synthesis methods allowing one to control the size and morphology of the catalytic MOF materials on the nanoscale as well as creation of MOF matrices with an hierarchical porous structure.

An efficient strategy will be also a search of the novel components for MOF based catalytic material (in addition to metal nanoparticles and polyoxometallates). For instance, enzyme/MOF hybrids (proposed by Chen et al. [36]) may provide an efficient cooperation between unique advantages of the pristine metal-organic matrices and the superior performance of enzymes in crude biomass transformation, such as carbohydrates depolymerization.

Finely, an important research focus remains the synthesis of highly porous and simultaneously robust and hydrothermally and chemically stable new MOF structures fitted for practical application fields, including catalytic biomass conversion, formulated by O. M. Yaghi twenty years ago.

A progress in achieving the formulated targets will allow one to consider MOF matrices and MOF-based materials with a greater confidence as heterogeneous catalytic systems of preferential choice of 21st century.

Funding:: This research received no external funding.

Acknowledgments: The work was carried out with a financial support from the Ministry of Education and Science of the Russian Federation in the framework of Increase Competitiveness Program of NUST «MISiS» (No. K2-2017-011) in the part related to catalytic aspects of the review and to Russian Science Foundation (grant no. 17-13-01526) in the part related to the structural aspects.

Conflicts of Interest: The authors declare no conflict of interest.

References

1. Bhanja, P.; Bhaumik, A. Porous nanomaterials as green catalyst for the conversion of biomass to bioenergy. *Fuel* **2016**, *185*, 432–441. [CrossRef]
2. De, S.; Saha, B.; Luque, R. Hydrodeoxygenation processes: Advances on catalytic transformations of biomass-derived platform chemical into hydrocarbon fuels. *Bioresour. Technol.* **2015**, *178*, 108–118. [CrossRef] [PubMed]

3. Jiang, Y.; Wang, X.; Cao, Q.; Dong, L.; Guan, J.; Mu, X. *Sustainable Production of Bulk Chemicals*; Xian, M., Ed.; Springer Science + Business Media: Dordrecht, The Netherlands, 2016. [CrossRef]

4. Climent, M.J.; Corma, A.; Iborra, S. Conversion of biomass platform molecules into fuel additives and liquid hydrocarbon fuels. *Green Chem.* **2014**, *16*, 516–547. [CrossRef]

5. Delidovich, I.; Leonhard, K.; Palkovits, R. Cellulose and hemicellulose valorization: An integrated challenge of catalysis and reaction engineering. *Energy Environ. Sci.* **2014**, *7*, 2803–2830. [CrossRef]

6. Chughtai, A.H.; Ahmad, N.; Younus, H.A.; Laypkov, A.; Verpoort, F. Metal–organic frameworks: Versatile heterogeneous catalysts for efficient catalytic organic transformations. *Chem. Soc. Rev.* **2015**, *44*, 6804–6849. [CrossRef] [PubMed]

7. Agarwala, B.; Kailasamb, K.; Sangwana, R.S.; Elumalai, S. Traversing the history of solid catalysts for heterogeneous synthesis of 5-hydroxymethylfurfural from carbohydrate sugars: A review. *Renew. Sustain. Energy Rev.* **2018**, *82*, 2408–2425. [CrossRef]

8. Sotelo-Boyas, R.; Trejo-Zarraga, F.; Hernandez-Loyo, F.d.J. Hydroconversion of Triglycerides into Green Liquid Fuels. In *Hydrogenation*; Karam, Y., Ed.; IntechOpen (Open Access), IntechOpen Limited: London, UK, 2012; pp. 188–216, Chapter 8, ISBN 978-953-51-0785-9.

9. Rodrigues, A.; Bordado, J.C.; Galhano dos Santos, R. Upgrading the glycerol from biodiesel production as a source of energy carriers and chemicals—A technological review for three chemical pathways. *Energies* **2017**, *10*, 1817. [CrossRef]

10. Melero, J.A.; Iglesias, J.; Morales, G.; Paniagua, M. Chemical routes for the conversion of cellulosic platform molecules into high-energy-density biofuels. Chapter 13. In *Handbook of Biofuels Production*, 2nd ed.; Carol, R.L., Karen, L., Clark, W.J., Eds.; Elsevier: Amsterdam, The Netherlands, 2016; pp. 359–388, ISBN 978-0-08-100455-5.

11. Sanders, J.P.M.; Clark, J.H.; Harmsen, G.J.; Heeres, H.J.; Heijnen, J.J.; Kersten, S.R.A.; van Swaaij, W.P.M.; Moulijn, J.A. Process intensification in the future production of base chemicals from biomass. *Chem. Eng. Process.* **2012**, *51*, 117–136. [CrossRef]

12. Besson, M.; Gallezot, P.; Pinel, C. Conversion of biomass into chemicals over metal catalysts. *Chem. Rev.* **2014**, *114*, 1827–1870. [CrossRef] [PubMed]

13. Herbst, A.; Janiak, C. MOF catalysts in biomass upgrading towards value-added fine chemicals. *CrystEngComm* **2017**, *19*, 4092–4117. [CrossRef]

14. Hara, M.; Nakajima, K.; Kamata, K. Recent progress in the development of solid catalysts for biomass conversion into high value-added chemicals. *Sci. Technol. Adv. Mater.* **2015**, *16*, 034903. [CrossRef] [PubMed]

15. Fechete, I.; Wang, Y.; Vedrine, J.C. The past, present and future of heterogeneous catalysis. *Catal. Today* **2012**, *189*, 2–27. [CrossRef]

16. Li, H.; Fang, Z.; Smith, R.L., Jr.; Yang, S. Efficient valorization of biomass to biofuels with bifunctional solid catalytic materials. *Prog. Energy Combust. Sci.* **2016**, *55*, 98–194. [CrossRef]

17. Farrusseng, D.; Aguado, S.; Pinel, C. Metal-organic frameworks: opportunities for catalysis. *Angew. Chem. Int. Ed.* **2009**, *48*, 7502–7513. [CrossRef] [PubMed]

18. Garcia-Garcia, P.; Müller, M.; Corma, A. MOF catalysis in relation to their homogeneous counterparts and conventional solid catalysts. *Chem. Sci.* **2014**, *5*, 2979–3007. [CrossRef]

19. Gascon, J.; Corma, A.; Kapteijn, F.; Llabres, I.; Xamena, F.X. Metal organic framework catalysis: Quo vadis? *ACS Catal.* **2014**, *4*, 361–378. [CrossRef]

20. Liu, J.; Chen, L.; Cui, H.; Zhang, J.; Zhang, L.; Su, C.Y. Applications of metal–organic frameworks in heterogeneous supramolecular catalysis. *Chem. Soc. Rev.* **2014**, *43*, 6011–6061. [CrossRef] [PubMed]

21. Zhu, L.; Liu, X.Q.; Jiang, H.L.; Sun, L.B. Metal–organic frameworks for heterogeneous basic catalysis. *Chem. Rev.* **2017**, *117*, 8129–8176. [CrossRef] [PubMed]

22. Huang, Y.B.; Liang, J.; Wang, X.S.; Cao, R. Multifunctional metal-organic framework catalysts: synergistic catalysis and tandem reactions. *Chem Soc. Rev.* **2017**, *46*, 126–157. [CrossRef] [PubMed]

23. Zhou, H.C.; Long, J.R.; Yaghi, O.M. Introduction to metal–organic frameworks. *Chem. Rev.* **2012**, *112*, 673–674. [CrossRef] [PubMed]

24. Férey, G. Hybrid porous solids: Past, present, future. *Chem. Soc. Rev.* **2008**, *37*, 191–214. [CrossRef] [PubMed]

25. Rowsell, J.L.C.; Yaghi, O.M. Metal-organic frameworks: A new class of porous materials. *Micropor. Mesopor. Mater.* **2004**, *73*, 3–14. [CrossRef]

26. Pintado-Sierra, M.; Rasero-Almansa, A.M.; Corma, A.; Iglesias, M.; Sanchez, F. Bifunctional iridium-(2-aminoterephthalate)–Zr-MOF chemoselective catalyst for the synthesis of secondary amines by one-pot three-step cascade reaction. *J. Catal.* **2013**, *299*, 137–145. [CrossRef]

27. Long, J.R.; Yaghi, O.M. The pervasive chemistry of metal–organic frameworks. *Chem. Soc. Rev.* **2009**, *38*, 1213–1214. [CrossRef] [PubMed]

28. Furukawa, H.; Cordova, K.E.; O'Keeffe, M.; Yaghi, O.M. The chemistry and applications of metal–organic frameworks. *Science* **2013**, *341*, 1230444. [CrossRef] [PubMed]

29. Lee, J.Y.; Farha, O.K.; Roberts, J.; Schedt, K.A.; Nguen, S.B.T.; Hupp, J.T. Metal–organic framework materials as catalysts. *Chem. Soc. Rev.* **2009**, *38*, 1450–1459. [CrossRef] [PubMed]

30. Isaeva, V.I.; Kustov, L.M. The application of metal–organic frameworks in catalysis. *Petrol. Chem.* **2010**, *50*, 167–180. [CrossRef]

31. Jiao, L.; Wang, Y.; Jiang, H.L.; Xu, Q. Metal–organic frameworks as platforms for catalytic applications. *Adv. Mater.* **2017**, 1703663. [CrossRef] [PubMed]

32. Xu, W.; Thapa, K.B.; Ju, Q.; Fang, Z.; Huang, W. Heterogeneous catalysts based on mesoporous metal–organic frameworks. *Coord. Chem Rev.* **2017**. [CrossRef]

33. Dhakshinamoorthy, A.; Asiri, A.M.; Garcia, H. Mixed-metal or mixed-linker metal organic frameworks as heterogeneous catalysts. *Catal. Sci. Technol.* **2016**, *6*, 5238–5261. [CrossRef]

34. Dhakshinamoorthy, A.; Asiri, A.M.; Garcia, H. Tuneable nature of metal organic frameworks as heterogeneous solid catalysts for alcohol oxidation. *Chem. Commun.* **2017**, *53*, 10851–10869. [CrossRef] [PubMed]

35. Corma, A.; Garcia, H.; Llabrés i Xamena, F.X. Engineering metal organic frameworks (MOFs) for heterogeneous catalysis. *Chem. Rev.* **2010**, *110*, 4606–4655. [CrossRef] [PubMed]

36. Chen, J.; Shen, K.; Li, Y. Greening the processes of metal–organic framework synthesis and their use in sustainable catalysis. *ChemSusChem* **2017**, *10*, 3165–3187. [CrossRef] [PubMed]

37. Wu, P.D.; Zhao, M. Incorporation of molecular catalysts in metal–organic frameworks for highly efficient heterogeneous catalysis. *Adv. Mater.* **2017**, *29*, 1605446. [CrossRef] [PubMed]

38. Zhou, T.; Du, Y.; Borgna, A.; Hong, J.; Wang, Y.; Han, J.; Zhang, W.; Xu, R. Post-synthesis modification of a metal–organic framework to construct a bifunctional photocatalyst for hydrogen production. *Energy Environ. Sci.* **2013**, *6*, 3229–3234. [CrossRef]

39. Lin, Y.; Kong, C.; Chen, L. Amine-functionalized metal–organic frameworks: Structure, synthesis and applications. *RSC Adv.* **2016**, *6*, 32598–32614. [CrossRef]

40. Dhakshinamoorthy, A.; Asiri, A.M.; Garcia, H. Organic frameworks as versatile hosts of Au nanoparticles in heterogeneous catalysis. *ACS Catal.* **2017**, *7*, 2896–2919. [CrossRef]

41. Falcaro, P.; Yazdi, A.; Imaz, I.; Furukawa, S.; Maspoch, D.; Ameloot, R.; Evans, J.D.; Doonan, C.J. Application of metal and metal oxide nanoparticles@MOFs. *Coord. Chem. Rev.* **2016**, *307*, 237–254. [CrossRef]

42. An, B.; Zeng, L.; Jia, M.; Li, Z.; Lin, Z.; Song, J.; Zhou, Y.; Cheng, J.; Cheng, W.; Lin, W. Molecular iridium complexes in metal–organic frameworks catalyze CO$_2$ hydrogenation via concerted proton and hydride transfer. *J. Am. Chem. Soc.* **2017**, *139*, 17747–17750. [CrossRef] [PubMed]

43. Corma, A. State of the art and future challenges of zeolites as catalysts. *J. Catal.* **2003**, *216*, 298–312. [CrossRef]

44. De Azevedo, D.C.S.; Cardoso, D.M.; Fraga, A.; Pastore, H.O. Zeolites for a sustainable world. *Microporous Mesoporous Mater.* **2017**, *254*, 1–2. [CrossRef]

45. Tafipolsky, M.; Amirjalayer, S.; Schmid, R. Atomistic theoretical models for nanoporous hybrid materials. *Microporous Mesoporous Mater.* **2010**, *129*, 304–318. [CrossRef]

46. Liang, J.; Liang, Z.; Zou, R.; Zhao, Y. Heterogeneous catalysis in zeolites, mesoporous silica, and metal–organic frameworks. *Adv. Mater.* **2017**, *29*, 1701139. [CrossRef] [PubMed]

47. Wang, Z.Q.; Cohen, S.M. Postsynthetic modification of metal–organic frameworks. *Chem. Soc. Rev.* **2009**, *38*, 1315–1329. [CrossRef] [PubMed]

48. Hwang, K.; Hong, D.Y.; Chang, J.S.; Jhung, S.H.; Seo, Y.K.; Kim, J.; Vimont, A.; Daturi, M.; Serre, C.; Férey, G. Amine grafting on coordinatively unsaturated metal centers of MOFs: Consequences for catalysis and metal encapsulation. *Angew. Chem. Int. Ed.* **2010**, *47*, 4144–4148. [CrossRef] [PubMed]

49. Müller, M.; Devaux, A.; Yang, C.H.; De Cola, L.; Fischer, R.A. Highly emissive metal–organic framework composites by host–guest chemistry. *Photochem. Photobiol. Sci.* **2010**, *9*, 846–853. [CrossRef] [PubMed]

50. Arnanz, A.; Pintado-Sierra, M.; Corma, A.; Iglesias, M.; Sanchez, F. Bifunctional metal organic framework catalysts for multistep reactions: MOF-Cu(BTC)-[Pd] catalyst for one-pot heteroannulation of acetylenic compounds. *Adv. Synth. Catal.* **2012**, *354*, 1347–1355. [CrossRef]

51. Ferey, G.; Mellot-Draznieks, C.; Serre, C.; Millange, F.; Dutour, J.; Surble, S.; Margiolaki, I. A chromium terephthalate-based solid with unusually large pore volumes and surface area. *Science* **2005**, *309*, 2040–2042. [CrossRef] [PubMed]

52. Hu, Z.; Zhao, D. Metal–organic frameworks with Lewis acidity: Synthesis, characterization, and catalytic applications. *CrystEngComm* **2017**, *19*, 4066–4081. [CrossRef]

53. Chowdhury, P.; Bikkina, C.; Meister, D.; Dreisbach, F.; Gumma, S. Comparison of adsorption isotherms on Cu-BTC metal organic frameworks synthesized from different routes. *Microporous Mesoporous Mater.* **2009**, *117*, 406–413. [CrossRef]

54. Peng, Y.; Krungleviciute, V.; Eryazici, I.; Hupp, J.T.; Farha, O.K.; Yildirim, T. Methane storage in metal–organic frameworks: Current records, surprise findings, and challenges. *J. Am. Chem. Soc.* **2013**, *135*, 11887–11894. [CrossRef] [PubMed]

55. Pérez-Mayoral, E.; Cějka, J. [Cu$_3$(BTC)$_2$]: A metal–organic framework catalyst for the friedländer reaction. *ChemCatChem* **2011**, *3*, 157–159. [CrossRef]

56. Yepez, R.; Garcia, S.; Schachat, P.; Sarnchez-Sarnchez, M.; Gonzarlez-Estefan, J.H.; Gonzarlez-Zamora, E.; Ibarra, I.A.; Aguilar-Pliego, J. Catalytic activity of HKUST-1 in the oxidation of trans-ferulic acid to vanillin. *New J. Chem.* **2015**, *39*, 5112–5115. [CrossRef]

57. Cavka, J.H.; Jakobsen, S.; Olsbye, U.; Guillou, N.; Lamberti, C.; Bordiga, S.; Lillerud, K.P. A new zirconium inorganic building brick forming metal organic frameworks with exceptional stability. *J. Am. Chem. Soc.* **2008**, *130*, 13850–13851. [CrossRef] [PubMed]

58. Chavan, S.; Vitillo, J.G.; Gianolio, D.; Zavorotynska, O.; Civalleri, B.; Jakobsen, S.; Nilsen, M.H.; Valenzano, L.; Lamberti, C.; Lillerud, K.P.; et al. H$_2$ storage in isostructural UiO-67 and UiO-66 MOFs. *Phys. Chem. Chem. Phys.* **2012**, *14*, 1614–1626. [CrossRef] [PubMed]

59. Caratelli, C.; Hajek, J.; Cirujano, F.G.; Francesc, M.W.; Llabrés i Xamena, X.; Van Speybroeck, V. Nature of active sites on UiO-66 and beneficial influence of water in the catalysis of Fischer esterification. *J. Catal.* **2017**, *352*, 401–414. [CrossRef]

60. Cirujano, F.G.; Corma, A.; Llabrés, I.; Xamena, F.X. Zirconium-containing metal organic frameworks as solid acid catalysts for the esterification of free fatty acids: Synthesis of bio diesel and other compounds of interest. *Catal. Today* **2015**, *257*, 213–220. [CrossRef]

61. Phan, A.C.; Doonan, J.; Uribe-Romo, F.J.; Knobler, C.B.; O'Keeffe, M.; Yaghi, O.M. Synthesis, structure, and carbon dioxide capture properties of zeolitic imidazolate frameworks. *Acc. Chem. Res.* **2010**, *43*, 58–67. [CrossRef] [PubMed]

62. Yin, H.; Lee, T.; Choi, J.; Yip, A.C.K. On the zeolitic imidazolate framework-8 (ZIF-8) membrane for hydrogen separation from simulated biomass-derived syngas. *Microporous Mesoporous Mater.* **2016**, *233*, 70–77. [CrossRef]

63. Banerjee, R.; Phan, A.; Wang, B.; Knobler, C.; Furukawa, H.; O'Keeffe, M.; Yaghi, O.M. High-throughput synthesis of zeolitic imidazolate frameworks and application to CO$_2$ capture. *Science* **2008**, *319*, 939–943. [CrossRef] [PubMed]

64. Gücüyener, C.; van den Bergh, J.; Gascon, J.; Kapteijn, F. Ethane/Ethene separation turned on its head: selective ethane adsorption on the metal–organic framework ZIF-7 through a gate-opening mechanism. *J. Am. Chem. Soc.* **2010**, *132*, 17704–17706. [CrossRef] [PubMed]

65. Kandiah, M.; Usseglio, S.; Svelle, S.; Olsbye, U.; Lillerud, K.P.; Tilset, M. Post-synthetic modification of the metal–organic framework compound UiO-66. *J. Mater. Chem.* **2010**, *20*, 9848–9851. [CrossRef]

66. Kaneti, Y.V.; Dutta, S.; Hossain, M.S.A.; Shiddiky, M.J.A.; Tung, K.L.; Shieh, F.K.; Tsung, C.K.; Wu, C.W.; Yamauchi, Y. Strategies for improving the functionality of zeolitic imidazolate frameworks: Tailoring nanoarchitectures for functional applications. *Adv. Mater.* **2017**, *29*, 1700213. [CrossRef] [PubMed]

67. Lu, X.; Wang, L.; Lu, X. Catalytic conversion of sugars to methyl lactate over Mg-MOF-74 in near-critical methanol solutions. *Catal. Commun.* **2018**, *110*, 23–27. [CrossRef]

68. Holm, M.S.; Pagán-Torres, Y.J.; Saravanamurugan, S.; Riisager, A.; Dumesic, J.A.; Taarning, E. Sn-beta catalysed conversion of hemicellulosic sugars. *Green Chem.* **2012**, *14*, 702–706. [CrossRef]

69. Cabello, C.P.; Gómez-Pozuelo, G.; Opanasenko, M.; Nachtigall, P.; Čejka, J. Metal–organic frameworks M-MOF-74 and M-MIL-100: Comparison of textural, acidic, and catalytic properties. *ChemPlusChem* **2016**, *81*, 828–835. [CrossRef]

70. Holm, M.S.; Saravanamurugan, S.; Taarning, E. Conversion of sugars to lactic acid derivatives using heterogeneous zeotype catalysts. *Science* **2010**, *328*, 602–605. [CrossRef] [PubMed]

71. Murillo, B.; Zornoza, B.; Iglesia, O.; Téllez, C.; Coronas, J. Chemocatalysis of sugars to produce lactic acid derivatives on zeolitic imidazolate frameworks. *J. Catal.* **2016**, *334*, 60–67. [CrossRef]

72. Kruger, J.S.; Nikolakis, V.; Vlachos, D.G. Carbohydrate dehydration using porous catalysts. *Curr. Opin. Chem. Eng.* **2012**, *1*, 312–320. [CrossRef]

73. Timofeeva, M.N.; Panchenko, V.N.; Khan, N.A.; Hasan, Z.; Prosvirin, I.P.; Tsybulya, S.V.; Jhung, S.H. Isostructural metal-carboxylates MIL-100(M) and MIL-53(M.) (M.: V., Al, Fe and Cr) as catalysts for condensation of glycerol with acetone. *Appl. Catal. A. Gen.* **2017**, *529*, 167–174. [CrossRef]

74. Zhang, B.; Hao, J.; Sha, Y.; Zhou, H.; Yang, K.; Song, Y.; Ban, Y.; He, R.; Liu, Q. Utilization of lignite derivatives to construct Zr-based catalysts for the conversion of biomass-derived ethyl levulinate. *Fuel* **2018**, *217*, 122–130. [CrossRef]

75. Cirujano, F.G.; Corma, A.; Llabrés, I.; Xamena, F.X. Conversion of levulinic acid into chemicals: Synthesis of biomass derived levulinate esters over Zr-containing MOFs. *Chem. Eng. Sci.* **2015**, *124*, 52–60. [CrossRef]

76. Védrine, J.C. Heterogeneous catalysis on metal oxides. *Catalysts* **2017**, *7*, 341. [CrossRef]

77. Hu, Z.; Peng, Y.; Gao, Y.; Qian, Y.; Ying, S.; Yuan, D.; Horike, S.; Ogiwara, N.; Babarao, R.; Wang, Y.; et al. Direct synthesis of hierarchically porous metal–organic frameworks with high stability and strong Brønsted acidity: The decisive role of hafnium in efficient and selective fructose dehydration. *Chem. Mater.* **2016**, *28*, 2659–2667. [CrossRef]

78. Hu, Z.G.; Peng, Y.W.; Tan, K.M.; Zhao, D. Enhanced catalytic activity of a hierarchical porous metal-organic framework CuBTC. *CrystEngComm* **2015**, *17*, 7124–7129. [CrossRef]

79. Wee, L.H.; Lescouet, T.; Fritsch, J.; Bonino, F.; Rose, M.; Sui, Z.; Garrier, E.; Packet, D.; Bordiga, S.; Kaskel, S.; et al. Synthesis of monoglycerides by esterification of oleic acid with glycerol in heterogeneous catalytic process using tin-organic framework catalyst. *Catal. Lett.* **2013**, *143*, 356–363. [CrossRef]

80. Su, Y.; Chang, G.; Zhang, Z.; Xing, H.; Su, B.; Yang, Q.; Ren, Q.; Yang, Y.; Bao, Z. Catalytic dehydration of glucose to 5-hydroxymethylfurfural with a bifunctional metal-organic framework. *AIChE J.* **2016**, *62*, 4403–4417. [CrossRef]

81. Akiyama, G.; Matsuda, R.; Sato, H.; Takata, M.; Kitagawa, S. Cellulose hydrolysis by a new porous coordination polymer decorated with sulfonic acid functional groups. *Adv. Mater.* **2011**, *23*, 294–3297. [CrossRef] [PubMed]

82. Herbst, A.; Janiak, C. Selective glucose conversion to 5-hydroxymethylfurfural (5-HMF) instead of levulinic acid with MIL-101Cr MOF-derivatives. *New J. Chem.* **2016**, *40*, 7958–7967. [CrossRef]

83. Liu, X.F.; Li, H.; Zhang, H.; Pan, H.; Huang, S.; Yang, K.L.; Yang, S. Efficient conversion of furfuryl alcohol to ethyl levulinate with sulfonic acid-functionalized MIL-101(Cr). *RSC Adv.* **2016**, *6*, 90232–90238. [CrossRef]

84. Jiang, J.; Yaghi, O.M. Brønsted acidity in metal–organic frameworks. *Chem. Rev.* **2015**, *115*, 6966–6997. [CrossRef] [PubMed]

85. Evans, J.D.; Sumby, C.J.; Doonan, C.J. Post-synthetic metalation of metal–organic frameworks. *Chem. Soc. Rev.* **2014**, *43*, 5933–5951. [CrossRef] [PubMed]

86. Juan-Alcanñiz, J.; Gascon, J.; Kapteijn, F. Metal-organic frameworks as scaffolds for the encapsulation of active species: State of the art and future perspectives. *J. Mater. Chem.* **2012**, *22*, 10102–10118. [CrossRef]

87. Chatterjee, A.; Hu, X.; FL-Yuk, L. A dual acidic hydrothermally stable MOF-composite for upgrading xylose to furfural. *Appl. Catal. A Gen.* **2018**. [CrossRef]

88. Luan, Y.; Zheng, N.; Qi, Y.; Yu, J.; Wang, G. Development of a SO$_3$H-functionalized UiO-66 metal–organic framework by postsynthetic modification and studies of its catalytic activities. *Eur. J. Inorg. Chem.* **2014**, *26*, 4268–4272. [CrossRef]

89. Yabushita, M.; Li, P.; Islamoglu, T.; Kobayashi, H.; Fukuoka, A.; Farha, O.K.; Katz, A. Selective Metal–Organic Framework Catalysis of Glucose to 5-Hydroxymethylfurfural Using Phosphate-Modified NU-1000. *Ind. Eng. Chem. Res.* **2017**, *56*, 7141–7148. [CrossRef]

90. Katz, M.J.; Mondloch, J.E.; Totten, R.K.; Park, J.K.; Nguyen, S.T.; Farha, O.K.; Hupp, J.T. Simple and compelling biomimetic metal-organic framework catalyst for the degradation of nerve agent simulants. *Angew. Chem. Int. Ed.* **2014**, *53*, 497–501. [CrossRef] [PubMed]

91. Liu, W.G.; Truhlar, D.G. Computational linker design for highly crystalline metal–organic framework NU-1000. *Chem. Mater.* **2017**, *29*, 8073–8081. [CrossRef]

92. Li, P.; Klet, R.C.; Moon, S.Y.; Wang, T.C.; Deria, P.; Peters, A.W.; Klahr, B.M.; Park, H.J.; Al-Juaid, S.S.; Hupp, J.T.; et al. Synthesis of nanocrystals of Zr-based metal–organic frameworks with csq-net: Significant enhancement in the degradation of a nerve agent simulant. *Chem. Commun.* **2015**, *51*, 10925–10928. [CrossRef] [PubMed]

93. Chatterjee, A.; Hu, X.; Lam, F.L.Y. Towards a recyclable MOF catalyst for efficient production of furfural. *Catal. Today* **2018**, *314*, 129–136. [CrossRef]

94. Chen, J.; Li, K.; Chen, L.; Liu, R.; Huang, X.; Ye, D. Conversion of fructose into 5-hydroxymethylfurfural catalyzed by recyclable sulfonic acid-functionalized metal–organic frameworks. *Green Chem.* **2014**, *16*, 2490–2499. [CrossRef]

95. Chiappe, C.; Rajamani, S. Synthesis of glycerol carbonate from glycerol and dimethyl carbonate in basic ionic liquids. *Pure Appl. Chem.* **2012**, *84*, 755–762. [CrossRef]

96. Climent, M.J.; Corma, A.; Frutos, P.D.; Iborra, S.; Noy, M.; Velty, A.; Concepción, P. Chemicals from biomass: Synthesis of GC by transesterification and carbonylation with urea with hydrotalcite catalysts. The role of acid-base pairs. *J. Catal.* **2010**, *269*, 140–149. [CrossRef]

97. Lee, S.D.; Park, G.A.; Kim, D.W.; Park, D.W. Catalytic performance of functionalized IRMOF-3 for the synthesis of glycerol carbonate from glycerol and urea. *J. Nanosci. Nanotechnol.* **2014**, *14*, 4551–4556. [CrossRef] [PubMed]

98. Xiang, W.; Zhang, Y.; Lin, H.; Liu, C.J. Nanoparticle/metal-organic framework composites for catalytic applications: Current status and perspective. *Molecules* **2017**, *22*, 2103. [CrossRef] [PubMed]

99. Yu, J.; Chao, M.; Yan, B.; Qin, X.; Shen, C.; Xue, H.; Pang, H. Nanoparticle/MOF composites: Preparations and applications. *Mater. Horiz.* **2017**, *4*, 557–569. [CrossRef]

100. Shen, K.; Chen, X.; Chen, J.; Li, Y. Development of MOF-derived carbon-based nanomaterials for efficient catalysis. *ACS Catal.* **2016**, *6*, 5887–5903. [CrossRef]

101. Li, P.; Zeng, H.C. Immobilization of metal–organic framework nanocrystals for advanced design of supported nanocatalysts. *Appl. Mater. Interfaces* **2016**, *43*, 29551–29564. [CrossRef] [PubMed]

102. Casas, N.; Schell, J.; Blom, R.; Maz, M. MOF and UiO-67/MCM-41 adsorbents for pre-combustion CO_2 capture by PSA: Breakthrough experiments and process design. *Sep. Purif. Technol.* **2013**, *111*, 34–48. [CrossRef]

103. Zhu, Q.L.; Qiang Xu, Q. Metal–organic framework composites. *Chem. Soc. Rev.* **2014**, *43*, 5468–5512. [CrossRef] [PubMed]

104. Yang, Q.; Xu, Q.; Jiang, H.L. Metal–organic frameworks meet metal nanoparticles: Synergistic effect for enhanced catalysis. *Chem. Soc. Rev.* **2017**, *46*, 4774–4808. [CrossRef] [PubMed]

105. Chen, L.; Huang, W.; Wang, X.; Chen, Z.; Yang, X.; Luque, R.; Li, Y. Catalytically active designer crown-jewel Pd-based nanostructures encapsulated in metal–organic frameworks. *Chem. Commun.* **2017**, *53*, 1184–1187. [CrossRef] [PubMed]

106. Juan-Alcanñiz, J.; Ramos-Fernandez, E.V.; Lafont, U.; Gascon, J.; Kapteijn, F. Building MOF bottles around phosphotungstic acid ships: One-pot synthesis of bifunctional polyoxometalate-MIL-101 catalysts. *J. Catal.* **2010**, *269*, 229–241. [CrossRef]

107. Juan-Alcanñiz, J.; Goesten, M.G.; Ramos-Fernandez, E.V.; Gascon, J.; Kapteijn, F. Towards efficient polyoxometalate encapsulationin MIL-100(Cr): Influence of synthesis conditions. *New J. Chem.* **2012**, *36*, 977–987. [CrossRef]

108. Wan, H.; Chen, C.; Wu, Z.; Que, Y.; Feng, Y.; Wang, W.; Liu, X. Encapsulation of Heteropolyanion-Based Ionic Liquid within the Metal−Organic Framework MIL-100(Fe) for Biodiesel Production. *ChemCatChem* **2015**, *7*, 441–449. [CrossRef]

109. Aijaz, A.; Zhu, Q.L.; Tsumori, N.; Akita, T.; Xu, Q. Surfactant-free Pd nanoparticles immobilized to a metal–organic framework with size- and location-dependent catalytic selectivity. *Chem. Commun.* **2015**, *51*, 2577–2580. [CrossRef] [PubMed]

110. Maksimchuk, N.V.; Kholdeeva, O.A.; Kovalenko, K.A.; Fedin, V.P. MIL-101 supported polyoxometalates: Synthesis, characterization, and catalytic applications in selective liquid-phase oxidation. *Isr. J. Chem.* **2011**, *51*, 281–289. [CrossRef]

111. Ren, Y.; Wang, M.; Chen, X.; Yue, B.; He, H. Heterogeneous catalysis of polyoxometalate based organic–inorganic hybrids. *Materials* **2015**, *8*, 1545–1567. [CrossRef] [PubMed]

112. Bohre, A.; Dutta, S.; Saha, B. Upgrading furfurals to drop-in biofuels: An overview. *ACS Sustain. Chem. Eng.* **2015**, *3*, 1263–1277. [CrossRef]

113. Salomon, W.; Yazigi, F.J.; Roch-Marchal, C.; Mialane, P.; Horcajada, P.; Serre, C.; Haouas, M.; Taulelle, F.; Dolbecq, A. Immobilization of Co-containing polyoxometalates in MIL-101(Cr): Structural integrity versus chemical transformation. *Dalton Trans.* **2014**, *43*, 12698–12705. [CrossRef] [PubMed]

114. Song, J.; Luo, Z.; Britt, D.K.; Furukawa, H.; Yaghi, O.M.; Hardcastle, K.I.; Hill, C.L. A multiunit catalyst with synergistic stability and reactivity: A polyoxometalate metal organic framework for aerobic decontamination. *J. Am. Chem. Soc.* **2011**, *133*, 16839–16846. [CrossRef] [PubMed]

115. Zhao, J.; Anjali, J.; Yan, Y.; Lee, J.M. Cr-MIL-101-Encapsulated Keggin Phosphomolybdic Acid as a Catalyst for the One-Pot Synthesis of 2,5-Diformylfuran from Fructose. *ChemCatChem* **2017**, *9*, 1187–1191. [CrossRef]

116. Zhang, V.; Degirmenci, Y.; Li, C.; Hensen, E.J.M. Phosphotungstic Acid Encapsulated in Metal–Organic Framework as Catalysts for Carbohydrate Dehydration to 5-Hydroxymethylfurfural. *ChemSusChem* **2011**, *4*, 59–64. [CrossRef] [PubMed]

117. Chen, D.; Liang, F.; Feng, D.; Xian, M.; Zhang, H.; Liu, H.; Du, F. An efficient route from reproducible glucose to 5-hydroxymethylfurfural catalyzed by porous coordination polymer heterogeneous catalysts. *Chem. Eng. J.* **2016**, *300*, 177–184. [CrossRef]

118. Wee, L.H.; Bajpe, S.R.; anssens, N.; Hermans, I.; Kristof Houthoofd, K.; Kirschhock, C.E.A.; Johan, A.; Martens, J.A. Convenient synthesis of $Cu_3(BTC)_2$ encapsulated Keggin heteropolyacid nanomaterial for application in catalysis. *Chem. Commun.* **2010**, *46*, 8186–8188. [CrossRef] [PubMed]

119. Bajpe, S.R.; Kirschhock, C.E.A.; Aerts, A.; Breynaert, E.; Absillis, G.; Parac-Vogt, T.N.; Giebeler, L.; Martens, J. Direct observation of molecular-level template action leading to self-assembly of a porous framework. *Chem. Eur. J.* **2010**, *16*, 3926–3932. [CrossRef] [PubMed]

120. Wang, Z.; Chen, Q. Conversion of 5-hydroxymethylfurfural into 5-ethoxymethylfurfural and ethyl levulinate catalyzed by MOF-based heteropolyacid materials. *Green Chem.* **2016**, *18*, 5884–5889. [CrossRef]

121. Zhu, J.; Wang, P.; Lu, M. Study on the one-pot oxidative esterification of glycerol with MOF supported polyoxometalates as catalyst. *Catal. Sci. Technol.* **2015**, *5*, 3383–3393. [CrossRef]

122. Guo, Q.; Ren, L.; Kumar, P.; Cybulskis, V.J.; Mkhoyan, M.K.A.; Davis, E.; Tsapatsis, M. A chromium Hydroxide/MIL-101(Cr) MOF composite catalyst and its use for the selective isomerization of glucose to fructose. *Angew. Chem. Int. Ed.* **2018**, *57*, 4926–4930. [CrossRef] [PubMed]

123. Luo, H.Y.; Lewis, J.D.; Roman-Leshkov, Y. Lewis acid zeolites for biomass conversion: Perspectives and challenges on reactivity, synthesis, and stability. *Annu. Rev. Chem. Biomol. Eng.* **2016**, *7*, 663–692. [CrossRef] [PubMed]

124. Wang, J.S.; Jin, F.Z.; Ma, H.C.; Li, X.B.; Liu, M.Y.; Kan, J.L.; Chen, G.J.; Dong, Y.B. Au@Cu(II)-MOF: Highly efficient bifunctional heterogeneous catalyst for successive oxidation–condensation reactions. *Inorg. Chem.* **2016**, *55*, 6685–6691. [CrossRef] [PubMed]

125. Zhao, M.; Yuan, K.; Wang, Y.; Li, G.D.; Gu, L.; Hu, W.P.; Zhao, H.J.; Tang, Z.Y. Metal-organic frameworks as selectivity regulators for hydrogenation reactions. *Nature* **2016**, *539*, 76–80. [CrossRef] [PubMed]

126. Fang, R.; Liu, H.; Luque, R.; Li, Y. Efficient and selective hydrogenation of biomass-derived furfural to cyclopentanone using Ru catalysts. *Green Chem.* **2015**, *17*, 4183–4188. [CrossRef]

127. Scognamiglio, J.; Jones, L.; Letizia, C.S.; Api, A.M. Fragrance material review on cyclopentanone. *Food Chem. Toxicol.* **2012**, *50*, S608–S612. [CrossRef] [PubMed]

128. Yin, D.; Ren, H.; Li, C.; Liu, J.; Liang, C. Highly selective hydrogenation of furfural to tetrahydrofurfuryl alcohol over MIL-101(Cr)-NH$_2$ supported Pd catalyst at low temperature. *Chin. J. Catal.* **2018**, *39*, 319–326. [CrossRef]

129. Chen, J.; Liu, R.; Guo, Y.; Chen, L.; Gao, H. Selective hydrogenation of biomass-based 5-Hydroxymethylfurfural over catalyst of palladium immobilized on amine-functionalized metal–organic frameworks. *ACS Catal.* **2015**, *5*, 722–733. [CrossRef]

130. Hester, P.; Xu, S.; Liang, W.; Al-Janabi, N.; Vakili, R.; Hill, P.; Muryn, C.A.; Chen, X.; Martin, P.A.; Fan, X. On thermal stability and catalytic reactivity of Zr-based metal–organic framework (UiO-67) encapsulated Pt catalysts. *J. Catal.* **2016**, *340*, 85–94. [CrossRef]

131. Lin, Z.; Cai, X.; Fu, Y.; Zhu, W.; Zhang, F. Cascade catalytic hydrogenation–cyclization of methyl levulinate to form g-valerolactone over Ru nanoparticles supported on a sulfonic acidfunctionalized UiO-66 catalyst. *RSC Adv.* **2017**, *7*, 44082–44088. [CrossRef]

132. Yang, J.; Ma, J.; Yuan, Q.; Zhang, P.; Guan, Y. Selective hydrogenation of furfural on Ru/Al-MIL-53: A comparative study on the effect of aromatic and aliphatic organic linkers. *RSC Adv.* **2016**, *6*, 92299–92304. [CrossRef]

133. Li, H.; Zhao, W.; Fang, Z. Hydrophobic Pd nanocatalysts for one-pot and high-yield production of liquid furanic biofuels at low temperatures. *Appl. Catal. B Environ.* **2017**, *215*, 18–27. [CrossRef]

134. Ning, L.; Liao, S.; Cui, H.; Yu, L.; Tong, X. Selective conversion of renewable furfural with ethanol to produce furan-2-acrolein mediated by Pt@MOF-5. *ACS Sustain. Chem. Eng.* **2018**, *6*, 135–142. [CrossRef]

135. Yuan, Q.; Zhan, D.; van Haandel, L.; Yea, F.; Xue, T.; Hensen, E.J.M.; Guan, Y. Selective liquid phase hydrogenation of furfural to furfuryl alcohol by Ru/Zr-MOFs. *J. Mol. Catal. A Chem.* **2015**, *406*, 58–64. [CrossRef]

136. Li, X.; Tjiptoputro, A.K.; Ding, J.; Xue, J.M.; Zhu, Y. Pd-Ce nanoparticles supported on functional Fe-MIL-101-NH$_2$: An efficient catalyst for selective glycerol oxidation. *Catal. Today* **2017**, *279*, 77–83. [CrossRef]

137. Chen, J.; Wang, S.; Huang, J.; Chen, L.; Ma, L.; Huang, X. Conversion of cellulose and cellobiose into sorbitol catalyzed by ruthenium supported on a polyoxometalate/metal–organic framework hybrid. *ChemSusChem* **2013**, *6*, 1545–1555. [CrossRef] [PubMed]

138. Chatterjee, A.; Hu, X.; Lam, F.L.Y. Catalytic activity of an economically sustainable fly-ash-metal-organic framework composite towards biomass valorization. *Catal. Today* **2018**, *314*, 137–146. [CrossRef]

139. Liu, X.W.; Sun, T.J.; Hu, S.D.; Wang, J.L. Composites of metal–organic frameworks and carbon-based materials: Preparations, functionalities and applications. *J. Mater. Chem. A* **2016**, *4*, 3584–3616. [CrossRef]

140. Canivet, J.; Fateeva, A.; Guo, Y.M.; Coasne, B.; Farrusseng, D. Water adsorption in MOFs: Fundamentals and applications. *Chem. Soc. Rev.* **2014**, *43*, 5594–5617. [CrossRef] [PubMed]

141. Liu, M.; Zhang, Y.; Zhu, E.; Jin, P.; Wang, K.; Zhao, J.; Li, C.; Yan, Y. Facile synthesis of halloysite nanotubes-supported acidic metal–organic frameworks with tunable acidity for efficient fructose dehydration to 5-Hydroxymethylfurfural. *ChemistrySelect* **2017**, *2*, 10413–10419. [CrossRef]

142. Guimarjes, L.; Enyashin, A.N.; Seifert, G.; Duarte, H.A. Structural, electronic, and mechanical properties of single-walled halloysite nanotube models. *J. Phys. Chem. C* **2010**, *114*, 11358–11363. [CrossRef]

143. Čelič, T.B.; Grilc, M.; Likozar, B.; Tušar, N.N. In-situ generation of Ni nanoparticles from metal–organic framework precursors and their use for biomass hydrodeoxygenation. *ChemSusChem* **2015**, *8*, 1703–1710. [CrossRef] [PubMed]

144. Insyani, R.; Verma, D.; Kim, S.M.; Kim, J. Direct one-pot conversion of monosaccharides into high-yield 2,5-dimethylfuran over a multifunctional Pd/Zr-based metal–organic framework@sulfonated graphene oxide catalyst. *Green Chem.* **2017**, *19*, 2482–2490. [CrossRef]

145. Yap, M.H.; Fow, K.L.; Chen, G.Z. Synthesis and applications of MOF-derived porous nanostructures. *Green Energy Environ.* **2017**, *2*, 218–245. [CrossRef]

146. Wang, Y.; Miao, Y.; Li, S.; Gao, L.; Xiao, G. Metal-organic frameworks derived bimetallic Cu-Co catalyst for efficient and selective hydrogenation of biomass-derived furfural to furfuryl alcohol. *Mol. Catal.* **2017**, *436*, 128–137. [CrossRef]

147. Fang, R.; Luque, R.; Li, Y. Selective aerobic oxidation of biomass-derived HMF to 2,5-diformylfuran using a MOF-derived magnetic hollow Fe–Co nanocatalyst. *Green Chem.* **2016**, *18*, 3152–3157. [CrossRef]

148. Fang, R.; Luque, R.; Li, Y. Efficient one-pot fructose to DFF conversion using sulfonated magnetically separable MOF-derived Fe$_3$O$_4$ (111) catalysts. *Green Chem.* **2017**, *19*, 647–655. [CrossRef]

149. Wang, Y.; Sang, S.; Zhu, W.; Gao, L.; Xiao, G. CuNi@C catalysts with high activity derived from metal–organic frameworks precursor for conversion of furfural to cyclopentanone. *Chem. Eng. J.* **2016**, *299*, 104–111. [CrossRef]

150. Lai, H.K.; Chou, Y.Z.; Lee, M.H.; Lin, K.Y.A. Coordination polymer-derived cobalt nanoparticle-embedded carbon nanocomposite as a magnetic multi-functional catalyst for energy generation and biomass conversion. *Chem. Eng. J.* **2018**, *332*, 717–726. [CrossRef]

151. Jin, P.; Zhang, Y.; Chen, Y.; Pan, J.; Dai, X.; Liu, M.; Yan, Y.; Li, C. Facile synthesis of hierarchical porous catalysts for enhanced conversion of fructose to 5-hydroxymethylfurfural. *J. Taiwan Inst. Chem. Eng.* **2017**, *75*, 59–69. [CrossRef]

152. Grilc, M.; Likozar, B.; Levec, J. Hydrodeoxygenation and hydrocracking of solvolysed lignocellulosic biomass by oxide, reduced and sulphide form of NiMo, Ni, Mo and Pd catalysts. *Appl. Catal. B* **2014**, *150–151*, 275–287. [CrossRef]

153. Kong, P.S.; Aroua, M.K.; Daud, W.M.A.W. Conversion of crude and pure glycerol into derivatives: A feasibility evaluation. *Renew. Sustain. Energy Rev.* **2016**, *63*, 533–555. [CrossRef]

154. Massa, M.; Andersson, A.; Finocchio, E.; Busca, G. Gas-phase dehydration of glycerol to acrolein over Al_2O_3-, SiO_2-, and TiO_2-supported Nb- and W-oxide catalysts. *J. Catal.* **2013**, *307*, 170–184. [CrossRef]

155. Huang, L.; Qin, F.; Huang, Z.; Zhuang, Y.; Ma, J.; Xu, H.; Shen, W. Metal–organic framework mediated synthesis of small-sized γ-Alumina as a highly active catalyst for the dehydration of glycerol to acrolein. *ChemCatChem* **2018**, *10*, 381–386. [CrossRef]

156. Zhu, H.Y.; Riches, J.D.; Barry, J.C. γ-Alumina nanofibers prepared from aluminum hydrate with poly(ethylene oxide) surfactant. *Chem. Mater.* **2002**, *14*, 2086–2093. [CrossRef]

157. Shen, S.; Ng, W.K.; Chia, L.S.O.; Dong, R.; Tan, B.H. Morphology controllable synthesis of nanostructured boehmite and γ-Alumina by facile dry gel conversion. *Cryst. Growth Des.* **2012**, *12*, 4987–4994. [CrossRef]

158. Zheng, L.; Li, X.; Du, W.; Shi, D.; Ning, W.; Lu, X.; Hou, Z. Metal-organic framework derived Cu/ZnO catalysts for continuous hydrogenolysis of glycerol. *Appl. Catal. B* **2017**, *203*, 146–153. [CrossRef]

159. Loiseau, T.; Serre, C.; Huguenard, C.; Fink, G.; Taulelle, F.; Henry, M.; Bataille, T.; Férey, G. A rationale for the large breathing of the porous aluminum terephthalate (MIL-53) upon hydration. *Chem. Eur. J.* **2004**, *10*, 1373–1382. [CrossRef] [PubMed]

160. Ortiz, A.U.; Springuel-Huet, M.A.; Coudert, F.X.; Fuchs, A.H.; Boutin, A. Predicting Mixture Coadsorption in Soft Porous Crystals: Experimental and Theoretical Study of CO_2/CH_4 in MIL-53(Al). *Langmuir* **2012**, *28*, 494–498. [CrossRef] [PubMed]

161. Millange, F.; Serre, C.; Férey, G. Synthesis, structure determination and properties of MIL-53as and MIL-53ht: The first Cr^{III} hybrid inorganic–organic microporous solids: $Cr^{III}(OH)\cdot\{O_2C–C_6H_4–CO_2\}\cdot\{HO_2C–C_6H_4–CO_2H\}_x$. *Chem. Commun.* **2002**, 822–823. [CrossRef]

162. Zi, G.; Yan, Z.; Wang, Y.; Chen, Y.; Guo, Y.; Yuan, F.; Gao, W.; Wang, Y.; Wang, J. Catalytic hydrothermal conversion of carboxymethyl cellulose to value-added chemicals over metal–organic framework MIL-53(Al). *Carbohydr. Polym.* **2015**, *115*, 146–151. [CrossRef] [PubMed]

163. Pérez-Ramírez, J.; Christensen, K.; Egeblad, C.H.; Christensen, C.H.; Groen, J.C. Hierarchical zeolites: Enhanced utilisation of microporous crystals in catalysis by advances in materials design. *Chem. Soc. Rev.* **2008**, *37*, 2530–2542. [CrossRef] [PubMed]

164. Wee, L.H.; Wiktor, C.; Turner, S.; Vanderlinden, W.; Janssens, N.; Bajpe, S.R.; Houthoofd, K.; Van Tendeloo, G.; De Feyter, S.; Kirschhock, C.E.A.; et al. Copper benzene tricarboxylate metal–organic framework with wide permanent mesopores stabilized by keggin polyoxometallate ions. *J. Am. Chem. Soc.* **2012**, *134*, 10911–10919. [CrossRef] [PubMed]

165. Cao, Y.; Ma, Y.; Wang, T.; Wang, X.; Huo, Q.; Liu, Y. Facile Fabricating Hierarchically Porous Metal–Organic Frameworks via a Template-Free Strategy. *Cryst. Growth Des.* **2016**, *16*, 504–510. [CrossRef]

166. Feng, L.; Yuan, S.; Zhang, L.L.; Tan, K.; Li, J.L.; Kirchon, A.; Liu, L.M.; Zhang, P.; Han, Y.; Chabal, Y.J.; Zhou, H.C. Creating hierarchical pores by controlled linker thermolysis in multivariate metal–organic frameworks. *Am. Chem. Soc.* **2018**, *140*, 2363–2372. [CrossRef] [PubMed]

167. Duan, C.; Li, F.; Zhang, H.; Li, J.; Wang, X.; Xi, H. Template synthesis of hierarchical porous metal–organic frameworks with tunable porosity. *RSC Adv.* **2017**, *7*, 52245–52251. [CrossRef]

168. Wee, L.H.; Meledina, M.; Turner, S.; Van Tendeloo, G.; Zhang, K.; Rodriguez-Albelo, L.M.; Masala, A.; Bordiga, S.; Jiang, J.; Navarro, J.A.R.; et al. 1D-2D-3D transformation synthesis of hierarchical metal–organic framework adsorbent for multicomponent alkane separation. *J. Am. Chem. Soc.* **2017**, *139*, 819–828. [CrossRef] [PubMed]

169. Wee, L.H.; Lescouet, T.; Ethiraj, J.; Bonino, F.; Vidruk, E.; Garrier, E.; Packet, D.; Bordiga, S.; Farrusseng, D.; Herskowitz, M.; et al. Zeolitic Imidazolate Framework-8 Catalyst for Monoglyceride Synthesis. *ChemCatChem* **2013**, *5*, 3562–3566. [CrossRef]

170. Guan, H.Y.; LeBlanc, R.J.; Xie, S.Y.; Yue, Y. Recent progress in the syntheses of mesoporous metal–organic framework materials. *Coord. Chem. Rev.* **2018**, *369*, 76–90. [CrossRef]

171. Junggeburth, S.C.; Schwinghammer, K.; Virdi, K.S.; Scheu, C.; Lotsch, B.V. Towards Mesostructured Zinc Imidazolate Frameworks. *Chem. Eur. J.* **2012**, *18*, 2143–2152. [CrossRef] [PubMed]

172. Park, J.; Wang, Z.U.; Sun, L.B.; Chen, Y.P.; Zhou, H.C. Introduction of functionalized mesopores to metal–organic frameworks via metal–ligand–fragment coassembly. *J. Am. Chem. Soc.* **2012**, *134*, 20110–20116. [CrossRef] [PubMed]

173. McNamara, N.D.; Hicks, J.C. Chelating Agent-Free, Vapor-Assisted Crystallization Method to Synthesize Hierarchical Microporous/Mesoporous MIL-125 (Ti). *ACS Appl. Mater. Interfaces* **2015**, *7*, 5338–5346. [CrossRef] [PubMed]

174. Yabushita, M.; Li, P.; Bernales, V.; Kobayashi, H.; Fukuoka, A.; Gagliardi, L.; Farha, O.K.; Katz, A. Unprecedented selectivity in molecular recognition of carbohydrates by a metal–organic framework. *Chem. Commun.* **2016**, *52*, 7094–7097. [CrossRef] [PubMed]

175. Stavila, V.; Parthasarathi, R.; Davis, R.W.; El Gabaly, F.; Sale, K.L.; Simmons, B.A.; Singh, S.; Allendorf, M.D. MOF-Based Catalysts for Selective Hydrogenolysis of Carbon-Oxygen Ether Bonds. *ACS Catal.* **2016**, *6*, 55–59. [CrossRef]

176. Dhakshinamoorthy, A.; Garcia, H. Cascade Reactions Catalyzed by Metal Organic Frameworks. *ChemSusChem* **2014**, *7*, 2392–2410. [CrossRef] [PubMed]

177. Nandasiri, M.I.; Jambovane, S.R.; McGrail, B.P.; Schaef, H.T.; Nune, S.K. Adsorption, separation, and catalytic properties of densified metal–organic frameworks. *Coord. Chem. Rev.* **2016**, *311*, 38–52. [CrossRef]

![catalysts logo] *catalysts*

MDPI

Review

Valorization of Biomass Derived Terpene Compounds by Catalytic Amination

Irina L. Simakova [1,2,*]**, Andrey V. Simakov** [3] **and Dmitry Yu. Murzin** [4]

[1]   Boreskov Institute of Catalysis, pr. Lavrentieva, 5, Novosibirsk 630090, Russia
[2]   Department of Natural Sciences, Novosibirsk State University, Pirogova 2, Novosibirsk 630090, Russia
[3]   Centro de Nanociencias y Nanotecnología, Universidad Nacional Autónoma de México, km. 107 Carretera
       Tijuana a Ensenada, Ensenada C.P. 22860, Baja California, Mexico; andrey@cnyn.unam.mx
[4]   Johan Gadolin Process Chemistry Centre, Åbo Akademi University, FI-20500 Turku/Åbo, Finland;
       dmurzin@abo.fi
*   Correspondence: simakova@catalysis.ru

Received: 30 July 2018; Accepted: 26 August 2018; Published: 29 August 2018

Abstract: This review fills an apparent gap existing in the literature by providing an overview of the readily available terpenes and existing catalytic protocols for preparation of terpene-derived amines. To address the role of solid catalysts in amination of terpenes the same reactions with homogeneous counterparts are also discussed. Such catalysts can be considered as a benchmark, which solid catalysts should match. Although catalytic systems based on transition metal complexes have been developed for synthesis of amines to a larger extent, there is an apparent need to reduce the production costs. Subsequently, homogenous systems based on cheaper metals operating by nucleophilic substitution (e.g., Ni, Co, Cu, Fe) with a possibility of easy recycling, as well as metal nanoparticles (e.g., Pd, Au) supported on amphoteric oxides should be developed. These catalysts will allow synthesis of amine derivatives of terpenes which have a broad range of applications as specialty chemicals (e.g., pesticides, surfactants, etc.) and pharmaceuticals. The review will be useful in selection and design of appropriate solid materials with tailored properties as efficient catalysts for amination of terpenes.

Keywords: terpenes; terpenoids; biomass; heterogeneous and homogeneous catalysts; amination; transition metals; supported metals

---

1. Introduction

A vast expansion in research activities on biomass derived compounds is clearly related to a growing interest in sustainable feedstock. The current review is focused on synthesis of various amines from biomass, namely terpenes. In general amine derivatives have found important applications as corrosion inhibitors, in cosmetics and toiletries, and color reprography to name but a few. Well known is also their utilization for production of different pesticides and dyes, such as azine, azo dyes, as well as indigo dyes [1]. Besides being important platform chemicals [2–4], they can be also applied in synthesis of pharmaceuticals in particular anticancer agents and DNA alkylators. Unfortunately, most of the industrially relevant aliphatic and aromatic amines, as well as aminoalcohols are currently manufactured from fossil resources [5,6]. For synthesis of shorter chain amines, (e.g., ethylene diamine [5], ethanolamines [6]) ammonia and respectively 1,2-dichloroethane and ethylene oxide are used. This is rather energy-intensive also resulting in significant CO_2 emissions and problems with corrosion when HCl is produced as a by-product. For such shorter chain amines apparently more sustainable reaction routes should be developed.

For longer chain amines there is a clear alternative relying on utilization of biomass feedstock, namely production of bio-based amines can be done from bio-derived alcohols obtained from carbohydrates, fats, oils, and lignins [4,7–10] (Figure 1). In particular the development of efficient heterogeneous catalysts for such syntheses starting from carbohydrates [11–14], lignin derived phenolics [4,15–21], fatty acid (esters) and glycerol from oleochemical sources [8,22], monomers from chitin [23], and amino acids from proteins, was comprehensively reviewed by Froidevaux et al. [10] and Pelckmans et al. [4].

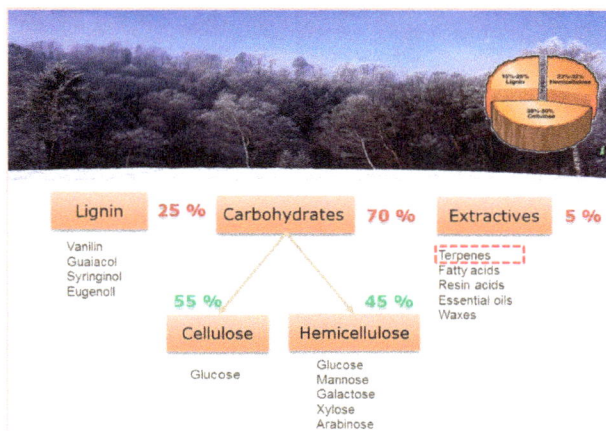

Figure 1. Chemistry of wood.

Another available biomass feedstock is the family of terpenes, being present in leaves, flowers, and fruits of many plants [24]. Distillation of turpentine, a byproduct in the pulp mills making cellulose, gives different terpenes. Apart from recent reviews [10,25] terpenes, have, however, not been considered in detail as a promising feedstock for biobased amines.

This review fills the apparent existing gap in the literature giving an overview of the readily available terpenes and describing the developed catalytic protocols for preparation of terpene-derived amines using homogeneous and heterogeneous catalysts. Bio-catalysis is beyond the scope of this review, while it should be mentioned that some interesting results have been reported [26–29] for intramolecular C–H amination of carbonazidate derivatives of menthol and borneol to corresponding five-membered cyclic compounds [30].

For some biomass derived compounds amination in the presence of heterogeneous catalysts has been extensively studied as described in detail in [4]. At the same time the same concept has been scarcely applied for so called extractives, constituting ca. 5% of lignocellulosic biomass. In particular, terpenes can be considered as very valuable components of biomass because of the potential industrial application of their derivatives ranging from basic and specialty chemicals to pharmaceuticals.

In order to address the role of solid catalysts in the amination of terpenes it was important to have an overview first of the same reactions occurring with the homogeneous counterparts. Such reactions can be considered as benchmarks, which solid catalysts should match. It should be mentioned in this connection, that it was recognized many years ago that at molecular level, there is little to distinguish between homogeneous and heterogeneous catalysis, while there are clear distinctions at the industrial level [31].

The subsequent sections consider respectively the significance of terpenes and their amine derivatives and main catalytic reactions for introduction of amine functionalities.

2. Terpenes Valorization into Valuable Amines

Terpenes are hydrocarbons consisting of isoprene (C5) basic units even if they are structurally very diverse. Terpenes or more precisely monoterpenes (C10), sesquiterpenes (C15), diterpenes (C20), sesterterpenes (C25), triterpenes (C30), and rubber (C5)n can be either branched or cyclic unsaturated molecules. Being typically extracted from the resins of coniferous trees they can be also present as acyclic or mono- to pentacyclic derivatives containing alcoxy, ether, carbonyl, keto, or ester ketone groups (i.e., "terpenoids"). These substrates present in various living species [32], particular in higher plants, are characteristic of a specific plant type.

The well-known application of natural (or even synthetic) resins of terpenes in perfumes and fragrances is related to their odor. In addition, synthesis of vitamins, insecticides, and pharmaceuticals also starts from terpenes [33–37]. Acyclic terpene amines are of special interest for production of insecticides, fungicides, and herbicides as well as in development of new pharmaceuticals [38–43]. Amino terpenes on the basis of (−)-menthol and (+)-3-carene were used for the preparation of potential inhibitors of γ-aminobutyric acid neuro-receptors for neurological applications [44,45]. Efficiency of limonene amino derivatives against in vitro cultures of the Leishmania (Viannia) braziliensis [46], egg hatchability, and mortality [47], as well as tobacco growth inhibitors was demonstrated [48,49]. Another interesting synthetic option is to use the amine group as a suitable protecting group, when there is a need to selectively hydrate some bonds in terpenes (e.g., myrcene) containing several double bonds. This strategy was applied in the synthesis of myrcenol, hydroxycitronellal [33] as well as terpenol [50]. It is also possible to use amino derivatives of terpenes as ligands in enantioselective reactions, such as catalytic asymmetric transfer hydrogenation of aromatic alkyl ketones [51] or enantioselective alkynyl zinc additions to aromatic and aliphatic aldehydes [52]. Amino terpenes on the basis of dihydromyrcenol were applied in synthesis of surfactants [53].

The history of plant terpenoids application in traditional herbal remedies is very extensive, therefore it is not surprising that they are currently under investigation due to their different therapeutic properties [54]. Even simple terpenes such as D-limonene, farnesol, and geraniol were reported to possess some chemotherapeutic activity against human cancer [55]. Carboranes with cinamyl, prenyl, and geranyl terpenoid fragments [55] were used to enhance boron delivery in boron neutron capture therapy [56–59]. Treatment with an alkyl halide of citronellal after amination with dimethyl or diethyl amines gives the corresponding chiral ionic liquids [60].

In general, a larger scale production of chemicals from plant extracts is limited, even if there are some examples when aminoterpenes play an important role in asymmetric and chemoselective catalysis. The Takasago Perfumery Company produces optically pure (−)-citronellal and pure (−)-menthol (1500 t/a) using *N,N*-diethylnerylamine [33]. SCM Corporation utilizes amination of myrcene for the synthesis of an insect repellent possessing insecticidal activity against the American flour beetle and the German cockroach [61].

This short overview illustrates a diverse scope of potential applications of terpenes-based amines in synthesis of valuable products including pharmaceuticals.

3. Possible Catalytic Tools for Synthesis of Terpene-Based Amines

This section is devoted to several reaction routes available to form C–N bonds in the terpenes of interest. Classical approaches for amine synthesis by a direct reaction of ammonia with alkyl halides or alternatively reduction of nitro or nitrile compounds are not considered here, instead the main focus is on hydroamination, hydroaminomethylation, reductive amination, and alcohols coupling with amines [1]. As mentioned above terpenes are highly functionalized molecules that contain double bonds, while their derivatives bear carbonyl and alkoxy or hydroxyl groups (Figure 2) that can be readily involved in various amination strategies.

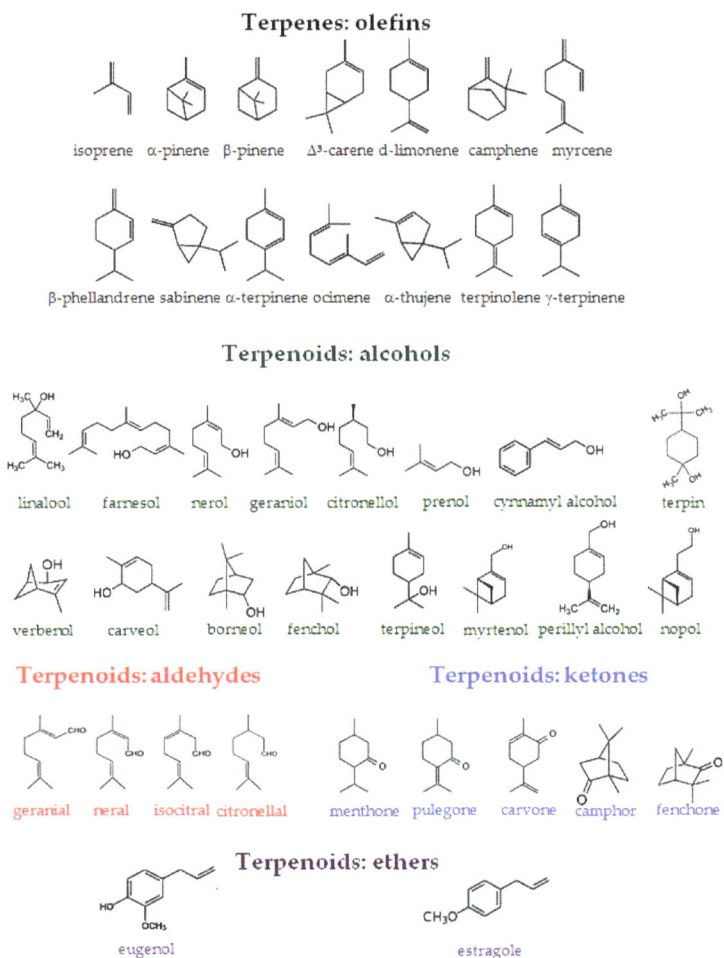

Figure 2. The main available terpenes and terpenoids.

Thus, five possible strategies in the formation of C–N bonds in terpenes and their derivatives were distinguished (Figure 3):

(1) reductive amination of aldehydes and ketones
(2) hydroaminomethylation
(3) hydroamination of double C=C bonds
(4) hydrogen borrowing methodology for amination of alcohols
(5) C–H amination of terpenes, which is a very specific case.

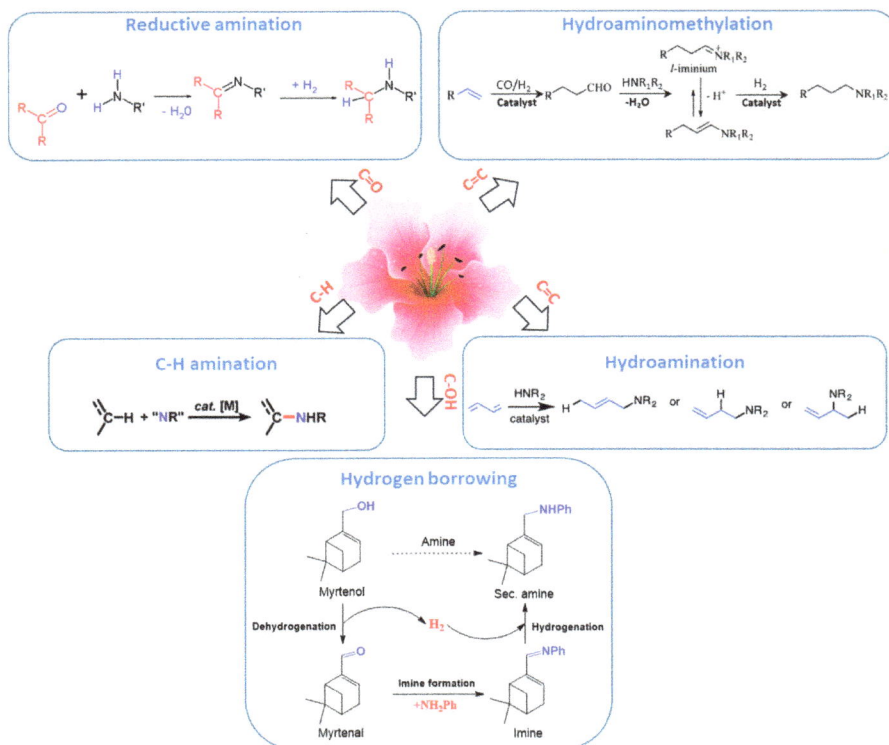

Figure 3. Overview of the available tools for amination of terpenoids.

3.1. Reductive Amination of Terpenes with Carbonyl Moiety

3.1.1. Reductive Amination of Aldehydes

As an example a particular terpenoid containing an aldehyde function is considered.

Citronellal (3,7-dimethyloct-6-en-1-al, **1**) (Figure 4) is well known as a flavoring agent and an insect repellent. The (*R*)-isomer of citronellal is typically found in citronella while the essential oil of kaffir lime contains the (*S*)-isomer. Citronellyl amine can be synthesized from the amide [62], oxime [62], and from geranylnitrile [63]. An issue related to reductive amination of aldehydes with ammonia using transition metals as catalysts is the need to suppress side reactions (Figure 4). Reductive amination of citronellal with aqueous ammonia giving primary amines was described by Behr et al. [62]. In this atom efficient method [Rh(cod)Cl]$_2$/TPPTS (TPPTS = 3,3′,3″-phosphanetriyl benzenesulfonic acid) as a homogenous catalyst was used in a biphasic solvent system. The organic compounds (substrate and product) are located in the apolar solvent phase of the biphasic solvent system, following an established concept applied for hydroformylation. This approach allowed a high yield of primary amines (**4, 5**) up to 87% effectively suppressing side reactions. These yields were obtained at 60 bar of hydrogen and 130 °C. Such high pressure was required as selectivity was seen to be dependent on pressure (Figure 5).

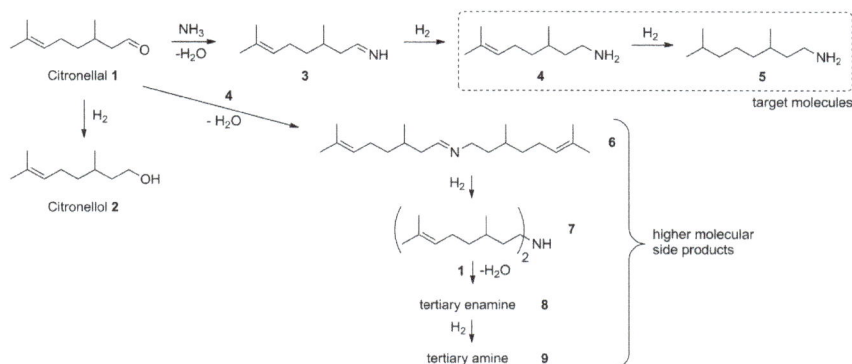

Figure 4. Reaction network in reductive amination of citronellal (**1**) with ammonia. Reprinted from [62] with permission from Elsevier.

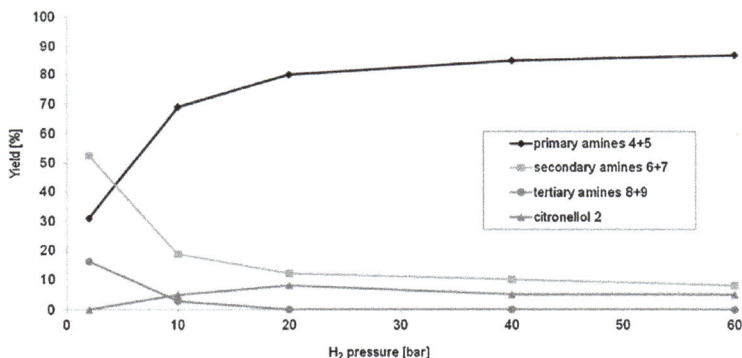

Figure 5. Influence of hydrogen pressure on the product distribution in the reductive amination of citronellal; data points are connected to show a better comparison. Reaction conditions: 6 mmol citronellal (**1**), 216 mmol NH_3, 0.5 mol% [Rh(cod)Cl]$_2$, 2.0 mol% TPPTS, 0.5 mol% CTAC, 5 mL toluene, 130 °C, 800 rpm, 6 h. Reprinted from [62] with permission from Elsevier.

All details of the experimental conditions are reported in [62]. The biphasic solvent inevitably requires efficient mass transfer, therefore surfactants, including ionic liquids or native cyclodextrins and their derivatives such as 1-decyl-3-methylimidazolium bromide ([DecMIM]Br) and methylcyclodextrin were applied [64]. Another option to increase selectivity towards the desired primary alcohols is to cleave the secondary imine formed as an undesired by-product [65].

3.1.2. Reductive Amination of Ketones

Reductive amination of d-fenchone to prepare fenchylamine, which are intermediates for some biologically active compounds [66] has been studied already long time ago [67] applying heterogeneous catalysts. In particular in the gas-phase amination of D-fenchone (**1**) with aliphatic nitriles (acetonitrile, acrylonitrile. or butyronitrile) performed at 220–260 °C under pressure of hydrogen ranging from 10 to 15 bar over copper on alumina modified with LiOH a mixture of isomeric endo-*N*-alkyl-l,3,3-trimethylbicyclo[2.2.l]hept-2-ylamines (**2**) and the corresponding exo compounds (**3**) with a ratio of 3:1 and yield of 50–60% was formed along with the intermediate *N*-fenchylidenalkylamine (**4**) (Figure 6). The main side products were α-fenchol (**5**) and β-fenchol (**6**).

$$RC{\equiv}N \longrightarrow RCH_2NH_2$$

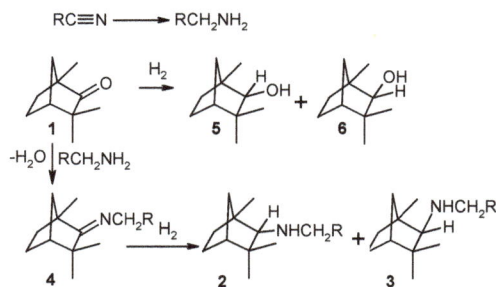

Figure 6. Scheme of reductive amination of D-fenchone (**1**) by aliphatic nitriles to prepare fenchylamine. Adapted from [67].

An interesting feature of this reaction is the generation of a primary amine from the initial nitrile on the metal sites. This primary amine then reacts with the substrate giving *N*-fenchylidenalkylamine (**4**) which is followed by hydrogenation into diastereomeric secondary amines (**2, 3**). This reaction competes with the intermolecular dehydration of the alcohols (**5, 6**). Formation of secondary amines is, however, predominant.

A systematic study on reductive amination of carbonyl terpenoids (camphor, carvone, hexahydropseudoionone, isocamphone) with different nitriles (acetonitrile, propionitrile, benzonitrile) over 15% Cu/Al$_2$O$_3$ modified with 2–6% LiOH resulting in both unsaturated and completely hydrogenated amines of diverse structure was conducted by Kozlov and co-workers [68].

Another example reported in the literature for reductive amination was related to camphor as a substrate. Influence of the heterogeneous catalysts type on the amination product yields in the reductive amination of camphor (**1**) with methylamine (Figure 7) was investigated. When Raney nickel was used as a catalyst, *N*-methylbornan-2-imine (**2**) (yield 82.8%) was predominantly formed, whereas the reaction over 5% Pd/C yielded a mixture of the imine (**2**) and *N*-methylbornan-2-ylamine (**3**) (30.4% and 65.7%, respectively). When platinum oxide was used as a catalyst, the yield of the amine **3** reached 92.7%.

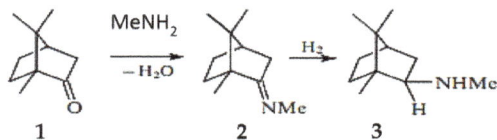

Figure 7. Reductive amination of camphor (**1**) with methylamine for synthesis of *N*-methylbornan-2-ylamine. Adapted from [68].

After reduction the promoted fused iron catalyst was applied for camphor (**1**) conversion (Figure 7) exhibiting high stereoselectivity to endobornan-2-ylamines (**3**), which is somewhat unusual for metal heterogeneous catalysts. In particular conversion of D,L-camphor (**1**) into endo- and exobornan-2-ylamines (**3**) during hydroamination reached 92%, with the endo to exo ratio being (1.4–1.8):1. Apparently, this stereoselectivity is due to the "imine-enamine" tautomerization occurring on the acid–base sites of the catalyst [68].

3.2. Hydroaminomethylation of Olefin Bonds in Terpenes

Somewhat related to reductive amination described above is hydroaminomethylation (HAM), which in fact is a tandem reaction consisting of hydroformylation followed by reductive amination [69].

This one-pot process proceeds on the same catalyst responsible for hydroformylation of C=C double bond making first an aldehyde followed by amination and hydrogenation of the imine/enamine intermediate [70–73] finally giving a secondary or a tertiary amine. The only by-product in this very efficient process with good atom economy is water. Apparently a careful choice of reaction conditions is needed to satisfy requirements for all reactions comprising a complex reaction network [70,74].

A few examples of hydroaminomethylation reaction with terpenes were reported including α-pinene [75], β-pinene [74], camphene [74], limonene [74,76,77], β-myrcene and β-farnesene [78] and naturally occurring allyl benzenes such as eugenol [79] and estragole [80]. Hydroformylation of the internal double bonds is much more difficult than the terminal bonds, thus it is not surprising that the examples mentioned above are related to isolated terminal double bonds. In the only reported hydroaminomethylation of a conjugated terpene [81] regioselectivity in hydroformylation was low along with low catalytic activity per se explained by formation of relatively stable η3-allyl-Rh complexes [78].

Hydroaminomethylation of limonene (**1**) with secondary amines (*n*- and *i*-propylamine, benzylamine), cyclic amines (piperidine, morpholine, piperazine,) aromatic amine (aniline) and diamines (ethylenediamine, propilenediamine, tetramethylenediamine) was reported by Graebin et al. [76]. The yields of products varied from 50% to 89% and are presented in Figure 8. In the case of diamines only isomerization products were obtained for tetramethylenediamine, while no products were formed when ethylenediamine was used. A plausible explanation could be inactivation of rhodium because of formation of stable chelated compounds of diamine with the catalyst.

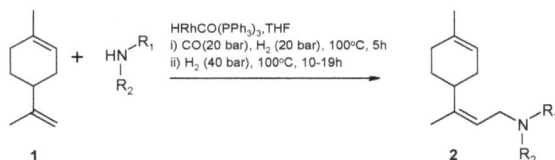

Figure 8. Hydroaminomethylation of limonene (**1**) with secondary amines, Y—yield of product: **2a**—R_1 = *n*-propyl, R_2 = H, Y = 85%; **2b**—R_1 = *i*-propyl, R_2 = H, Y = 50%; **2c**—R_1 = benzyl, R_2 = H, Y = 44%; **2d**—R_1 = R_2 = piperidine, Y = 80%; **2e**—R_1 = R_2 = morpholine, Y = 79%; **2f**—R_1 = R_2 = piperazine, Y = 89%; **2g**—R_1 = phenyl, R_2 = H, Y = 50%. Reprinted from [76] with permission from Elsevier.

Olefin hydroformylation was selective towards the linear aldehydes compared to the branched ones [76] which can be ascribed to the catalyst itself along with the steric hindrance of the terpene isopropenyl group [82–85].

Amination of aldehydes is more difficult with ammonia than with primary amines. Nevertheless Behr et al. [77] applied ammonia in HAM of limonene (**1**) (Figure 9). The reaction proceeds through hydroformylation of limonene (**1**) to the corresponding aldehyde (**2**) in the first step followed by condensation with ammonia giving an aldehyde observed experimentally and subsequent hydrogenation of the latter to a primary amine (**3**).

The desired amine (**3**) reacted also with the aldehyde 2 resulting in formation of secondary and tertiary amines. Moreover, limonene (**1**) underwent isomerization to its isomer isoterpinolene (**4**) [25,77].

[Rh(cod)(μ-OMe)]₂ as a pre-catalyst in the presence or absence of triphenylphosphine or tribenzylphosphine as ligands was applied in HAM of *R*-(+)-limonene (**1**) (Figure 10), camphene (**5**) (Figure 11), and (−)-β-pinene (**9**) (Figure 12) with di-*n*-butylamine, *n*-butylamine, morpholine, triphenylphosphine, and tribenzylphosphine using toluene as a solvent [74]. The reaction giving moderate to good yields (75–94%) was performed at 100 °C and 60 bar with an equimolar mixture of CO and H_2.

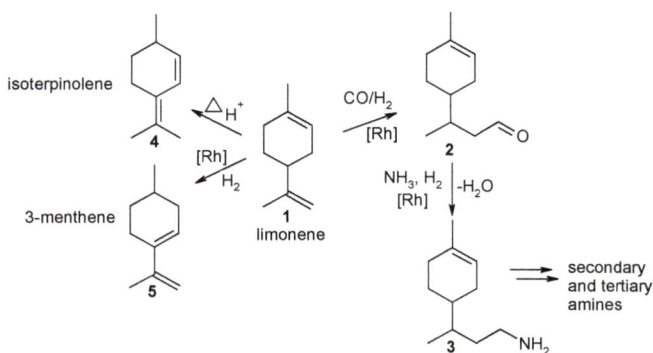

Figure 9. Reaction scheme of the hydroaminomethylation of limonene (**1**) with ammonia using [Rh(cod)Cl]$_2$ catalyst. Reprinted from [77] with permission from Willey.

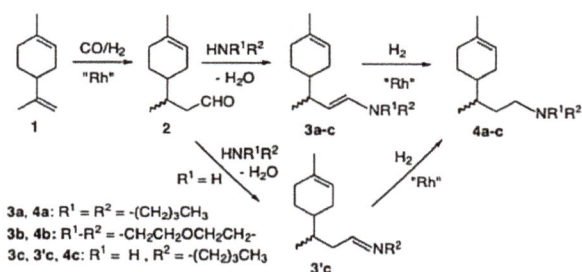

Figure 10. Hydroaminomethylation of limonene (**1**). Reprinted from [74] with permission from Elsevier.

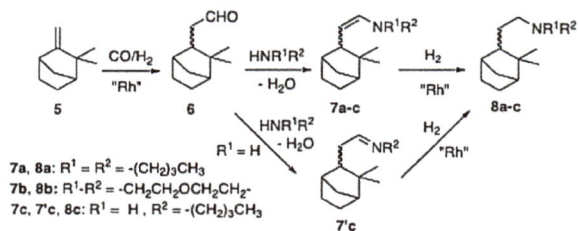

Figure 11. Hydroaminomethylation of camphene (**5**). Reprinted from [74] with permission from Elsevier.

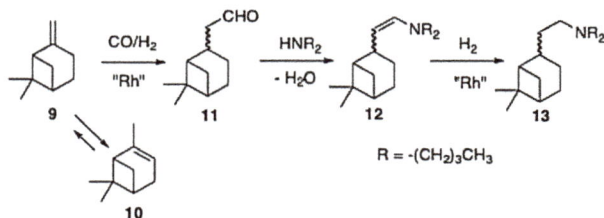

Figure 12. Hydroaminomethylation of β-pinene (**9**). Reprinted from [74] with permission from Elsevier.

Hydroaminomethylation of estragole (**1**) (Figure 13), a bio-renewable starting material, with di-*n*-butylamine was studied in [80]. Estragole being a primary constituent of essential oil of tarragon (60–75%) is also present in other sources, such as pine oil, turpentine, fennel or anise (2%) [86]. HAM consists of alkene hydroformylation followed by reductive amination of aldehydes. Different ligands were used with rhodium(I) catalysts including phosphine, phosphites, and phospholes. The latter were the most efficient not only in hydroformylation, but also in reductive amination. Three isomeric amines (**9–11**) were generated as final products (Figure 13). Along with these imines aldehydes (**3–5**) and enamines (**6–8**) were observed depending on conditions. Side reactions included for example aldol condensation. Some other hydrogenation products as well as unidentified products were also formed.

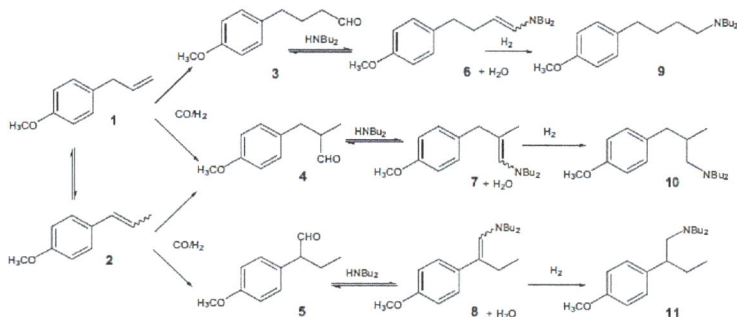

Figure 13. Hydroaminomethylation of estragole (**1**). Reprinted from [80] with permission from Elsevier.

Hydroaminomethylation of eugenol (**1**) with di-*n*-butylamine (Figure 14) involved bis[(1,5-ciclooctadiene)(μ-methoxy)rhodium(I)] as a pre-catalyst [79]. The presence of phosphines was needed to improve chemoselectivity in hydroformylation, being detrimental for hydrogenation of enamine intermediates. Similar to the cases described above mainly linear aldehyde was obtained in hydroformylation. Efficiency of HAM could be also improved by addition of triflic acid as a promoter [79].

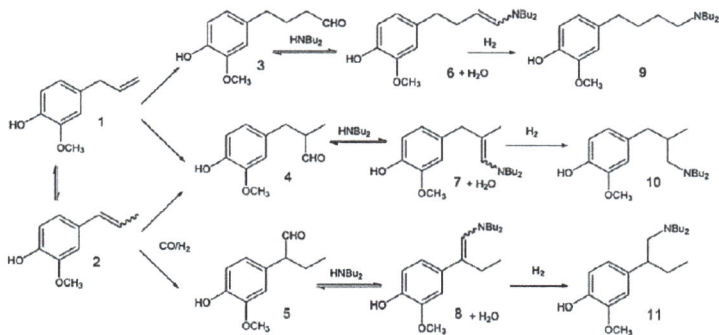

Figure 14. Hydroaminomethylation of eugenol with di-*n*-butylamine. Reprinted from [79] with permission from Elsevier.

As can be seen from Figure 14 hydroaminomethylation of eugenol with di-*n*-butylamine gives three isomeric amines (**9–11**) of which compound **9** is predominant. Similar to estragole the intermediate aldehydes (**3–5**) and enamines (**6–8**) were also observed. Tables 1 and 2 contain the results for HAM of eugenol for different catalysts and reaction conditions.

Table 1. Hydroaminomethylation of eugenol (**1**, Figure 14) with di-*n*-butylamine: ligand effect [a]. Adapted from [79].

Ligand	P/Rh [b]	Con. (%)	Product Distribution (%)				Regioselectivity (%) [c]		
			2	Aldehydes	Enamines	Amines	9	10	11
None	0	100 [d]	4	1	0	90	61	33	6
PPh₃	2	100	0	24	3	73	96	4	0
PPh₃	10	100	0	32	10	58	94	5	1
NAPHOS	2	34	3	34	56	7	>99	0	0
NAPHOS [e]	2	100	10	21	5	64	97	3	0

[a] Conditions: **1** (10 mmol); di-*n*-butylamine (10 mmol); [Rh(cod)(μ-OMe)]₂ (5.0 × 10⁻³ mmol), toluene (30 mL), 4.0 MPa (CO:H₂ = 1:3), 100 °C, 24 h. For products, the value "zero" means not observed or <0.5%. [b] Phosphorus/Rhodium atomic ratio. [c] Related to the sum of amines (**9** + **10** + **11**). [d] 5% of hydrogenation of substrate. [e] 120 °C.

Table 2. Hydroaminomethylation of eugenol (**1**, Figure 14) with di-*n*-butylamine: effect of added acid [a]. Adapted from [79].

Ligand	P/Rh [b]	Acid	T (°C)	Con. (%)	Product Distribution (%)				Regioselectivity (%) [c]		
					2	Aldehydes	Enamines	Amines	9	10	11
PPh₃	2	None	120	100 [d]	0	17	0	83	96	4	0
PPh₃	2	H₂SO₄ [d]	120	86	13	76	9	2	94	6	0
PPh₃	2	HOTs [d]	120	100	0	23	0	75	89	11	0
PPh₃	2	HOTf [d]	120	100	1	11	0	88	84	14	2
NAPHOS	2	HOTf [e]	100	100	5	9	0	86	99	1	0
NAPHOS	20	HOTf [e]	100	100	2	5	0	93	99	1	0
NAPHOS	40	HOTf [e]	100	100	2	3	0	95	98	2	0

[a] Conditions: **1** (10 mmol); di-*n*-butylamine (10 mmol); [Rh(cod)(μ-OMe)]₂ (5.0 × 10⁻³ mmol), toluene (30 mL), 4.0 MPa (CO:H₂ = 1:3), 24 h. For products, the value "zero" means not observed or <0.5%. [b] Phosphorus/Rhodium atomic ratio. [c] Related to the sum of amines (**9** + **10** + **11**). [d] 1.0 mmol. [e] 2.0 mmol.

Triflic acid, being more stable than HBF₄, was reported to be an efficient promoter for eugenol hydroaminomethylation in the presence of phosphines as ancillaries [79].

Rh/1,2-bis(diphenylphosphino)ethane was used [78] for hydroaminomethylation of industrially available β-myrcene and β-farnesene. The reaction network is basically the same as for other terpenes presented above (Figure 15) with also high regioselectivity towards the linear aldehyde. Such efficient synthetic protocol allows preparation of environmentally friendly biobased surfactants.

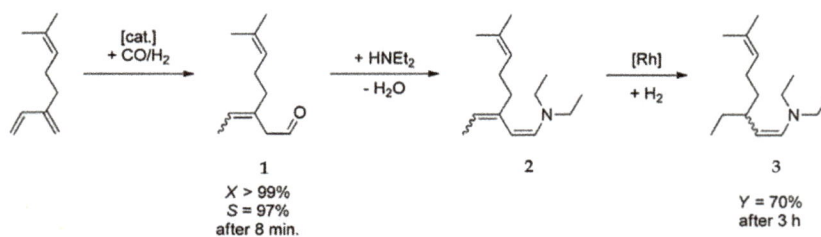

Figure 15. Hydroaminomethylation of myrcene. GC based yields (Y) and conversions (X) are given in % of reacted myrcene. Reproduced from [78] with permission from Willey.

The final example of this section is related to a two-step synthesis protocol when hydroformylation of a-pinene was done either using rhodium or cobalt based catalysts. Chiral aminomethyl pinane was prepared in 100 g scale [75] with rhodium as a catalyst giving (+)-3-formylpinane and subsequent reductive amination with ammonia. On the contrary Co₂(CO)₈ led to (−)-2-formylborane.

3.3. Hydroamination on Olefin Bonds of Terpenes

Regioselective hydroamination of alkenes is more challenging than hydroamination of aldehydes. In particular a product of anti-Markovnikov addition is desired, while known synthetic protocols are mainly selective towards the product of Markovnikov addition [87–89].

Pd(cod)Cl$_2$ in combination with bis(2-diphenylphosphinophenyl) ether (DPEphos) [89] afforded selective hydroamination of acyclic and cyclic dienes with several aromatic and aliphatic amines not requiring presence of any additive. Significant efforts in [89] were concentrated on transformations of isoprene. More relevant in the context of this review is hydroamination of myrcene catalyzed either by alkali metals or transition metals [33]. Following the same pattern as for other 1,3-dienes, predominantly 1,4-amines were formed (Figure 16).

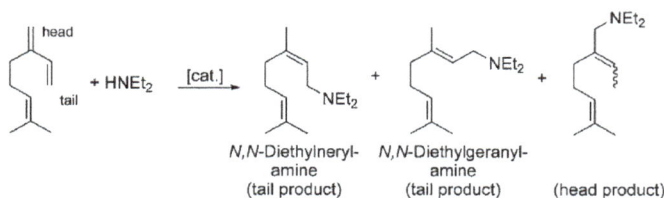

Figure 16. Identification of tail products in hydroamination of myrcene. Reproduced from [33] with permission from Willey.

The reactive double bond is the terminal one (denoted as tail in Figure 16) resulting in linear amines, which in fact are the desired products. Such a method is rather efficient compared to alternatives, which rely on the corresponding acetates as the starting compounds and allylic amination catalyzed by palladium complexes [90–93].

Superior atom efficiency of hydroamination (100%) is a reason for a plentitude of studies on addition of amines, in particular diethylamine, to myrcene (Table 3).

Table 3. Overview of hydroamination investigations *. Adapted from [33].

Catalyst/Conditions [a]	Amine	Geranylamine Selectivity (%)	Amine Yield (%)	TON [c]	TOF [d] (h^{-1})
20.0 mol% BuLi, 50 °C, 4 h	HNEt$_2$	95	77–83	4	1.04
33.3 mol% Na/16.7 mol% naphthalene, 20 °C, 1 h	HNEt$_2$	80	53	2	1.60
36.0 mol% Li, 50 °C, 20 h	HN*i*Pr$_2$	n.m.	80	2	0.11
100 mol% Li, 50 mol% naphthalene, 100 mol% TMDAP, 20 °C, 4 h	HNEt$_2$, morpholine, pyrrolidine, piperidine, HN*n*Bu$_2$	geranylamine 88 [b]	72 [b]	1 [b]	0.18 [b]
10 mol% BuLi, 50 °C, 2.5 h	HNMe$_2$	88	79	8	3.18
1.7 mol% Na, 50 °C, 4 h	HNMe$_2$, HNEt$_2$, HN*i*Pr$_2$, HN*n*Bu$_2$	90 [b]	83 [b]	49 [b]	12.21 [b]
3.6 mol% Li, 25 °C, 72 h	HNMe$_2$, HNEt$_2$, HN*i*Pr$_2$, HN*n*Bu$_2$, piperidine, 2-methylpiperidine, 2,6-dimethyl-piperidine	96 [b]	87 [b]	24 [b]	0.34 [b]
36.0 mol% Li, 55 °C, 5 h	HNEt$_2$	92	74–77	2	0.43
15.0 mol% BuLi, 55 °C, 12 h	HNEt$_2$	95	85	6	0.47
0.1 mol% [RhCl(cod)]$_2$, 2.3 mol% TPPTS, 100 °C, 21.5 h (biphasic)	morpholine	53	59	590	27.44
0.2 mol% Pd(CF$_3$CO$_2$)$_2$, 1.6 mol% DPPB, 100 °C, 4 h	morpholine	57	98	490	123.50

[a] TMDAP: N,N,N′,N″-tetramethyldiaminopropane, TPPTS: triphenylphosphine trisulfonate. [b] According to diethylamine. [c] TON: turn over number. [d] TOF: turn over frequency. * References to original papers are provided in [33].

N,N-Diethylgeranyl- and nerylamine are valuable starting compounds for synthesis of a variety of terpenoids (Figure 17) including (−)-menthol, myrcenol, hydroxycitronellol, nerol, geraniol, linalool, (+)-citronellal, and (+)-citronellol.

The hydroamination reaction is mostly catalyzed by alkali metal-based systems such as sodium or lithium (Table 3, entries 1–9). More expensive transition metals, nevertheless provide higher TON values being effectively recycled. Nowadays two routes are known using transition metal catalysts (Rh and Pd complexes) developed by Rhone–Poulenc [94] and Berh [33] with TOF values of 27 and 124 h^{-1}, respectively (Table 3, entries 10–11).

Preparation of optically pure (−)-citronellal and (−)-menthol (route A) developed by Takasago Perfumery Company is the largest application of asymmetric catalysis. It includes enantioselective isomerization to an optically active enamine with high enantioselectivity (ee = 95–99%) [95]. Subsequent hydrolysis results in (+)-citronellal, which then undergoes cyclization in the presence of $ZnBr_2$ and forms (−)-isopulegol. This reaction can also be performed over heterogeneous catalysts [96]. Hydrogenation of isopulegol gives finally (−)-menthol.

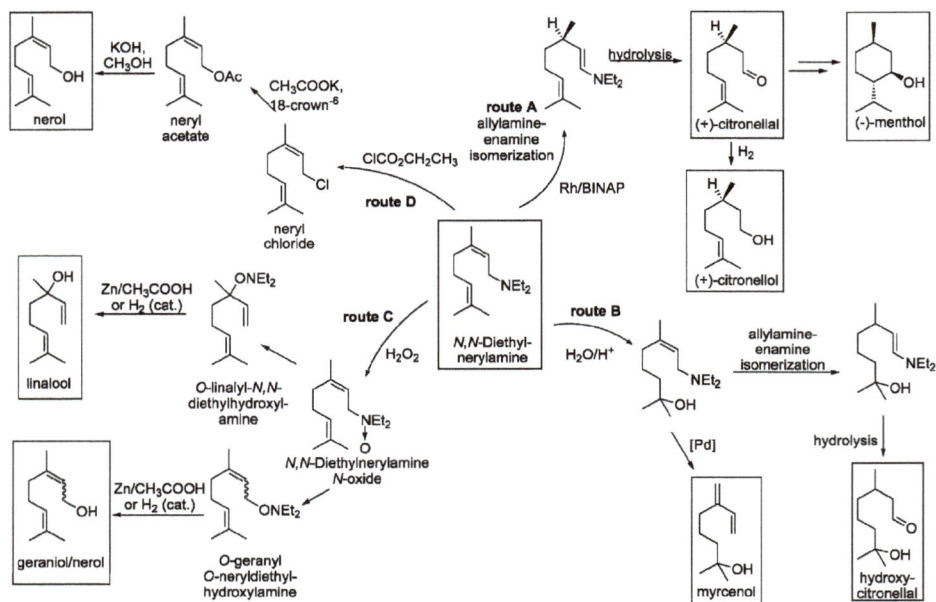

Figure 17. Product network based on *N,N*-diethylnerylamine. Reproduced from [33] with permission from Willey.

In route B (Figure 17) the amine group serves as a suitable protecting group allowing selective hydration of the isolated double bond of myrcene. Further transformations using Pd complexes [97–99] or sodium hydride [100] give respectively myrcenol and hydroxycitronellal.

Oxidation of *N,N*-diethylnerylamine with hydrogen peroxide (route C in Figure 17) is the first step in the synthesis of geraniol, nerol, or linalool [101].

Amination of myrcene with 2-amino-2-methyl-1-propanol developed by SCM Corporation leads to products acting as repellents against the American flour beetle and the German cockroach [61].

Hydroamination of myrcene into diethylgeranylamine using palladium complexes with bidentate ligands such as bis(diphenylphosphino)butane (DPPB) or bis(2-diphenylphosphinophenyl) ether (DPEphos) [102] raises an issue of palladium recovery combining the advantages of homogeneous catalysis (high selectivity and activity) with those of heterogeneous catalysis (catalyst reuse and simple separation). Application of thermomorphic solvent systems, such as for example dimethylformamide/heptane and acetonitrile (ACN)/heptane, allowed simple catalyst separation. ACN/heptane turned out to be a more suitable solvent mixture, permitting efficient product extraction with negligible catalyst leaching. This method is based on the temperature dependent miscibility gap of the solvent system components. Thus, when a mixture of two liquid components, immiscible at room temperature, is heated to a higher reaction temperature, a single liquid phase is formed. In this state no mass transfer limitations occur. Cooling down under the critical solution temperature leads to a biphasic system, from which the catalyst phase can be simply separated from the extract phase providing easy catalyst recycling. Overall 90% yield of diethylgeranylamine was reported [102].

3.4. Amination of Terpene Alcohols

Hydrogen borrowing amination of terpene alcohols has attracted a lot of attention generating only water as the byproduct [7,103–107]. While this method has been adapted industrially [108] for production of low alkyl chain amines such as *N*-methyl-, *N,N*-dimethyl-, and *N,N,N*-trimethyl-amines in the presence of Bronsted and Lewis acid catalysts, rather high temperatures are needed exceeding 300 °C and moreover mixtures of *N*-substituted amines are often produced.

In this section, we discuss recent progress in the development of efficient homogeneous and heterogeneous catalysts, capable of carrying out selective synthesis of desirable amines, especially taking into account that selective amination of rather labile terpene alcohols requires milder conditions and thus more efficient catalytic systems. Along with the hydrogen borrowing reactions, less atomic efficient homogeneous transition metal-catalyzed allylic substitution reactions with functionalized allylic terpene alcohols are considered.

3.4.1. Homogeneous Catalysts

A broad range of homogeneous catalysts based on transition metal complexes was applied for *N*-alkylation, including Rh, Pd, Au, Ag, Pt, Os, and Re. Moreover, even some systems with non-noble metals (Ni, Cu, Fe, and Co) have been proven to be efficient catalysts in *N*-alkylation [109].

In particular contribution of Milstein and co-workers in development of direct homogeneous catalytic amination of primary alcohols with ammonia should be acknowledged [104]. Transformations of secondary alcohols to primary amines with ammonia following the so-called hydrogen borrowing methodology were described extensively in the literature [7,64,105,110–122].

Specific applications of this approach to transformations of terpene derivatives were also reported. For instance, Pingen et al. [113] utilized [Ru₃(CO)₁₂] and different phosphor containing ligands to selectively convert various primary and secondary terpene alcohols to primary and secondary amines. This particular catalyst was considered to be an exception, as in fact the hydrogen borrowing approach rarely results in synthesis of amines and diamines with optimal yields using homogeneous catalysts based on Ru and Ir complexes combined with P-ligands. Some examples of reasonable yields of primary and secondary terpene amines exceeding 80% were reported for such primary and secondary terpene alcohols as myrtenol [123,124], citronellol, nerol, geraniol, farnesol, and fenchol [9,125].

Some examples of *N*-alkylation of terpene alcohols with ammonia are presented in Tables 4 and 5. Screening of different P-ligands showed [9] that the acridine-based diphosphine was the only bidentate ligand affording excellent results. This ligand was used in couple with [RuHCl(CO)(PPh₃)₃] for the amination of primary alcohols with ammonia. This combination, however, appeared to be inactive for secondary amines under the same conditions [104,125].

Table 4. Secondary terpene alcohol amination with [Ru$_3$(CO)$_{12}$]/L9 [a]. Adapted from [125].

Substrate	Time (h)	T (°C)	Conversion [b] (%)	Selectivity [c,d] (%)	Ketones (%)	Product
Menthol	21	150	25.3	18.0	82	Menthylamine
	21	170	31.6	70.8	29.2	
	63	150	32.7	24.3	75.7	
Carveol	21	150	67.3	74.7		Carvylamine
	21	170	99.1	55.2		
	65.5	150	82.0	56		
	42	170	99.3	50.4		
Verbenol	21	150	93.5	39.7	44.2	Verbenylamine
	21	170	96.4	27.7	62.2	
	63	170	98.8	25.0	47.7	
Borneol	21	150	58.5	73.2		Bornylamine
	21	170	76.8	83.9		
	65.5	150	72.2	78.7		
	63	170	79.1	81.2		
Fenchol	21	150	50.6	0	34.8	Fenchylamine
	21	170	59.2	0	46.0	
	63	150	54.5	0	38.2	
	117	170	63.9	0	44.3	

[a] [Ru$_3$(CO)$_{12}$] (0.066 mmol), L9 (2 mmol), substrate (10 mmol), *tert*-amyl alcohol (13.3 mL), NH3 (6 mL, 234 mmol), 150 and 170 °C. [b] Conversion determined from GC analysis on the basis of alcohol consumption and amine production. [c] Total of all primary amines. [d] Main byproducts are primary imines, amides, and ketones.

Table 5. Primary terpene alcohol amination with [Ru$_3$(CO)$_{12}$]/L9 [a]. Adapted from [125].

Substrate	Time (h)	T (°C)	Conversion [b] (%)	Selectivity [c,d] (%)	Product
Citronellol	21	150	33.8	64.8	Citronellylamine
	21	170	47.4	61.2	
	63	150	47.9	71.4	
	152	170	82.9	65.9	
Geraniol	21	150	98.9	79.9	Geranylamine
	21	170	99.3	71	
	63	150	99.7	70.3	
	42	170	99.4	67.4	
Nerol	21	150	66.7	92.9	Nerylamine
	21	170	85.5	75.1	
	120	150	85.8	87	
	49	170	96	69.8	
Farnesol	21	150	83.8	77.4	Farnesylamine
	21	170	94.8	73.8	
	95	150	93,2	72.5	
	45	170	99.4	76.1	

[a] [Ru$_3$(CO)$_{12}$] (0.066 mmol), L9 (0.2 mmol), substrate (10 mmol), *tert*-amyl alcohol (13.3 mL), NH3 (6 mL, 234 mmol), 150 and 170 °C. [b] Conversion determined from GC analysis on the basis of alcohol consumption and amine production. [c] Total of all primary amines. [d] Main byproducts are primary imines, amides, and ketones.

While palladium-catalyzed amination of allylic functionalized simple alkenes has been studied extensively, application of natural terpenic alkenes derivatives is much less common [91,92]. In this context an interesting example is the synthesis of *N,N*-diethylgeranylamine (**2**) and *N,N*-diethylneranylamine (**4**) from diethylamine and linalyl acetate (**1**) or nerolidyl acetate (**3**) (Figure 18) using Pd(PPh$_3$)$_4$ [91].

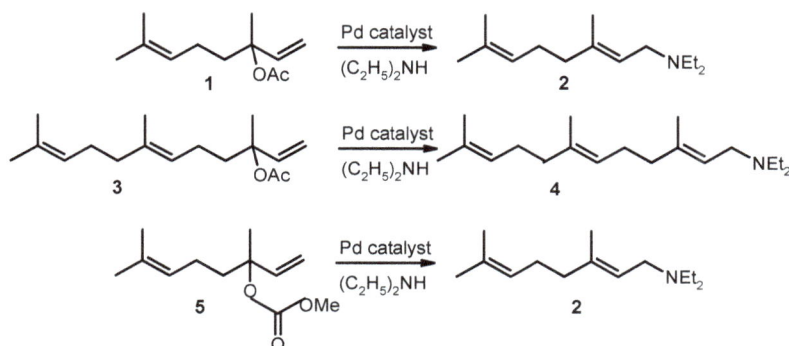

Figure 18. Pd catalyzed allylic amination of linalyl acetate (**1**), nerolidyl acetate (**3**), and linalyl methylcarbonate (**5**). Adapted from [91].

Besides diethylamine other amines can be applied for such transformations. For example, *N*-geranylaniline was obtained by reaction of aniline with linalyl acetate (**1**) displaying 100% stereoselectivity.

Analogously nerolidyl acetate (**3**) was transformed into amine derivatives [91]. Dependence of the yield and selectivity (E to Z ratio) for nerolidyl acetate (**3**) is presented in Table 6.

Table 6. Pd(0)-catalyzed amination of nerolidyl acetate (**3**). Adapted from [91].

Amine	Solvent	Ligand	T (°C)	Molar Ratio Amine/(3)	Yield of (4) (%) [a]	E:Z
(C$_2$H$_5$)$_2$NH	THF	PPh$_3$	20	2.0	40	60:40
(C$_2$H$_5$)$_2$NH	THF	PPh$_3$	50	2.0	14	62:38
(C$_2$H$_5$)$_2$NH	IPE	PPh$_3$	20	2.0	48	60:40
(C$_2$H$_5$)$_2$NH	IPE	PPh$_3$	50	2.0	8	62:38
(C$_2$H$_5$)$_2$NH	Benzene	PPh$_3$	20	2.0	35	76:24
(C$_2$H$_5$)$_2$NH	DMF	PPh$_3$	20	2.0	70	52:48
(C$_2$H$_5$)$_2$NH	CH$_2$Cl$_2$	PPh$_3$	20	2.0	52	54:46
(C$_2$H$_5$)$_2$NH	(C$_2$H$_5$)$_2$NH	PPh$_3$	20	Excess	57	59:41

Reaction conditions: nerolidyl acetate (**3**) 4 mmol; catalyst 0.02 mmol (5 mol%); solvent 12 mL. [a] Yields of (**4**) were isolated yields.

Similarly to the example above for linalyl acetate utilization of linalyl methylcarbonate with the same Pd(0) catalyst and diethylamine results also in *N,N*-diethylgeranylamine (Table 7).

Reaction of linalyl methylcarbonate (**5**) with aniline or morpholine gives selectively (*E*)-isomers (E:Z = 100:0) (Table 7).

Table 7. Pd(0)-catalyzed amination of linalyl methylcarbonate (**5**). Adapted from [91].

Amine	Solvent	Ligand	T (°C)	Molar Ratio Amine/(5)	Yield of (2) (%) [a]	E:Z
$(C_2H_5)_2NH$	THF	PPh_3	20	2.0	20	91:9
$(C_2H_5)_2NH$	IPE	PPh_3	20	2.0	21	92:8
$(C_2H_5)_2NH$	Benzene	PPh_3	20	2.0	15	94:6
$(C_2H_5)_2NH$	DMF	PPh_3	20	2.0	42	86:14
$(C_2H_5)_2NH$	$1,2\text{-}C_2H_4Cl_2$	PPh_3	20	2.0	30	86:14
$(C_2H_5)_2NH$	$(C_2H_5)_2NH$	PPh_3	20	Excess	48	91:9
Aniline	THF	PPh_3	20	2.0	52	100:0 [b]
Aniline	IPE	PPh_3	20	2.0	60	100:0
N-Me-Aniline	THF	PPh_3	20	2.0	50	100:0 [c]
N-Me-Aniline	IPE	PPh_3	20	2.0	59	100:0
Morpholine	THF	PPh_3	20	2.0	47	100:0 [d]
Morpholine	IPE	PPh_3	20	2.0	45	100:0
Pyrrolidine	THF	PPh_3	20	2.0	32	85:15 [e]
Pyrrolidine	IPE	PPh_3	20	2.0	49	94:6

Reaction conditions: linalyl merhylcarbonate (**5**) 5 mmol, catalyst 0.25 mmol (5 mol%), solvent 12 mL. [a] Yields of (**2**) were isolated yields. [b] The product is *N*-geranyl aniline. [c] The product is *N*-methyl-*N*-geranyl aniline. [d] The product is geranyl morpholine. [e] The product is geranyl pyrrolidine.

Pd(0)-catalyzed amination of allylic natural functionalized terpenes—myrtenyl acetate (**1**), perillyl acetate (**3**), geranyl acetate (**5**), mertynyl alkyl carbonate (**9**), perillyl alkyl carbonate (**10**), geranyl alkyl carbonate (**11**)—was studied in [124]. The reaction scheme is shown in Figure 19, while Table 8 illustrates the catalytic results for myrtenyl acetate with and without Et_3N.

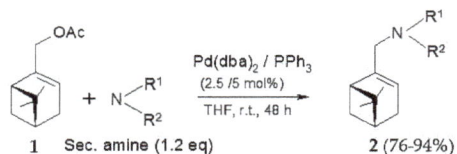

Figure 19. A general reaction amination of myrtenyl acetate (**1**) with different secondary amines (morpholine, pyrrolidine, and dimethylamine). Reproduced from [124] under the terms of the Creative Commons Licenses.

Table 8. Catalytic amination of myrtenyl acetate [a] (**1**). Adapted from [124].

Secondary Amine	Product [b]	$[\alpha]_D$ [c]	With 2.5 Equiv of Et_3N		Without Et_3N	
			Yield [d] (%)	Conv. (%)	Yield [d] (%)	Conv. [e] (%)
Pyrrolidine	**2a**	−4.0	86	94	76	93
Morpholine	**2b**	−18.4	90	96	91	95
Me_2NH	**2c**	−3.5	94	100	92	96

[a] The reaction performed in THF with 2.5 mol % of $Pd(dba)_2$, 5 mol% PPh-and 1.3 equivalents of nucleophile under nitrogen atmosphere at room temperature. [b] Experiments monitored until no further evolution was encountered. [c] $[\alpha]_D$ determined at: c = 2, $CHCl_3$ for (**2a**); c = 1.9, $CHCl_3$ for (**2b**); c = 2, $CHCl_3$ for (**2c**). [d] Isolated yield. [e] Conversions were determined by GC and based on (**1**).

This substrate was used in combination with different secondary amines (pyrrolidine, morpholine, and Me$_2$NH) and the Pd(dba)$_2$/PPh$_3$ catalytic system (Figure 19). As seen from Table 8 high yields could be reached under mild conditions. Application of an acceptor base, e.g., Et$_3$N was beneficial for pyrrolidine as the amine source. The nature of the ligand in case of morpholine as nucleophile and Et$_3$N as the acceptor base influenced significantly the catalytic behavior (Table 9). Presence of ligands was essential as in their absence there was practically no reaction.

Table 9. Ligand influence on the amination of myrtenyl acetate [a] (**1**) with morpholine in the presence of Et$_3$N. Adapted from [124].

Ligand [b]	Yield [c] (%)	Conversion [d] (%)
None	-	2
PPh$_3$	90	96
dppe	88	94
P(o-tolyl)$_3$	13	22
2,2′-dipyridyl	5	14

[a] The reaction mixture in THF with 2.5 mol% of Pd(dba)$_2$, 5 mol% of ligand and 1.2 equivalents of morpholine under nitrogen atmosphere at room temperature. [b] Experiments monitored by GC until no further evolution was encountered. [c] Isolated yield. [d] Conversions were determined by GC and based on (**1**).

Some other examples of Pd(0)-catalyzed amination of terpenic allylic esters, namely perillyl acetate (**3**), geranyl acetate (**5**), mertynyl alkyl carbonate (**9**), perillyl alkyl carbonate (**10**), and geranyl alkyl carbonate (**11**), are summarized in Table 10 [124].

Table 10. Catalytic amination of terpenic allylic esters [a]. Adapted from [124].

Terpenic Allylic Esters	Secondary Amine	Product	Yield [b] (%)	Conv [c] (%)
Allyl acetate				
3	Pyrrolidine	**4a**	48	60
	Morpholine	**4b**	48	56
				62
	Me$_2$NH	**4c**	54	
5	Pyrrolidine	**6a**	39	58
	Morpholine	**6b**	40	56
	Me$_2$NH	**6c**	42	62
Allyl carbonate				
9	Pyrrolidine	**2a**	94	96
	Morpholine	**2b**	94	96
				98
	Me$_2$NH	**2c**	96	
10	Pyrrolidine	**4a**	50	68
	Morpholine	**4b**	52	62
				68
	Me$_2$NH	**4c**	62	
11	Pyrrolidine	**6a**	43	64
	Morpholine	**6b**	44	62
	Me$_2$NH	**6c**	47	66

[a] Reaction performed in THF with 2.5 mmol% of Pd(dba)$_2$, 5 mol% dppe and 1.2 equivalents of nucleophile under nitrogen atmosphere at room temperature. [b] Isolated yield. [c] Conversions were determined by GC.

Lyubimov et al. [55] conducted Pd-catalyzed amination of allylic carbonates of terpenoids: (*E*)-cinnamyl ethyl carbonate (**1**), ethyl prenyl carbonate (**2**), ethyl geranyl carbonate (**3**), in supercritical CO_2 with *N*-(ortho-carboran-3-yl)-*N*-methylamine (**4**) (Figure 20). Corborane amines, which are promising agents for cancer therapy, were obtained at complete conversion with excellent regioselectivity using sodium bicarbonate as the acceptor base [55].

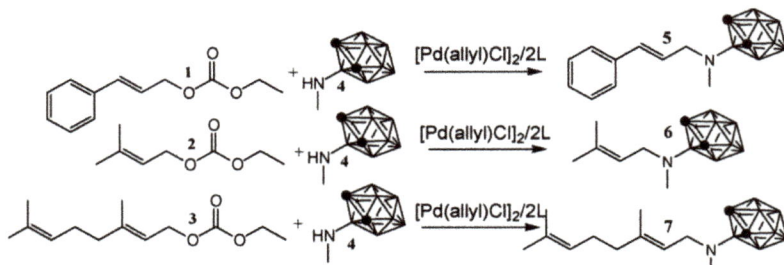

Figure 20. Allylic amination with *N*-(ortho-carboran-3-yl)-*N*-methylamine: *N*-Cinnamyl-*N*-(ortho-carboran-3-yl)methylamine (**5**), *N*-(3-Methylbut-2-enyl)-*N*-(ortho-carboran-3-yl)methylamine (**6**), (*E*)-*N*-(3,7-Dimethylocta-2,6-dienyl)-*N*-(ortho-carboran-3-yl)methyl amine (**7**). Reprinted from [55] with permission from Elsevier.

3.4.2. Heterogeneous Catalysts

Although various transition-metal complexes can provide reasonable selectivity to the desired products, application of expensive noble metal catalysts results in complicated reaction systems. Homogeneous transition metal catalysts based on noble metals are typically expensive and toxic, not stable and prone to contaminate the products and finally difficult to recover. Moreover, application of sophisticated and unpredictable organic ligands has obvious disadvantages, such as difficulties in their synthesis and recovery. Not surprisingly to increase the probability of industrial implementation significant efforts were devoted to heterogenization of homogeneous catalysts [126–129].

One example is related to introduction of a chiral ferrocenyl-based ligand to the surface of a mesoporous material MCM-41 [130]. The active catalytic species (Figure 21) were used in allylic amination of cinnamyl acetate with benzylamine with the aim of achieving high yields of the branched product (reaching 50%) with the highest possible *ee*. In fact, the enantiomeric excess was close to 100%, while the homogeneous catalyst gave exclusively a straight chain product with no *ee* [131].

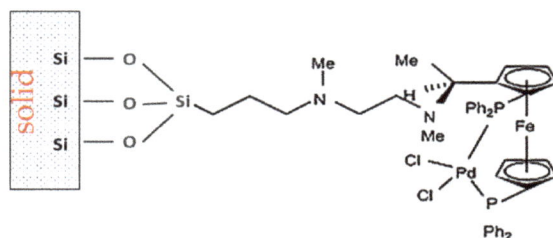

Figure 21. Amination catalysts used in allylic amination of cinnamyl acetate with benzylamine. Adapted from [130].

Even though immobilization of organic ligands or metal complexes by covalent binding [131,132] has been a subject of extensive research there are a number of technological challenges including complexity of the synthesis procedure and eventual leaching of the immobilized species. Advantages of heterogeneous catalysts in product and catalyst isolation, and catalyst reuse prompted their application in amination [133,134]. Specificity of heterogeneous catalysts sometimes displaying activity for only a particular substrate restricts their more widespread application. As an example sulfonamide alkylation with alcohols over Ru on iron oxide can be mentioned of such specificity, as the catalysts are not active in alkylation of carboxamides or amines [135]. Therefore, it is required to develop more active and general heterogeneous catalysts for the terpene alcohols amination without addition of organic ligands.

N-alkylation follows the hydrogen borrowing mechanism with first dehydrogenation of an alcohol to the corresponding aldehyde followed by condensation with an amine giving an imine. The latter is then hydrogenated by hydrogen "stored" on the catalyst resulting in the final alkylated amine products (Figure 22) [109]. The method is thus mainly suitable for primary amines and alcohols. According to this general mechanism, hydrogen transfer from alcohols to catalysts and then from adsorbed hydrogen to intermediate imines are the typical key processes of the methods. It should also be pointed out that inorganic bases were usually required in hydrogen autotransfer reactions for deprotonation of the alcohols to facilitate their coordination with transition metal catalysts. Since bases were usually used in large excess amounts in the early transition metal-free methods, they were used only as additives but not as catalysts in transition metal-catalyzed reactions, even though in many cases they were also used in minor (well below stoichiometry) amounts [109].

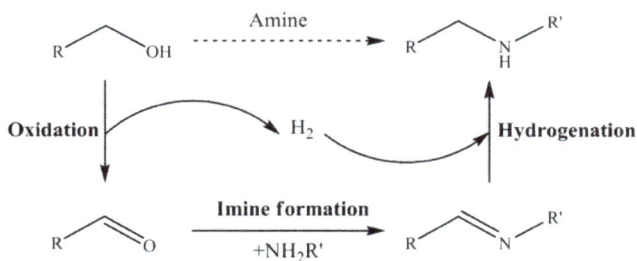

Figure 22. Schematic view of the one-pot alcohols amination applying "hydrogen borrowing approach", where oxidation in fact stands for dehydrogenation. Reprinted from [123] with permission from Elsevier.

Hydrogen transfer steps (dehydrogenation of alcohols and hydrogenation of imines) are considered as the rate controlling steps. Primary amines and alcohols are thus preferred displaying better results in these key steps. Inorganic bases as mentioned above are typically used as additives to facilitate deprotonation of alcohols and coordination to transition metals [109].

First application of typical heterogeneous catalysts in direct terpene alcohol amination was demonstrated in a series of recent studies by Simakova and co-workers [123,136–140]. Liquid-phase amination of myrtenol [123,136–139], nopol, and perillyl alcohol [140] was carried out over supported Au catalysts (Au 1.4 mol% to substrate) in toluene at 180 °C under 9 bar nitrogen pressure using equimolar amounts of substrates without any bases as additives. The reaction network for myrtenol amination is presented in Figure 23. In addition to the expected products, myrtenal and the corresponding imines (**1** and **2**), also myrtanol as well as myrtanal with the saturated C–C bond were formed subsequently resulting also in imines (**3**) and (**4**). It should be noted that prior to the work using heterogeneous catalysts only amine (**2**) was obtained by interacting myrtenol with PBr$_3$ giving a bromide which was then put in contact with aniline leading to amine (**2**) [141].

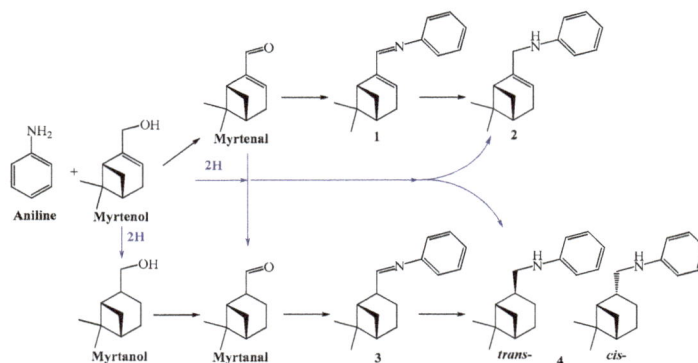

Figure 23. Myrtenol amination with aniline over Au catalysts. Reprinted from [138] with permission from Elsevier.

The authors showed that the support played a crucial role in this reaction. The product distribution during the reaction was found to depend strongly on the type of the support (Figure 24). A nearly complete conversion of myrtenol was achieved only in the presence of Au/ZrO_2 and Au/Al_2O_3 catalysts.

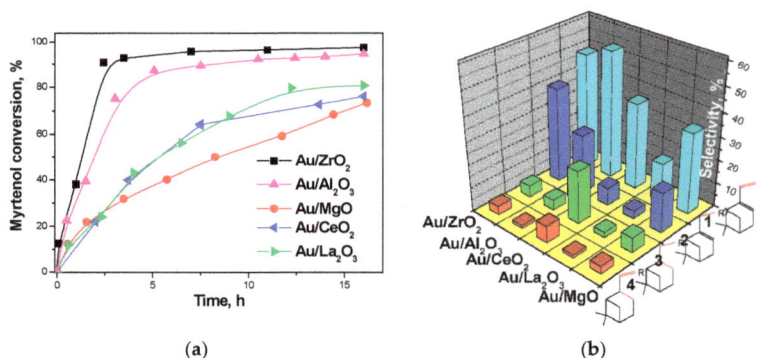

(a)   (b)

Figure 24. Myrtenol conversion (**a**) and the selectivity (**b**) to the corresponding products **1** (blue bars), **2** (dark blue bars), **3** (green bars), and **4** (red bars) at the same myrtenol conversion (74%) for myrtenol amination in the presence of gold supported on ZrO_2, Al_2O_3, CeO_2, La_2O_3, and MgO. The reaction conditions: T = 180 °C, $p(N_2)$ = 9 bar, myrtenol 1 mmol, aniline 1 mmol, toluene 10 mL, catalyst 1.4 mol% Au, R = aniline. Reprinted from [123] with permission from Elsevier.

Gold supported on ceria, magnesia, and lanthana showed a relatively high alcohol conversion, even if the reaction rate was lower than for gold on alumina and zirconia. The non-basic supports Au/Al_2O_3 and Au/ZrO_2 promoted much faster interactions of the aldehyde with aniline as well as hydrogen transfer [123].

The authors underlined the necessity of a certain balance between the different acid–base sites of the metal oxide for efficient alcohol amination. The initial myrtenol activation was shown to require the presence of the basic sites on metal oxide surfaces whereas availability of the protons on the support surface was suggested to be important for the target amine formation. The highest activity in one-pot myrtenol amination among the tested catalysts was obtained over Au/ZrO_2 with both acidic and basic sites.

Moreover, the complex influence of redox treatment on catalytic behavior of Au/ZrO$_2$, Au/CeO$_2$, and Au/La$_2$O$_3$ was studied (Figure 25) elucidating formation of the active species. Au/ZrO$_2$ treated under an oxidizing atmosphere was shown to be more effective in terms of the target secondary amine yield [136].

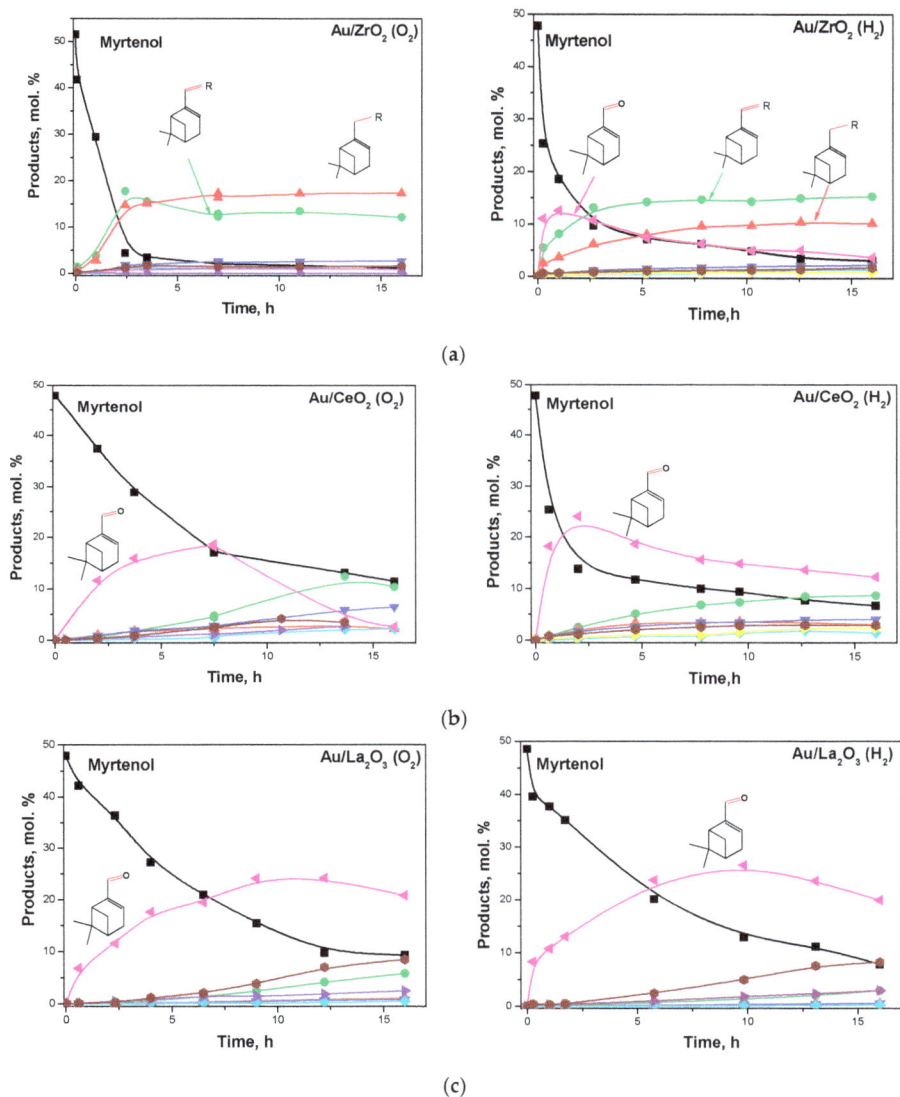

Figure 25. The products distribution as a function of the reaction time during myrtenol amination over Au/ZrO$_2$ (**a**), Au/CeO$_2$ (**b**), Au/La$_2$O$_3$ (**c**) pre-treated in O$_2$ (left) or in H$_2$ (right). The reaction products: myrtenol (■), myrtenal (◀), myrtanol (●), myrtanal (▶), 1 (●), 2(▲), 3 (◆), 4 (▼). The reaction conditions: T = 180 °C, myrtenol 1 mmol, aniline 1 mmol, toluene 10 mL, catalyst 1.4 mol% Au. Reprinted from [137] with permission from Elsevier.

Demidova et al. [123] found that myrtenal condensation with aniline per se was non-catalytic being, however, noticeably accelerated in the presence of a catalyst. The first step of alcohol deprotonation was concluded to be promoted at the basic sites of the support giving an alkoxide intermediate on the support surface. This is followed by β-hydride elimination catalyzed by Au to form myrtenal. The adsorbed myrtenal and aniline interact to form hemiaminal, which then undergoes an attack by the hydride ion from Au nanoparticles and a proton from the support surface resulting in production of the final amine. Formally in the last step the H^+ and H^- transfer to the hemiaminal is accompanied by dehydration. According to the literature data in the case of homogeneous catalysts a cooperative mechanism of a coordinately unsaturated metal center and adjacent acid/base center is widely accepted [123]. H^- in the metal hydrides and H^+ of OH or NH groups of the ligand are transferred to carbon and nitrogen of the C–N bond, respectively. Taking into account this model as well as the regularities obtained for other heterogeneous catalysts [106], the authors proposed [123] that the hemiaminal undergoes dehydration and subsequent H^+ addition resulting in formation of an immonium cation, which is then attacked by hydride ion from Au nanoparticles to form the final product [106,142]. The mechanism is illustrated in Figure 26.

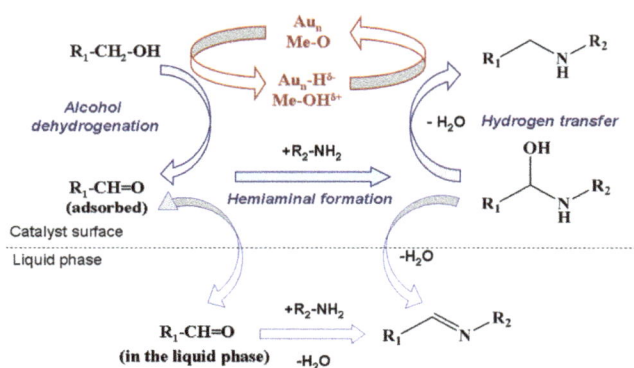

Figure 26. Proposed catalytic mechanism for myrtenol amination with aniline over Au on metal oxide MeO. Reprinted from [123] with permission from Elsevier.

It was found that Au on metal oxides was slightly deactivated during amination predominantly due to imine adsorption. Therefore, kinetic modelling of myrtenol amination was done for the mechanism which also incorporated the catalyst deactivation step [123,136].

Based on the kinetic data of myrtenol amination with aniline published in [123] it was proposed that introducing additional hydrogen is beneficial for the overall process by improving in particular hydrogenation of the imine. Such approach is of general interest for the so-called hydrogen borrowing reactions, when hydrogen generated in the dehydrogenation step is transferred to an intermediate imine.

In [138] the effect of hydrogen addition was thus explored to increase selectivity. Hydrogen addition timing depending on myrtenol conversion and hydrogenation temperature affected selectivity for the reaction products. Hydrogen addition (1 bar) after an almost complete myrtenol conversion at 100 °C increased the yield to amine up to 68% preserving the C=C bond in the initial myrtenol structure. Hydrogen addition at 180 °C independent on the level of myrtenol conversion promoted reduction of both C=C and C=N bonds with formation of two diastereomers (yield up to 93%). Formation of the trans-isomer was preferred when hydrogen was added at almost complete myrtenol conversion. As a result it was shown in [138], that in the presence of a gold catalyst controlled hydrogenation of competitive C=C and C=N groups can be performed during one-pot alcohol amination by regulating the reaction conditions (Table 11).

Table 11. Effect of reaction conditions on the content of amine (**4**) diastereomers and diastereomeric excess value at complete myrtenol conversion. Adapted from [138].

Conditions	*trans*-(%)	*cis*-(%)	*de*(%)
1 bar H$_2$, 8 bar N$_2$, 180 °C	65	35	30
2 bar H$_2$, 7 bar N$_2$, 180 °C	62	38	24
9 bar N$_2$, 180 °C/1 bar H$_2$, 8 bar N$_2$, 180 °C	80	20	60

The same authors studied the application of the more safe hydrogen sources and often more readily available in the fine chemical industry than molecular hydrogen, namely alcohols (methanol, 2-propanol) or formic acid [139]. It was found that small amounts of 2-propanol or formic acid (additive/myrtenol molar ratio equal to 0.5 and 0.25, respectively) helped to improve the yield of the target amine elevating it to 68% and 65%, respectively, compared to 52% amine yield in the absence of additives. However, a further increase of the additive amount decreased amine formation because 2-propanol itself reacted with aniline competing with myrtenol. Introduction of formic acid into the reaction mixture suppressed activity of the Au/ZrO$_2$ catalyst due to a strong adsorption of formic acid and decomposition products on the support basic sites required for activation of the initial alcohol. In comparison with other hydrogen additives methanol was found not to be as effective for one-pot alcohol amination. The catalytic activity and selectivity to the reaction products obtained using different external hydrogen sources are compared in Figure 27.

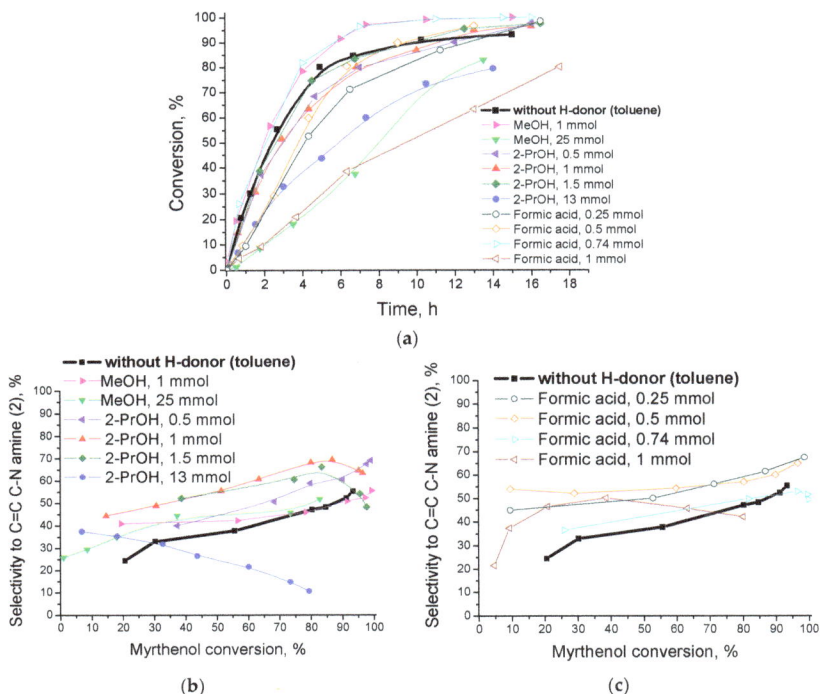

Figure 27. The effect of H-donors addition and its amount on myrtenol conversion (**a**) and selectivity to amine (**2**) (**b**,**c**). The reaction conditions: T = 180 °C, myrtenol 1 mmol, aniline 1 mmol, toluene 10 mL, Au/ZrO$_2$ catalyst 92 mg, 9 bar N$_2$. Reprinted from [139] with permission from Elsevier.

Finally, the structure effect of terpene alcohols selected based on their pharmaceutical relevance was studied over Au/ZrO_2 [140] (Tables 12 and 13). Primary bicyclic (myrtenol and nopol, bearing an unconjugated –OH group) and monocyclic (perillyl) alcohols were chosen for aniline amination under comparable conditions. The rate of alcohol dehydrogenation decreased 10-fold using nopol which after dehydrogenation gave an unconjugated aldehyde group. Selectivity to the desired amine in the latter case increased via selective hydrogen transfer to C=N bonds. Monocyclic perillyl alcohol was more reactive than myrtenol, giving a complex product mixture at 180 °C with the amine present only in small amounts. A decrease of the reaction temperature resulted in a more controlled transformation of perillyl alcohol to imines and amines, with predominant hydrogenation of the C=C bond (Table 12).

Table 12. Effect of the monoterpene alcohol structure on catalytic behavior of one-pot alcohol amination with aniline. The reaction conditions: T = 180 °C, monoterpene alcohol 1 mmol, aniline 1 mmol, toluene 10 mL, Au/ZrO_2 catalyst 92 mg, 9 bar N_2. Adapted from [140].

Alcohol	R^a $(mol \cdot L^{-1} \cdot h^{-1})$	Time (h)	Alcohol Conversion (%)	Sec. Amine		Imine		Aldehyde [b]
				1^c	2^d	1^c	2^d	
Myrtenol	2.1×10^{-2}	2	44	39	5	50	3	0
		8	87	52	6	34	2	0
		16	98	53	7	33	2	0
Nopol	2.4×10^{-3}	8	10	74	0	22	0	0
		16	40	76	0	19	0	0
Perillyl alcohol	3.1×10^{-2}	2	70	Complicated mixture of the products				
		8	99					
Perillyl alcohol [e]	1.3×10^{-2}	8	98	1	19	28	11	2

[a] Initial reaction rate of alcohol transformation, calculated within the linear part of the kinetic curves. [b] Selectivity to intermediate aldehyde formed from the corresponding primary or secondary alcohol, respectively. [c] Selectivity to the corresponding product with unsaturated C=C bond in alcohol structure (bold). [d] Selectivity to the corresponding product with saturated C–C bond in alcohol structure (italic). [e] The reaction temperature is 160 °C.

Table 13. Effect of amine structure on catalytic behavior of one-pot myrtenol amination. The reaction conditions: T = 180 °C, myrtenol 1 mmol, amine 1 mmol, toluene 10 mL, Au/ZrO_2 catalyst 92 mg, 9 bar N_2. Adapted from [140].

Amine	R^a $(mol \cdot L^{-1} \cdot h^{-1})$	Time (h)	Alcohol Conversion (%)	Sec. Amine		Imine		Myrtenal/Myrtanal
				1^b	2^c	1^b	2^c	
Aniline	2.1×10^{-2}	2	44	39	5	50	3	0/0
		8	87	52	6	34	2	0/0
		16	98	53	7	33	2	0/0
Aniline [d]	2.0×10^{-2}	16	98	69	9	10	2	5/2
4-Methylaniline	1.6×10^{-2}	2	34	19	1	73	0	$4/9 \times 10^{-1}$
		8	89	37	1	53	0	8/2
		16	98	43	2	49	0	5/1
4-Methylaniline [e]	1.4×10^{-2}	16	95	51	4	30	2	7/2
4-Bromoaniline	3.4×10^{-2}	2	60	0	14	0	43	14/0
		8	99	0	22	0	49	13/0
Benzylamine	2.3×10^{-2}	2	48	36	13	26	14	$1 \times 10^{-1}/0$
		8	98	3	35	0	49	0/0
Phenethylamine	3.3×10^{-3}	8	40	63	0	36	0	0/0
		16	53	51	0	47	0	0/0
		24	70	46	1	52	0	0/0
Phenethylamine [e]	3.0×10^{-3}	16	48	69	3	20	0	3/2

Table 13. *Cont.*

Amine	R [a] $(\text{mol·L}^{-1}\cdot\text{h}^{-1})$	Time (h)	Alcohol Conversion (%)	Selectivity (%)				Myrtenal/Myrtanal
				Sec. Amine		Imine		
				1 [b]	2 [c]	1 [b]	2 [c]	
3,4-Dimethoxy-phenethylamine	7.0×10^{-4}	16	13	72	0	27	0	0/0
3,4-Dimethoxy-phenethylamine [e]	6.9×10^{-4}	16	12	80	0	15	0	2/1
3-Aminopyridine	1.9×10^{-3}	8	19	21	0	74	0	2/1
		16	31	20	1	60	0	11/3
		24	44	23	2	58	1	12/3
3-Aminopyridine [e]	2.0×10^{-3}	16	33	30	2	49	0	14/2

[a] Initial reaction rate of alcohol transformation, calculated within the linear part of the kinetic curves. [b] Selectivity to the corresponding product with unsaturated C=C bond in myrtenol structure (bold). [c] Selectivity to the corresponding product with saturated C–C bond in myrtenol structure (italic). [d] The result obtained in [140] adding 0.5 mmol of 2-propanol into initial reaction mixture. [e] 0.5 mmol of 2-propanol was added to initial reaction mixture.

Myrtenol amination was also investigated for a range of aliphatic and aromatic amines, showing that the primary amine structure affected both the initial dehydrogenation rate and the selectivity to the desired amine (Table 13). Important issues to be considered were the substrate accumulation on the catalyst surface as well as the reactivity of the substrates and intermediates. Stronger adsorption of more basic amines on the cationic gold species can retard dehydrogenation of an alcohol, which is the first step of the overall process. The nature of the substituent and the reactivity of intermediates is even more important. This was confirmed, as a good correlation was found between the substrate structure and reactivity using the Hammett equation [140].

These data indicated that while an increase of hydrogen availability is an evident method of improving the yields, fine tuning of the amines reactivity can be an even more efficient tool.

3.5. C–H Amination of Terpenes

The final section of this review is devoted to C–H amination. Sulfonimidoylnitrene species (**1**) (Figure 28) allow the intermolecular chemoselective C–H amination of various complex molecules [143] giving a possibility to synthesize enantiopure aminated derivatives not easily accessible by classical organic synthesis. Allylic methylene units of terpenes and enol ethers have been efficiently aminated with the yields close to 98% and very high diastereomeric ratios (up to ca. 200 to 1) in the presence of a rhodium catalyst Rh$_2$(S-nta)$_4$ (nta = N-(1,8-naphthoyl)-alanine) (Figure 29) [143]. More importantly, the combination of steric, inductive, and conformational factors leads to chemoselective functionalization in the allylic positions allowing site-selective amination with yields of up to 88%. Steric and stereoelectronic effects, in this case, favor amination of tertiary equatorial C-H bonds.

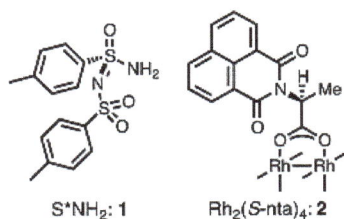

Figure 28. Sulfonimidamide S*NH$_2$ (**1**) and the rhodium catalyst Rh$_2$(S-nta)$_4$ (**2**). Reproduced with permission from [144]. Copyright American Chemical Society.

Catalytic C–H amination of terpene substrates was highly efficient allowing isolation of C–H aminated products (**3a–3e**) in high yields (Table 14). In Table 14 compounds (**3a**), (**3b**) stand for derivatives of α-pinene, while (**3c–3d**), and (**3e**) correspond to derivatives of respectively limonene, nopol trichloroacetate, and carene. An important feature of amination is chemoselectivity in activation of the C–H bond, while the double bond is inert.

Table 14. Allylic amination of terpenes with sulfonimidamide S*NH$_2$ (**1** in Figure 28) [a]. Adapted from [143].

Product	Yield	d.r. [b]	Product	Yield	d.r. [b]
3a	91	39:1	3f	85	
3b	71	49:1	3g	34(63) [c]	
3c	73	49:1	3h: R=Ac / 3I: R=COCCl$_3$	90 (3h) / 89 (3I)	19:1 / >20:1
3d R=COCCl$_3$	98	>20:1	3j: R=3,5-NO$_2$Bz	81	5:1
3e	27	10:1			

[a] Reaction conditions: terpene (0.2 mmol) in a 3:1 mixture of 1,1,2,2-tetrachloroethane/MeOH at −35 °C.
[b] The diastereomeric ratios have been determined by ^{1}H NMR or HPLC. [c] Yield in parentheses obtained using 5 equiv of substrate. (**3a**, **3b**)—enantiomers of α-pinene, (**3c**)—limonene, (**3d**)—nopol trichloroacetate, (**3e**)—carene.

4. Learning from Homogeneous Catalysis and Future Outlook for Heterogeneous Catalysis

The current review focuses on terpene amine synthesis in the presence of solid catalysts rather than with homogeneous catalysts even if the latter are also discussed including immobilized ones.

Terpene alcohol amination represents an interesting example of where catalytic synthesis might reflect different mechanistic views: the hydrogen borrowing pathway in general and the allylic substitution in a particular case of functionalized allylic alcohols substrates. The hydrogen borrowing pathway is a highly atom efficient approach matching green chemistry requirements and providing selective C–N bond formation while keeping the initial terpene moiety. In this connection close attention in the current review was paid to this very promising approach realized over homogeneous and heterogeneous catalysts. Amination of myrtenol over supported gold catalysts was reliably documented to proceed through hydrogen borrowing methodology [123]. Amination of various other allylic and non-allylic terpene alcohols with homogeneous Ru complexes was shown to occur via a hydrogen borrowing pathway as well [125]. Thus [Ru$_3$(CO)$_{12}$]/L9 catalyzed amination of secondary and primary terpene alcohols (Tables 4 and 5 in the manuscript) was reported to proceed through intermediate carbonyl compounds indicating a hydrogen borrowing pathway rather than an allylic substitution. The corresponding amine compounds were formed both in the case of substrates with an allylic –OH group (carveol, verbenol, geraniol, nerol, farnesol) and a non-allylic OH group (menthol, borneol, fenchol, citronellol). Reactivity of terpene alcohols depends rather on the presence of steric hindered substituents than on the conjugation with the double bond. Thus in the case of substrates with bulky substituents such as menthol, verbenol, and fenchol the intermediate ketones were the major products regardless of OH group conjugation with the C=C bond.

Along with hydrogen borrowing reactions less atomic efficient catalytic methodologies were also demonstrated in the review. In particular, homogeneous transition metal-catalyzed allylic substitution reactions with functionalized allylic terpene alcohols, discussed in the review (Figures 18–20, Tables 6–10) [55,92,124], resulted in stoichiometric by-products formation. Allylic substitution reactions typically utilize an activated allylic substrate (i.e., an allylic alcohol protected as an acetate or ester acting as a leaving group), a transition metal catalyst (usually palladium), and a nucleophile. In a very good recent review [145], which unfortunately did not present examples of terpene allylic alcohols amination, it was demonstrated that allylic substitution in general is possible for unactivated allylic alcohols. In the current review, allylic substitution type of transformations are related to amination of several functionalized allylic alcohols catalyzed by Pd complexes, namely linalyl acetate (Figure 18), nerolidyl acetate (Figure 18, Table 6), linalyl methylcarbonate (Figure 18, Table 7), myrtenyl acetate (Figure 19, Tables 8 and 9), perillyl acetate, geranyl acetate, mertynyl alkyl carbonate, perillyl alkyl carbonate, and geranyl alkyl carbonate (Table 10) [92,124], as well as cinnamyl ethyl carbonate, ethyl prenyl carbonate, and ethyl geranyl carbonate (Figure 20) [55].

Mechanistic aspects of myrtenol amination in the presence of supported gold catalysts were discussed in [123] suggesting an important role of the hydride ion. In this context it is interesting to find a common denominator for heterogeneous and homogeneous catalytic amination.

Palladium catalyzed allylic amination typically involves formation of neutral or cationic palladium π-allyl complexes via S_N2 reaction. A soft nucleophile attacks from the back side of the metal allowing retention of configuration in the product [144]. According to DFT calculations for palladium-catalyzed allylation of primary amines by allylic alcohols [146] one potential pathway involves formation of cationic Pd hydride species while in the second option decomplexation of the coordinated allylammonium can occur. In [124] it was assumed that both amination of allyl acetates and carbonates involves generation of a (π-allyl)-palladium complex (**2**) (Figure 29). Experimentally observed formation of the racemic product (**3**) was thus rationalized considering that the nucleophile can attack both allylic positions of the (π-allyl)-palladium complex (**2**).

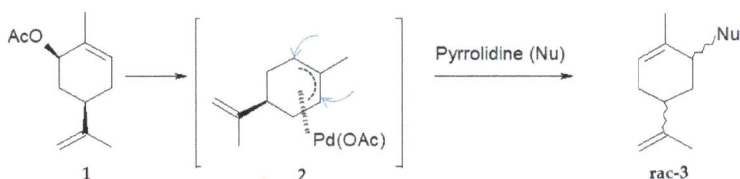

Figure 29. Proposed mechanism of Pd(0)-allylic amination of *cis*-carvyl acetate (**1**). Reproduced from [124] under the terms of the Creative Commons Licenses.

Formation of cationic π-allyl-Pd-complex intermediate **B** was supposed [89] to proceed through the initial formation of transient Pd–H species **A**, followed by their reaction with the diene. The nucleophilic attack of aniline on the less-substituted carbon of the intermediate species **B** (Figure 30) [89] results in a regioselective 1,4-hydroamination product.

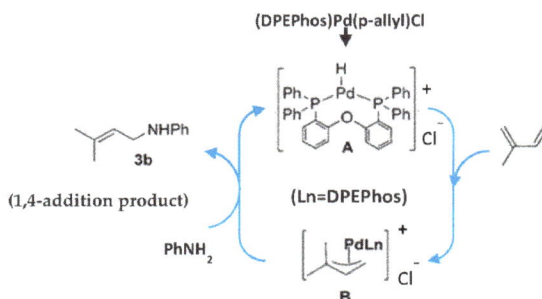

Figure 30. Proposed mechanism for the intermolecular hydroamination of 1,3-dienes with aniline. Reproduced from [89] by permission of The Royal Society of Chemistry.

As mentioned above analysis of available literature shows that there are just a few examples when terpene amines were synthesized using heterogeneous systems, comprising reductive amination over Ni Raney, Pt/C, Pd/C, copper on alumina modified with LiOH, hydroamination on alkali metals, and hydrogen borrowing reactions over Au and AuPd. In fact, there is a clear trend in the more widespread application of heterogeneous catalysts. It is interesting that complexes of precious metals are mainly applied as homogeneous catalysts, while despite utilization of noble heterogeneous catalysts (e.g., carbon supported Pt and Pd in amination of camphor), other metals such as supported Cu and Au were considered to be efficient. Moreover, while addition of Pd to heterogeneous catalysts deteriorates the overall performance by decreasing selectivity towards complex amines at the expense of hydrogenation, similar behavior was not observed for homogeneous catalysts. This is even more striking as according to the available data the mechanisms of amination in the presence of transition metal complexes discussed above and heterogeneous catalysts are similar. In particular, catalytically active species are formed by generation of the metal hydrides in the case of Pd–H from homogeneous Pd chloride complexes. Similar suggestions follow from the work of the authors of this review on amination of myrtenol with aniline over gold catalysts. Obviously, there is a need for more detailed studies of the nature of active sites in homogeneous catalysts to fully explore this knowledge in the design of heterogeneous systems. Alternatively if the mechanisms of homogeneous and heterogeneous catalysis are different, a significant amount of work should be devoted to heterogeneous catalysts in a quest for understanding the reaction mechanism. This and many other questions do not have clear answers at the moment urging on one hand more in depth and on the other more broader studies on amination of terpenes over heterogeneous catalysts.

5. Conclusions

Although amination of terpenoids has been extensively studied since the early decades of the last century, industrial implementation of biomass-based terpenes as starting materials is still in its infancy. Catalytic systems based on transition metals complexes have been developed for performing such reactions. However, to reduce the production costs, either easily recoverable homogeneous systems based on cheaper metals operating by nucleophilic substitution, as well as supported metal nanoparticles (Ni, Co, Cu, Pd, Au) on low alkaline supports should be developed. These catalysts will provide synthesis of amine derivatives of terpenes having a broad range of applications as specialty chemicals, surfactants, pharmaceuticals, etc.

Author Contributions: I.L.S. performed the literature search and drafted the manuscript; A.V.S. and D.Y.M. contributed to the writing and editing of the manuscript. All the authors revised and approved the manuscript.

Acknowledgments: This research was supported by RFBR Grant # 18-53-45013 IND_a.

Conflicts of Interest: The authors declare no conflict of interest.

References

1. Lawrence, S.A. *Amines: Synthesis, Properties and Applications*; Cambridge University Press: Cambridge, UK, 2004; p. 372.
2. Shi, F.; Cui, X. *Catalytic Amination for N-Alkylamine Synthesis*; Academic Press: Cambridge, MA, USA, 2018.
3. Maxwell, G.R. *Synthetic Nitrogen Products: A Practical Guide to the Products and Processes*; Kluwer Academic Publishers: New York, NY, USA, 2004.
4. Pelckmans, M.; Renders, T.; Van de Vyver, S.; Sels, B.F. Bio-based amines through sustainable heterogeneous catalysis. *Green Chem.* **2017**, *19*, 5303–5331. [CrossRef]
5. Eller, K.; Henkes, E.; Rossbacher, R.; Höke, H. Amines, aliphatic. In *Ullmann's Encyclopedia of Industrial Chemistry*; Wiley: Hoboken, NJ, USA, 2012; pp. 647–698.
6. Frauenkron, M.; Melder, J.-P.; Ruider, G.; Rossbacher, R.; Höke, H. Ethanolamines and Propanolamines. In *Ullmann's Encyclopedia of Industrial Chemistry*; Wiley: Hoboken, NJ, USA, 2012; pp. 405–431.
7. Bähn, S.; Imm, S.; Neubert, L.; Zhang, M.; Neumann, H.; Beller, M. The catalytic amination of alcohols. *ChemCatChem* **2011**, *3*, 1853–1864. [CrossRef]
8. Kimura, H. Progress in one-step amination of long-chain fatty alcohols with dimethylamine: Development of key technologies for industrial applications, innovations, and future outlook. *Catal. Rev. Sci. Eng.* **2011**, *53*, 1–90. [CrossRef]
9. Pera-Titus, M.; Shi, F. Catalytic amination of biomass-based alcohols. *ChemSusChem* **2014**, *7*, 720–722. [CrossRef] [PubMed]
10. Froidevaux, V.; Negrell, C.; Caillol, S.; Pascault, J.-P.; Boutevin, B. Biobased amines: From synthesis to polymers; present and future. *Chem. Rev.* **2016**, *116*, 14181–14224. [CrossRef] [PubMed]
11. Geboers, J.; Van de Vyver, S.; Carpentier, K.; de Blochouse, K.; Jacobs, P.; Sels, B. Efficient catalytic conversion of concentrated cellulose feeds to hexitols with heteropoly acids and Ru on carbon. *Chem. Commun.* **2010**, *46*, 3577–3579. [CrossRef] [PubMed]
12. Geboers, J.; Van de Vyver, S.; Carpentier, K.; Jacobs, P.; Sels, B. Efficient hydrolytic hydrogenation of cellulose in the presence of Ru-loaded zeolites and trace amounts of mineral acid. *Chem. Commun.* **2011**, *47*, 5590–5592. [CrossRef] [PubMed]
13. Geboers, J.A.; Van de Vyver, S.; Ooms, R.; Op de Beeck, B.; Jacobs, P.A.; Sels, B.F. Chemocatalytic conversion of cellulose: Opportunities, advances and pitfalls. *Catal. Sci. Technol.* **2011**, *1*, 714–726. [CrossRef]
14. Van de Vyver, S.; Geboers, J.; Jacobs, P.A.; Sels, B.F. Recent Advances in the Catalytic Conversion of Cellulose. *ChemCatChem* **2011**, *3*, 82–94. [CrossRef]
15. Tinikul, R.; Chenprakhon, P.; Maenpuen, S.; Chaiyen, P. Biotransformation of plant-derived phenolic acids. *Biotechnol. J.* **2018**, *13*. [CrossRef] [PubMed]
16. Du, X.; Li, J.; Lindström, M.E. Modification of industrial softwood kraft lignin using Mannich reaction with and without phenolation pretreatment. *Ind. Crop. Prod.* **2014**, *52*, 729–735. [CrossRef]
17. Li, C.; Zhao, X.; Wang, A.; Huber, G.W.; Zhang, T. Catalytic transformation of lignin for the production of chemicals and fuels. *Chem. Rev.* **2015**, *115*, 11559–11624. [CrossRef] [PubMed]
18. Van den Bosch, S.; Schutyser, W.; Vanholme, R.; Driessen, T.; Koelewijn, S.F.; Renders, T.; De Meester, B.; Huijgen, W.J.J.; Dehaen, W.; Courtin, C.M.; et al. Reductive lignocellulose fractionation into soluble lignin-derived phenolic monomers and dimers and processable carbohydrate pulps. *Energy Environ. Sci.* **2015**, *8*, 1748–1763. [CrossRef]
19. Zakzeski, J.; Bruijnincx, P.C.A.; Jongerius, A.L.; Weckhuysen, B.M. The catalytic valorization of lignin for the production of renewable chemicals. *Chem. Rev.* **2010**, *110*, 3552–3599. [CrossRef] [PubMed]
20. Wang, B.; Chena, T.-Y.; Wanga, H.-M.; Li, H.-Y.; Liub, C.-F.; Wena, J.-L. Amination of biorefinery technical lignins using Mannich reaction synergy with subcritical ethanol depolymerization. *Int. J. Biol. Macromol.* **2018**, *107*, 426–435. [CrossRef] [PubMed]
21. Pan, H.; Sun, G.; Zhao, T. Synthesis and characterization of aminated lignin. *Int. J. Biol. Macromol.* **2013**, *59*, 221–226. [CrossRef] [PubMed]
22. Fukuda, H.; Kondo, A.; Noda, H. Biodiesel fuel production by transesterification of oils. *J. Biosci. Bioeng.* **2001**, *92*, 405–416. [CrossRef]
23. Zargar, V.; Asghari, M.; Dashti, A. A review on chitin and chitosan polymers: Structure, *chemistry*, solubility, derivatives. *ChemBioEng. Rev.* **2015**, *2*, 204–226. [CrossRef]

24. Isikgor, F.H.; Becer, C.R. Lignocellulosic biomass: A Sustainable platform for production of bio-based chemicals and polymers. *Polym. Chem.* **2015**, *6*, 4497–4559. [CrossRef]
25. Behr, A.; Wintzer, A. From terpenoids to amines: A critical review. In *New Developments in Terpenes Research*; Hu, J., Ed.; Nova Science Publishers: New York, NY, USA, 2014; Chapter 6; pp. 113–134.
26. Kroutil, W.; Fischereder, E.-M.; Fuchs, C.S.; Lechner, H.; Mutti, F.G.; Pressnitz, D.; Rajagopalan, A.; Sattler, J.H.; Simon, R.C.; Siirola, E. Asymmetric preparation of prim-, sec-, and tert-amines employing selected biocatalysts. *Org. Process Res. Dev.* **2013**, *17*, 751–759. [CrossRef] [PubMed]
27. Turner, N.J.; Truppo, M.D. Biocatalytic routes to nonracemic chiral amines. In *Chiral Amine Synthesis: Methods, Developments and Applications*, 2nd ed.; Nugent, T.C., Ed.; Wiley: Hoboken, NJ, USA, 2010; p. 523.
28. Schrewe, M.; Ladkau, N.; Buehler, B.; Schmid, A. Direct terminal alkylamino-functionalization via multistep biocatalysis in one recombinant whole-cell catalyst. *Adv. Synth. Catal.* **2013**, *355*, 1693–1697. [CrossRef]
29. Song, J.-W.; Lee, J.-H.; Bornscheuer, U.T.; Park, J.-B. Microbial synthesis of medium-chain α,ω-dicarboxylic acids and ω-aminocarboxylic acids from renewable long-chain fatty acids. *Adv. Synth. Catal.* **2014**, *356*, 178–1788. [CrossRef]
30. Singh, R.; Kolev, J.N.; Sutera, P.A.; Fasan, R. Enzymatic C(sp3)-H amination: P450-catalyzed conversion of carbonazidates into oxazolidinones. *ACS Catal.* **2015**, *5*, 1685–1691. [CrossRef] [PubMed]
31. Deutschmann, O.; Knozinger, H.; Kochloefl, K.; Turek, T. Heterogeneous catalysis and solid catalysts. In *Ullmann's Encyclopedia of Industrial Chemistry*; Wiley: Hoboken, NJ, USA, 2009; pp. 1–110.
32. Harrewijn, P.; van Oosten, A.M.; Piron, P.G.M. *Natural Terpenoids as Messengers. A Multidisciplinary Study of their Production, Biological Functions and Practical Applications*; Kluwer Academic Publishers: South Holland, The Netherlands, 2001.
33. Behr, A.; Johnen, L. Myrcene as a natural base chemical in sustainable chemistry: A critical review. *ChemSusChem* **2009**, *2*, 1072–1095. [CrossRef] [PubMed]
34. Mäki-Arvela, P.; Simakova, I.L.; Salmi, T.; Murzin, D.Y. Catalytic transformations of extractives. In *Catalytic Process Development for Renewable Materials*; Hardcover Handbook; Wiley: Weinheim, Germany, 2013; Chapter 13; 450p.
35. Murzin, D.Y.; Simakova, I.L. Catalysis in biomass conversion. In *Comprehensive Inorganic Chemistry II*; Schlogl, R., Niemantsverdriet, J.W., Eds.; Elsevier: New York, NY, USA, 2013; Chapter 7.27; pp. 2–32.
36. Murzin, D.Y.; Simakova, I.L. Catalysis in biomass processing. *Catal. Ind.* **2011**, *3*, 218–249. [CrossRef]
37. Murzin, D.Y.; Demidova, Y.; Hasse, B.; Etzold, B.; Simakova, I.L. Synthesis of fine chemicals using catalytic nanomaterials: Structure sensitivity. In *Producing Fuels and Fine Chemicals from Biomass Using Nanomaterials*; Luque, R., Balu, A.M., Eds.; CRC Press: Boca Raton, FL, USA, 2013; pp. 267–281.
38. Salakhutdinov, N.F.; Volcho, K.P.; Yarovaya, O.I. Monoterpenes as a renewable source of biologically active compounds. *Pure Appl. Chem.* **2017**, *89*, 1105–1118. [CrossRef]
39. Kapitsa, I.G.; Suslov, E.V.; Teplov, G.V.; Korchagina, D.V.; Komarova, N.I.; Volcho, K.P.; Voronina, T.A.; Shevela, A.I.; Salakhutdinov, N.F. Synthesis and anxiolytic activity of 2-aminoadamantane derivatives containing monoterpene fragments. *Pharm. Chem. J.* **2012**, *46*, 263–265. [CrossRef]
40. Teplov, G.V.; Suslov, E.V.; Zarubaev, V.V.; Shtro, A.A.; Karpinskaya, L.A.; Rogachev, A.D.; Korchagina, D.V.; Volcho, K.P.; Salakhutdinov, N.F.; Kisilev, O.I. Synthesis of new compounds combining adamantanamine and monoterpene fragments and their antiviral activity against influenza virus A(H1N1)pdm09. *Lett. Drug Des. Discov.* **2013**, *10*, 477–485. [CrossRef]
41. Volcho, K.P.; Laev, S.S.; Ashraf, G.M.; Aliev, G.; Salakhutdinov, N.F. Application of monoterpenoids and their derivatives for treatment of neurodegenerative disorders. *Curr. Med. Chem.* **2017**, *24*, 3283–3309.
42. Silva, R.O.; Salvadori, M.S.; Sousa, F.B.M.; Santos, M.S.; Carvalho, N.S.; Sousa, D.P.; Gomes, B.S.; Oliveira, F.A.; Barbosa, A.L.R.; Frietas, R.M.; et al. Evaluation of the anti-inflammatory and antinociceptive effects of myrtenol, a plant-derived monoterpene alcohol, in mice. *Flavour Fragr. J.* **2014**, *29*, 184–192. [CrossRef]
43. Sarmento-Neto, J.F.; do Nascimento, L.G.; Felipe, C.F.B.; de Sousa, D.P. Analgesic potential of essential oils. *Molecules* **2016**, *21*, 20. [CrossRef] [PubMed]
44. Lochynski, S.; Kuldo, J.; Frackowiak, B.; Holband, J.; Wojcik, G. Stereochemistry of terpene derivatives. Part 2: Synthesis of new chiral amino acids with potential neuroactivity. *Tetrahedron Asymmetry* **2000**, *11*, 1295–1302. [CrossRef]

45. Gajcy, K.; Pekala, J.; Frackowiak-Wojtasek, B.; Librowski, T.; Lochynski, S. Stereochemistry of terpene derivatives. Part 7: Novel rigidified amino acids from (+)-3-carene designed as chiral GABA analogues. *Tetrahedron Asymmetry* **2010**, *21*, 2015–2020. [CrossRef]

46. Ferrarini, S.R.; Graebin, C.S.; Limberger, J.; Canto, R.F.S.; Dias, D.O.; da Rosa, R.G.; Madeira, M.D.F.; Eifler-Lima, V.L. Synthesis of limonene β-amino alcohol derivatives in support of new antileishmanial therapies. *Mem. Inst. Oswaldo Cruz* **2008**, *103*, 773–777. [CrossRef] [PubMed]

47. Ferrarini, S.R.; Duarte, M.O.; da Rosa, R.G.; Rolim, V.; Eifler-Lima, V.L.; von Poser, G.; Ribeiro, V.L.S. Acaricidal activity of limonene, limonene oxide and β-amino alcohol derivatives on Rhipicephalus (Boophilus) microplus. *Vet. Parasitol.* **2008**, *157*, 149–153. [CrossRef] [PubMed]

48. Strong, J. *N*-[3-(4-methyl-3-cyclohexenyl)butyl]amines. U.S. Patent 3,890,384, 27 January 1975.

49. Strong, J. *N*-[3-(4-methyl-3-cyclohexenyl)butyl]amines and Their Use as Plant Growth Regulators. U.S. Patent 4,030,908, 27 January 1972.

50. Keim, W.; Kurtz, K.R.; Roeper, M. Palladium catalyzed telomerization of isoprene with secondary amines and conversion of the resulting terpene amines to terpenols. *J. Mol. Catal.* **1983**, *20*, 129–138. [CrossRef]

51. Watts, C.C.; Thoniyot, P.; Cappuccio, F.; Verhagen, J.; Gallagher, B.; Singaram, B. Catalytic asymmetric transfer hydrogenation of ketones using terpene-based chiral β-amino alcohols. *Tetrahedron Asymmetry* **2006**, *17*, 1301–1307. [CrossRef]

52. Watts, C.C.; Thoniyot, P.; Hirayama, L.C.; Romano, T.; Singaram, B. Enantioselective alkynylations of aromatic and aliphatic aldehydes catalyzed by terpene derived chiral amino alcohols. *Tetrahedron Asymmetry* **2005**, *16*, 1829–1835. [CrossRef]

53. Alves, M.-H.; Sfeir, H.; Tranchant, J.-F.; Gombart, E.; Sagorin, G.; Caillol, S.; Billon, L.; Save, M. Terpene and dextran renewable resources for the synthesis of amphiphilic biopolymers. *Biomacromolecules* **2014**, *15*, 242–251. [CrossRef] [PubMed]

54. Ajikumar, P.K.; Tyo, K.; Carlsen, S.; Mucha, O.; Phon, T.H.; Stephanopoulos, G. Terpenoids: Opportunities for biosynthesis of natural product drugs using engineered microorganisms. *Mol. Pharm.* **2008**, *5*, 167–190. [CrossRef] [PubMed]

55. Lyubimov, S.E.; Kuchurov, I.V.; Verbitskaya, T.A.; Rastorguev, E.A.; Kalinin, V.N.; Zlotin, S.G.; Davankov, V.A. Pd-catalyzed allylic amination in supercritical carbon dioxide: Synthesis of carborane-containing terpenoids. *J. Supercrit. Fluids* **2010**, *54*, 218–221. [CrossRef]

56. Armstrong, A.F.; Valliant, J.F. The bioinorganic and medicinal chemistry of carboranes: From new drug discovery to molecular imaging and therapy. *Dalton Trans.* **2007**, 4240–4251. [CrossRef] [PubMed]

57. Tietze, L.F.; Griesbach, U.; Bothe, U.; Nakamura, H.; Yamamoto, Y. Novel carboranes with a DNA binding unit for the treatment of cancer by boron neutron capture therapy. *ChemBioChem* **2002**, *3*, 219–225. [CrossRef]

58. Di Meo, C.; Panza, L.; Capitani, D.; Mannina, L.; Banzato, A.; Rondina, M.; Renier, D.; Rosato, A.; Crescenzi, V. Hyaluronan as carrier of carboranes for tumor targeting in boron neutron capture therapy. *Biomacromolecules* **2007**, *8*, 552–559. [CrossRef] [PubMed]

59. Barth, R.F.; Coderre, J.A.; Vicente, M.G.H.; Blue, T.E. Boron neutron capture therapy of cancer: Current status and future prospects. *Clin. Cancer Res.* **2005**, *11*, 3987–4002. [CrossRef] [PubMed]

60. Nageshwar, D.; Rao, D.M.; Acharyulu, P.V.R. Terpenes to ionic liquids: Synthesis and characterization of citronellal-based chiral ionic liquids. *Synth. Commun.* **2009**, *39*, 3357–3368. [CrossRef]

61. Bordenca, C.; Dorschner, K.P.; Johnson, R.P. Insect Repellent Compositions and Process Having an N-substituted Hydroxyalkyl Amine as an Active Ingredient. U.S. Patent 3,933,915, 23 June 1972.

62. Behr, A.; Wintzer, A.; Lübke, C.; Müller, M. Synthesis of primary amines from the renewable compound citronellal via biphasic reductive amination. *J. Mol. Catal. A Chem.* **2015**, *404–405*, 74–82. [CrossRef]

63. Kukula, P.; Koprivova, K. Structure-selectivity relationship in the chemoselective hydrogenation of unsaturated nitriles. *J. Catal.* **2005**, *234*, 161–171. [CrossRef]

64. Bahn, S.; Imm, S.; Neubert, L.; Zhang, M.; Neumann, H.; Beller, M. Selective ruthenium-catalyzed alkylation of indoles by using amines. *Chem. Eur. J.* **2011**, *17*, 4705–4708. [PubMed]

65. Fuchs, S.; Rösler, T.; Grabe, B.; Kampwerth, A.; Meier, G.; Strutz, H.; Behr, A.; Vorholt, A.J. Synthesis of primary amines via linkage of hydroaminomethylation of olefins and splitting of secondary amines. *Appl. Cat. A Gen.* **2018**, *550*, 198–205. [CrossRef]

66. Donetti, A.; Casadio, S.; Bonardi, G.; Omodei-Sale, A. Terpene compounds as drugs. 13. o-Terpenylaminomethylphenols and their N-methyl derivatives. *J. Med. Chem.* **1972**, *15*, 1089–1091. [CrossRef] [PubMed]

67. Kozlov, N.G.; Kalechits, G.V.; Vyalimyae, T.K. Terpene amines. IV. Synthesis and study of the structure of amines from d-fenchone. *Khimiya Prirodnych Soedinenii (Chem. Nat. Comp.)* **1983**, *4*, 483–488. [CrossRef]

68. Tarasevich, V.A.; Kozlov, N.G. Reductive amination of oxygen-containing organic compounds. *Russ. Chem. Rev.* **1999**, *68*, 55–72. [CrossRef]

69. Kalck, P.; Urrutigoïty, M. Tandem hydroaminomethylation reaction to synthesize amines from alkenes. *Chem. Rev.* **2018**, *118*, 3833–3861. [CrossRef] [PubMed]

70. Eilbracht, P.; Barfacker, L.; Buss, C.; Hollmann, C.; Kitsos-Rzychon, B.E.; Kranemann, C.L.; Rische, T.; Roggenbuck, R.; Schmidt, A. Tandem reaction sequences under hydroformylation conditions: new synthetic applications of transition metal catalysis. *Chem. Rev.* **1999**, *99*, 3329–3366. [CrossRef] [PubMed]

71. Ahmed, M.; Seayad, A.M.; Jackstell, R.; Beller, M. Highly selective synthesis of enamines from olefins. *Angew. Chem. Int. Ed.* **2003**, *42*, 5615–5619. [CrossRef] [PubMed]

72. Ahmed, M.; Seayad, A.M.; Jackstell, R.; Beller, M. Amines made easily: A highly selective hydroaminomethylation of olefins. *J. Am. Chem. Soc.* **2003**, *125*, 10311–10318. [CrossRef] [PubMed]

73. Fogg, D.E.; dos Santos, E.N. Tandem catalysis: A taxonomy and illustrative review. *Coord. Chem. Rev.* **2004**, *248*, 2365–2379. [CrossRef]

74. Melo, D.S.; Pereira-Júniora, S.S.; dos Santosa, E.N. An efficient method for the transformation of naturally occurring monoterpenes into amines through rhodium-catalyzed hydroaminomethylation. *Appl. Catal. A Gen.* **2012**, *411–412*, 70–76. [CrossRef]

75. Börner, A.; Franke, R. *Hydroformylation: Fundamentals, Processes, and Applications in Organic Synthesis*; Wiley: Hoboken, NJ, USA, 2016; 736p.

76. Graebin, C.S.; Eifler-Lima, V.L.; da Rosa, R.G. One-pot synthesis of secondary and tertiary amines from R(+)-limonene by tandem hydroformylation/reductive amination (hydroaminomethylation). *Catal. Commun.* **2008**, *9*, 1066–1070. [CrossRef]

77. Behr, A.; Wintzer, A. Hydroaminomethylation of the renewable limonene with ammonia in an aqueous biphasic solvent system. *Chem. Eng. Technol.* **2015**, *38*, 2299–2304. [CrossRef]

78. Faßbach, T.A.; Gaide, T.; Terhorst, M.; Behr, A.; Vorholt, A.J. Renewable surfactants through the hydroaminomethylation of terpenes. *ChemCatChem* **2017**, *9*, 1359–1362. [CrossRef]

79. Oliveira, K.C.B.; Santos, A.G.; dos Santos, E.N. Hydroaminomethylation of eugenol with di-*n*-butylamine catalyzed by rhodium complexes: Bringing light on the promoting effect of Bronsted acids. *Appl. Catal. A Gen.* **2012**, *445–446*, 204–208. [CrossRef]

80. Oliveira, K.C.B.; Carvalho, S.N.; Duarte, M.F.; Gusevskaya, E.V.; dos Santos, E.N.; Karroumi, J.E.; Gouygou, M.; Urrutigoïty, M. Phospholes as efficient ancillaries for the rhodium-catalyzed hydroformylation and hydroaminomethylation of estragole. *Appl. Catal. A Gen.* **2015**, *497*, 10–16. [CrossRef]

81. Behr, A.; Reyer, S.; Manz, V. Hydroaminomethylation of isoprene: Recycling of the homogeneous rhodium catalyst in aqueous biphasic systems. *Chem. Ing. Tech.* **2012**, *84*, 108–113. [CrossRef]

82. Sirol, S.; Kalck, P. Hydroformylation of optically pure monoterpenes catalyzed by dinuclear thiolato-bridged rhodium complexes. *New J. Chem.* **1997**, *21*, 1129–1137.

83. Foca, C.M.; Barros, H.J.V.; dos Santos, E.N.; Gusevskaya, E.V.; Bayon, J.C. Hydroformylation of myrcene: Metal and ligand effects in the hydroformylation of conjugated dienes. *New J. Chem.* **2003**, *27*, 533–539. [CrossRef]

84. Halligudi, S.B.; Bhatt, K.N.; Venkatasubramanian, K. Hydroformylation of olefins catalyzed by rhodium complex anchored on clay matrices. *React. Kinet. Catal. Lett.* **1993**, *51*, 459–464. [CrossRef]

85. Barros, H.J.V.; Ospina, M.L.; Arguello, E.; Rocha, W.R.; Gusevskaya, E.V.; dos Santos, E.N. Rhodium catalyzed hydroformylation of β-pinene and camphene: Effect of phosphorous ligands and reaction conditions on diastereoselectivity. *J. Organomet. Chem.* **2003**, *671*, 150–157. [CrossRef]

86. Estragol. Available online: https://en.wikipedia.org/wiki/Estragole#cite_note-2 (accessed on 28 June 2018).

87. Haggin, J. Chemists seek greater recognition for catalysis. *Chem. Eng. News* **1993**, *71*, 23–27. [CrossRef]

88. Beller, M.; Seavad, J.; Tillack, A.; Jiao, H. Catalytic Markovnikov and anti-Markovnikov functionalization of alkenes and alkynes: Recent developments and trends. *Angew. Chem. Int. Ed.* **2004**, *43*, 3368–3398. [CrossRef] [PubMed]

89. Banerjee, D.; Junge, K.; Beller, M. Palladium-catalysed regioselective hydroamination of 1,3-dienes: Synthesis of allylic amines. *Org. Chem. Front.* **2014**, *1*, 368–372. [CrossRef]

90. Watson, I.D.G.; Yudin, A.K. New insights into the mechanism of palladium-catalyzed allylic amination. *J. Am. Chem. Soc.* **2005**, *127*, 17516–17529. [CrossRef] [PubMed]

91. Watanabe, S.; Fujita, T.; Sakamoto, M.; Haga, T.; Kuramochi, T. Palladium-catalyzed addition of dialkylamines to linalyl acetate and related compounds. *J. Essent. Oil Res.* **1994**, *9*, 441–445. [CrossRef]

92. Nguyen, D.H.; Urrutigoïty, M.; Fihri, A.; Hierso, J.-C.; Meunier, P.; Kalck, P. Efficient palladium-ferrocenylphosphine catalytic systems for allylic amination of monoterpene derivatives. *Appl. Organomet. Chem.* **2006**, *20*, 845–850. [CrossRef]

93. Fihri, A.; Meunier, P.; Hierso, J.-C. Performances of symmetrical achiral ferrocenylphosphine ligands in palladium-catalyzed cross-coupling reactions: A review of syntheses, catalytic applications and structural properties. *Coord. Chem. Rev.* **2007**, *251*, 2017–2055. [CrossRef]

94. Mignani, G.; Morel, D. Processes Amination of Conjugated Dienes. Patent FR 2,569,403, 23 August 1986.

95. Akutagawa, S. Asymmetric synthesis by metal BINAP catalysts. *Appl. Catal. A Gen.* **1995**, *128*, 171–207. [CrossRef]

96. Mäki-Arvela, P.; Kumar, N.; Nieminen, V.; Sjöholm, R.; Salmi, T.; Murzin, D.Y. Cyclization of citronellal over zeolites and mesoporous materials for production of isopulegol. *J. Catal.* **2004**, *225*, 155–169. [CrossRef]

97. Kumobayashi, H.; Mitsuhashi, S.; Akutagawa, S.; Ohtsuka, S. A practical synthesis of myrcenol by palladium complex-catalyzed elimination reaction. *Chem. Lett.* **1986**, *15*, 157–160. [CrossRef]

98. Chalk, A.J.; Magennis, S.A.; Wertheimer, V.S.; Naipawer, R.E. Process for the Catalytic Synthesis of Conjugated Dienes from Dialkylallylamines. U.S. Patent 4,467,118, 21 August 1984.

99. Chalk, A.J.; Wertheimer, V.; Magennis, S.A. A new palladium catalyzed equivalent of hofmann elimination for allylic amines. *J. Mol. Catal.* **1983**, *19*, 189–200. [CrossRef]

100. Hata, G.; Tanaka, M. Terpene Hydrocarbons. JP Patent 50,123,605, 29 September 1975.

101. Murata, A.; Tsuchiya, S.; Suzuki, H.; Ikeda, H. A Method for Producing of Chain Terpene Alcohols. DE Patent 2720839, 14 May 1977.

102. Behr, A.; Johnen, L.; Rentmeister, N. Novel Palladium-catalysed hydroamination of myrcene and catalyst separation by thermomorphic solvent systems. *Adv. Synth. Catal.* **2010**, *352*, 2062–2072. [CrossRef]

103. Hamid, M.H.S.A.; Slatford, P.A.; Williams, J.M.J. Borrowing hydrogen in the activation of alcohols. *Adv. Synth. Catal.* **2007**, *349*, 1555–1575. [CrossRef]

104. Gunanathan, C.; Milstein, D. Applications of acceptorless dehydrogenation and related transformations in chemical synthesis. *Science* **2013**, *341*, 1229712. [CrossRef] [PubMed]

105. Guillena, G.; Ramon, D.J.; Yus, M. Hydrogen autotransfer in the N-alkylation of amines and related compounds using alcohols and amines as electrophiles. *Chem. Rev.* **2010**, *110*, 1611–1641. [CrossRef] [PubMed]

106. Shimizu, K.-I.; Kon, K.; Onodera, W.; Yamazaki, H.; Kondo, J.N. Heterogeneous Ni catalyst for direct synthesis of primary amines from alcohols and ammonia. *ACS Catal.* **2013**, *3*, 112–117. [CrossRef]

107. Dang, T.T.; Ramalingam, B.; Shan, S.P.; Seayad, A.M. Reductive N-Alkylation of nitro compounds to N-alkyl and N,N-dialkyl amines with glycerol as the hydrogen source. *ACS Catal.* **2013**, *3*, 2536–2540. [CrossRef]

108. Murzin, D.Y. *Chemical Reaction Technology*; De Gruyter: Berlin, Germany, 2015; 428p.

109. Ma, X.; Su, C.; Xu, Q. N-Alkylation by hydrogen autotransfer reactions. In *Hydrogen Transfer Reactions: Reductions and Beyond*; Guillena, G., Ramon, D.J., Eds.; Springer International Publishing: Basel, Switzerland, 2016; pp. 291–364.

110. Imm, S.; Bahn, S.; Neubert, L.; Neumann, H.; Beller, M. An efficient and general synthesis of primary amines by ruthenium catalyzed amination of secondary alcohols with ammonia. *Angew. Chem. Int. Ed.* **2010**, *49*, 8126–8129. [CrossRef] [PubMed]

111. Imm, S.; Bahn, S.; Zhang, M.; Neubert, L.; Neumann, H.; Klasovsky, F.; Pfeffer, J.; Haas, T.; Beller, M. Improved ruthenium-catalyzed amination of alcohols with ammonia: Synthesis of diamines and amino esters. *Angew. Chem. Int. Ed.* **2011**, *50*, 7599–7603. [CrossRef] [PubMed]

112. Lamb, G.W.; Williams, J.M.J. Borrowing hydrogen-C-N bond formation from alcohols. *Chim. Oggi-Chem. Today* **2008**, *26*, 17–19.

113. Pingen, D.; Muller, C.; Vogt, D. Direct amination of secondary alcohols using ammonia. *Angew. Chem. Int. Ed.* **2010**, *49*, 8130–8133. [CrossRef] [PubMed]

114. Corma, A.; Navas, J.; Sabater, M.J. Advances in one-pot synthesis through borrowing hydrogen catalysis. *Chem. Rev.* **2018**, *118*, 1410–1459. [CrossRef] [PubMed]

115. Watson, A.J.A.; Maxwell, A.C.; Williams, J.M.J. Borrowing hydrogen methodology for amine synthesis under solvent-free microwave conditions. *J. Org. Chem.* **2011**, *76*, 2328–2331. [CrossRef] [PubMed]

116. Saidi, O.; Blacker, A.J.; Farah, M.M.; Marsden, S.P.; Williams, J.M.J. Iridium-catalysed amine alkylation with alcohols in water. *Chem. Commun.* **2010**, *46*, 1541–1543. [CrossRef] [PubMed]

117. Kawahara, R.; Fujita, K.; Yamaguchi, R. N-Alkylation of Amines with Alcohols catalyzed by a water soluble Cp*Iridium complex: An efficient method for the synthesis of amines in aqueous media. *Adv. Synth. Catal.* **2011**, *353*, 1161–1168. [CrossRef]

118. Hollmann, D.; Tillack, A.; Michalik, D.; Jackstell, R.; Beller, M. An improved ruthenium catalyst for the environmentally benign amination of primary and secondary alcohols. *Chem. Asian J.* **2007**, *2*, 403–410. [CrossRef] [PubMed]

119. Martinez-Asencio, A.; Ramon, D.J.; Yus, M. N-alkylation of poor nucleophilic amines and derivatives with alcohols by a hydrogen autotransfer process catalyzed by copper (II) acetate: Scope and mechanistic considerations. *Tetrahedron* **2011**, *67*, 3140–3149. [CrossRef]

120. Martinez-Asencio, A.; Yus, M.; Ramon, D.J. Palladium (II) acetate as a catalyst for the N-alkylation of aromatic amines, sulfonamides and related nitrogenated compounds with alcohols by a hydrogen autotransfer process. *Synthesis* **2011**, 3730–3740. [CrossRef]

121. Blank, B.; Kempe, R. Catalytic alkylation of methyl-N-heteroaromatics with alcohols. *J. Am. Chem. Soc.* **2010**, *132*, 924–925. [CrossRef] [PubMed]

122. Michlik, S.; Hille, T.; Kempe, R. The Iridium-catalyzed synthesis of symmetrically and unsymmetrically alkylated diamines under mild reaction conditions. *Adv. Synth. Catal.* **2012**, *354*, 847–862. [CrossRef]

123. Demidova, Y.S.; Simakova, I.L.; Estrada, M.; Beloshapkin, S.; Suslov, E.V.; Korchagina, D.V.; Volcho, K.P.; Salakhutdinov, N.F.; Simakov, A.V.; Murzin, D.Y. One-pot myrtenol amination over Au nanoparticles supported on different metal oxides. *Appl. Catal. A Gen.* **2013**, *464–465*, 348–356. [CrossRef]

124. Houssame, S.E.; Anane, H.; Firdoussi, L.E.; Karim, A. Palladium(0)-catalyzed amination of allylic natural terpenic functionalized olefins. *Cent. Eur. J. Chem.* **2008**, *6*, 470–476. [CrossRef]

125. Pingen, D.; Diebolt, O.; Vogt, D. Direct amination of bio-alcohols using ammonia. *ChemCatChem* **2013**, *5*, 2905–2912. [CrossRef]

126. Valkenberg, M.H.; Holderich, W.F. Preparation and use of hybrid organic–inorganic catalysts. *Catal. Rev.* **2002**, *44*, 321–374. [CrossRef]

127. Huang, X.; Wu, H.; Liao, X.P.; Shi, B. Liquid phase hydrogenation of olefins using heterogenized ruthenium complexes as high active and reusable catalyst. *Catal. Commun.* **2010**, *11*, 487–492. [CrossRef]

128. Cao, Y.; Hu, J.C.; Yang, P.; Dai, W.L.; Fan, K.N. CuCl catalyst heterogenized on diamide immobilized SBA-15 for efficient oxidative carbonylation of methanol to dimethylcarbonate. *Chem. Commun.* **2003**, 908–909. [CrossRef]

129. Mukhopadhyay, K.; Chaudhari, R.V. Heterogenized HRh(CO)(PPh₃)₃ on zeolite Y using phosphotungstic acid as tethering agent: A novel hydroformylation catalyst. *J. Catal.* **2003**, *213*, 73–77. [CrossRef]

130. Johnson, B.F.G.; Raynor, S.A.; Shephard, D.S.; Mashmeyer, T.; Thomas, J.M.; Sankar, G.; Bromley, S.; Oldroyd, R.; Gladden, L.; Mantle, M.D. Superior performance of a chiral catalyst confined within mesoporous silica. *Chem. Commun.* **1999**, 1167–1168. [CrossRef]

131. Dyal, A.; Loos, K.; Noto, M.; Chang, S.W.; Spagnoli, C.; Shafi, K.V.P.M.; Ulman, A.; Cowman, M.; Gross, R.A. Activity of Candida rugose lipase immobilized on γ-Fe₂O₃ Magnetic Nanoparticles. *J. Am. Chem. Soc.* **2003**, *125*, 1684–1685. [CrossRef] [PubMed]

132. Wang, W.; Xu, Y.; Wang, D.I.C.; Li, Z. Recyclable nanobiocatalyst for enantioselective sulfoxidation: Facile fabrication and high performance of chloroperoxidase-coated magnetic nanoparticles with iron oxide core and polymer shell. *J. Am. Chem. Soc.* **2009**, *131*, 12892–12893. [CrossRef] [PubMed]

133. Shylesh, S.; Schuenemann, V.; Thiel, W.R. Magnetically separable nanocatalysts: Bridges between homogeneous and heterogeneous catalysis. *Angew. Chem. Int. Ed.* **2010**, *49*, 3428–3459. [CrossRef] [PubMed]

134. Campelo, M.; Luna, D.; Luque, R.; Marinas, J.M.; Romero, A.A. Sustainable preparation of supported metal nanoparticles and their applications in catalysis. *ChemSusChem* **2009**, *2*, 18–45. [CrossRef] [PubMed]

135. Shi, F.; Tse, M.; Zhou, S.; Pohl, M.-M.; Radnik, J.; Hubner, S.; Jahnisch, K.; Bruckner, A.; Beller, M. Green and efficient synthesis of sulfonamides catalyzed by nano-Ru/Fe$_3$O$_4$. *J. Am. Chem. Soc.* **2009**, *131*, 1775–1779. [CrossRef] [PubMed]

136. Demidova, Y.S.; Simakova, I.L.; Warne, J.; Simakov, A.; Murzin, D.Y. Kinetic modeling of one-pot myrtenol amination over Au/ZrO$_2$ catalyst. *Chem. Eng. J.* **2014**, *238*, 164–171. [CrossRef]

137. Simakova, I.L.; Demidova, Y.S.; Estrada, M.; Beloshapkin, S.; Suslov, E.V.; Volcho, K.P.; Salakhutdinov, N.F.; Murzin, D.Y.; Simakov, A. Gold catalyzed one-pot myrtenol amination: Effect of catalyst redox activation. *Catal. Today* **2017**, *279*, 63–70. [CrossRef]

138. Demidova, Y.S.; Suslov, E.V.; Simakova, I.L.; Korchagina, D.V.; Mozhajcev, E.S.; Volcho, K.P.; Salakhutdinov, N.F.; Simakov, A.; Murzin, D.Y. Selectivity control in one-pot amination of Au/ZrO$_2$ by molecular hydrogen addition. *J. Mol. Catal. A Chem.* **2017**, *426*, 60–67. [CrossRef]

139. Demidova, Y.S.; Suslov, E.V.; Simakova, I.L.; Volcho, K.P.; Salakhutdinov, N.F.; Simakov, A.; Murzin, D.Y. Promoting effect of alcohols and formic acid on Au-catalyzed one-pot alcohol amination. *Mol. Catal.* **2017**, *433*, 414–419. [CrossRef]

140. Demidova, Y.S.; Suslov, E.V.; Simakova, I.L.; Korchagina, D.V.; Mozhajcev, E.S.; Volcho, K.P.; Salakhutdinov, N.F.; Simakov, A.; Murzin, D.Y. One-pot monoterpene alcohol amination over Au/ZrO$_2$ catalyst: Effect of the substrates structure. *J. Catal.* **2018**, *360*, 127–134. [CrossRef]

141. Cherng, Y.-J.; Fang, J.-M.; Lu, T.-J. A new pinane-type tridentate modifier for asymmetric reduction of ketones with lithium aluminum hydride. *Tetrahedron Asymmetry* **1995**, *6*, 89–92. [CrossRef]

142. Ishida, T.; Takamura, R.; Takei, T.; Akita, T.; Haruta, M. Support effects of metal oxides on gold-catalyzed one-pot *N*-alkylation of amine with alcohol. *Appl. Catal. A Gen.* **2012**, *413–414*, 261–266. [CrossRef]

143. Lescot, C.; Darses, B.; Collet, F.; Retailleau, P.; Dauban, P. Intermolecular C–H amination of complex molecules: Insights into the factors governing the selectivity. *J. Org. Chem.* **2012**, *77*, 7232–7240. [CrossRef] [PubMed]

144. Trost, B.M.; van Vranken, D.L. Asymmetric transition metal-catalyzed allylic alkylations. *Chem. Rev.* **1996**, *96*, 395–422. [CrossRef] [PubMed]

145. Butt, N.A.; Zhang, W. Transition metal-catalyzed allylic substitution reactions with unactivated allylic substrates. *Chem. Soc. Rev.* **2015**, *44*, 7929–7967. [CrossRef] [PubMed]

146. Piechaczyk, O.; Thoumazet, C.; Jean, Y.; Le Floch, P. DFT study on the palladium-catalyzed allylation of primary amines by allylic alcohol. *J. Am. Chem. Soc.* **2006**, *128*, 14306–14317. [CrossRef] [PubMed]

catalysts

MDPI

Article

Brønsted and Lewis Solid Acid Catalysts in the Valorization of Citronellal

Federica Zaccheria [1], Federica Santoro [1], Elvina Dhiaul Iftitah [2] and Nicoletta Ravasio [1,*]

[1] CNR Institute of Molecular Science and Technology, Via Golgi 19, 20133 Milano, Italy; f.zaccheria@istm.cnr.it (F.Z.); f.santoro@istm.cnr.it (F.S.)

[2] Chemistry Department Faculty of Science, Brawijaya University Jl. Veteran, Malang 65145, Indonesia; vin_iftitah@ub.ac.id

* Correspondence: n.ravasio@istm.cnr.it; Tel.: +39-02-5031-4382

Received: 30 July 2018; Accepted: 12 September 2018; Published: 22 September 2018

Abstract: Terpenes are valuable starting materials for the synthesis of molecules that are of interest to the flavor, fragrance, and pharmaceutical industries. However, most processes involve the use of mineral acids or homogeneous Lewis acid catalysts. Here, we report results obtained in the liquid-phase reaction of citronellal with anilines under heterogeneous catalysis conditions to give tricyclic compounds with interesting pharmacological activity. The terpenic aldehyde could be converted into octahydroacridines with a 92% yield through an intramolecular imino Diels–Alder reaction of the imine initially formed in the presence of an acidic clay such as Montmorillonite KSF. Selectivity to the desired product strongly depended on the acid sites distribution, with Brønsted acids favoring selectivity to octahydroacridine and formation of the *cis* isomer. Pure Lewis acids such as silica–alumina with a very low amount of alumina gave excellent results with electron-rich anilines like toluidine and p-anisidine. This protocol can be applied starting directly from essential oils such as kaffir lime oil, which has a high citronellal content.

Keywords: solid acids; acidic clays; terpenes; citronellal; octahydroacridines; heterogeneous catalysis

1. Introduction

Terpenes represent a class of natural compounds suited for the synthesis of several types of molecules useful for the industrial production of intermediates for fragrances, flavors, and pharmaceuticals [1]. α-pinene, e.g., one of the main components of turpentine, can be used as a starting material for the synthesis of β-santalol and sandalwood fragrances. These are valuable alternatives to toxic nitro-musks and low-biodegradability polycyclic musks that are among the commonly used fragrances in European laundry detergents, fabric softeners, cleaning agents, and cosmetic products, therefore ubiquitously present in the aquatic environment [2].

In previous years, the use of terpenes has been widely investigated in the polymer industry because of the strong need for renewable and biodegradable materials in this sector [3,4].

Among terpenic molecules, citronellal, citral, limonene, carene, and pinene are common because of their large occurrence in essential oils and in oils derived from agro-industrial residues. Citronellal and citral offer noteworthy opportunities in synthetic strategies because of their condensation and addition reactions by virtue of their unsaturated aldehydic structure. The abundance of these two molecules in essential oils such as kaffir lime, citronella, lemongrass, and krangean oil, promotes their use as raw materials for the sustainable synthesis of different chemicals, specifically N-containing ones. A Schiff base with antibacterial activity was recently synthesized through acid catalysis from citronellal, [5] while benzimidazole derivatives of both aldehydes were obtained by using microwave irradiation [6]. In this case, the aldehyde reacts with *o*-phenylenediamine to give the benzimidazol moiety, but when an aromatic monoamine reacts with an unsaturated aldehyde it gives an N-aryl imine that can further react with the electron-rich alkene moiety through the Povarov reaction, which is an intramolecular imino Diels–Alder reaction [7] (Scheme 1).

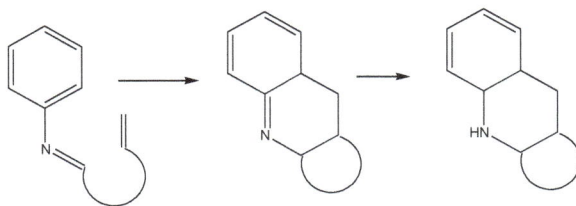

Scheme 1. General intramolecular Povarov reaction.

This class of organic reactions has attracted interest as a useful route of access to N-containing polycyclic structures, such as substituted tetrahydroquinoline and octahydroacridine (OHAs) derivatives. Octahydroacridine systems are an interesting class of biologically active molecules in the field of drugs and pharmaceuticals, acting as gastric acid secretion inhibitors [8] in the prevention of senile dementia [9].

Despite several varying methods available offering access to the octahydroacridine skeleton, such as the Beckmann rearrangement of oxime sulfonate, [10] the catalytic hydrogenation of acridine, [11] and Friedel–Crafts acylation, [12] the simplest one remains the acid catalyzed imino Diels–Alder reaction of 2-azadienes. Because of their poor reactivity in the Povarov reaction, 2-azadienes need to be activated through coordination with acidic catalysts that enhance their electron-deficient character. Both Lewis and Brønsted acids have been used, but, despite their effectiveness, many of these catalysts show disadvantages, such as multistep procedures, long reaction times, use of inert atmosphere, expensiveness, and tedious work-up. Therefore, developing green and efficient catalysts for this reaction is an important challenge.

An efficient synthesis of OHAs starts from citronellal and N-arylamines [13,14] (Scheme 2). This Lewis-acid catalyzed imino Diels–Alder reaction is the most atom-economic way to OHAs, with high yields and, in some cases, high stereoselectivity.

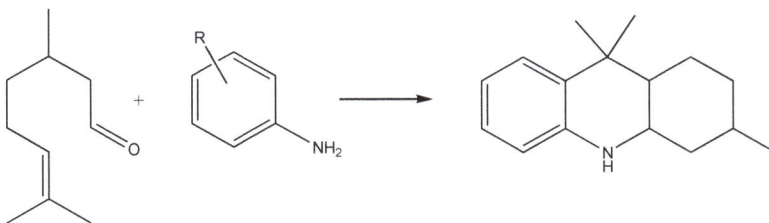

Scheme 2. Reaction of citronellal and anilines into the corresponding octahydroacridine.

During our recent studies on reductive amination of ketones, [15] the attempts to synthesize amine derivatives of C=O-containing terpenes led us to obtain OHAs structures. Herein, we report our results about the synthesis of OHAs from citronellal promoted by solid acid catalysts.

2. Results and Discussion

Previously, we have reported on the use of pre-reduced Cu/SiO_2 (Cu/Si) catalysts in the Direct Reductive Amination (DRA) of aromatic ketones [15]. This is a significant reaction allowing one to obtain secondary amines in one step starting from a ketone, which is still carried out in the presence of unfavorable reagents such as $NaBH_3CN$ or $NaBH(OAc)_3$. The use of a heterogeneous catalyst based on a non-noble metal can yield up to 98% of the amine at 100 °C in a few hours. Unfortunately, under the same conditions, aliphatic terpenic ketones, in particular menthone, a major component of dementholized mint oil, react slowly, requiring a longer time span, although selectivity to the amine is excellent (Scheme 3).

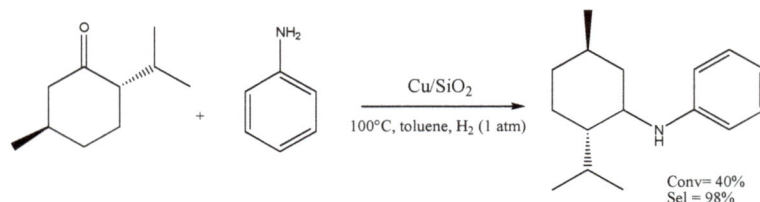

Scheme 3. Amination of menthone with aniline (Conv = Conversion; Sel = Selectivity).

This prompted us to test the reaction of terpenic molecules with a greater number of reactive aldehyde groups.

Table 1 reports selected results obtained in the reaction of citronellal under varying conditions and in the presence of several catalytic materials.

Surprisingly, the reaction of citronellal with aniline under DRA conditions, that is, in the presence of pre-reduced Cu catalyst and molecular H_2, was quick and gave the corresponding OHA as the only product in high yield (Table 1, entry 3). A blank test carried out without a catalyst led to the formation of the intermediate imine, without observing the formation of the Diels–Alder product, showing that, in this specific reaction, a catalyst is only necessary for the second step, and a reductive environment is not required (Table 1, entries 1–2). Therefore, both pre-reduction of the catalyst and molecular H_2 could be avoided, and the unreduced CuO/SiO_2 (CuO/Si) catalyst was used at room temperature in the presence of air, giving comparable results.

This reactivity is ascribed to the Lewis acid character of CuO/SiO_2. Thus, the chemisorption–hydrolysis technique used in the preparation of this catalyst allows us to reach a high dispersion of CuO on the silica surface [16]. The CuO particles are small and defective, thus giving an account of the Lewis acidity of this material, able to promote, e.g., the ring opening of epoxides with alcohols [17]. After the reduction pre-treatment, highly defective metal nanoparticles are formed showing Lewis acid activity in a wide range of reactions, particularly in the one-pot production of valeric esters from γ-valerolactone (GVL) and alcohols [18].

Therefore, other solid acids can be used in the synthesis of OHA. Because of our experience in the use of amorphous mixed oxides, we tested some of them in this reaction, namely, a silica–alumina cracking catalyst with a 13% content of Al_2O_3 (SiAl 13) and a 0.6% Alumina on silica (SiAl0.6). Furthermore, we also tested two commercial acid-treated clays, namely, Montmorillonite K10 and KSF. Clays are versatile materials widely used for various applications, including catalytic ones. Thus, increasingly stringent environmental issues and process optimization call for the substitution of liquid acids by more sustainable solid materials. Clays have both Brønsted (B) and Lewis (L) sites, the amount and the strength of which can be modified by acid treatment. Commercial acid-treated clays are widely used industrially as acid catalysts, therefore they are well-defined and reliable materials. K10 is commonly used in the formation of a C–N bond as an alternative to HCl, e.g., in the synthesis of alkylquinolines [19].

It should be emphasized that, under acidic conditions, citronellal can undergo the *ene* reaction to give isopulegol (Scheme 4). Although this is an interesting reaction, as isopulegol can be hydrogenated to menthol, in the present case this is a secondary reaction that lowers the yield in OHA and, therefore, a fine-tuning of the acidity is mandatory in order to reach a high selectivity to the desired product.

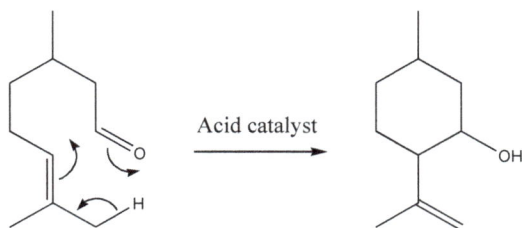

Scheme 4. *Ene* reaction of citronellal into isopulegol.

Results reported in Table 1 show significant differences in selectivity. Considering the reaction between citronellal and aniline, the two amorphous silica–aluminas and Montmorillonite K10 gave a low yield in OHA due to a significant formation of isopulegols (Table 1, entries 6,11,13). This agrees with our previous findings [20], showing that these three catalysts are highly active in the cyclization of citronellal to isopulegol. However, moving from aniline to electron-rich amines, the selectivity to OHA increased, reaching a remarkable 96% yield in the case of p-anisidine over the catalyst with a very low amount of alumina (Table 1, entry 9).

Table 1. Synthesis of octahydroacridines starting from (±)-Citronellal and anilines.

Entry	Catalyst	Amine	Conditions	T h	Conv. %	OHAs Sel %	Cis/Trans.	Imine Sel %	Isopulegols %
1	None	aniline	Toluene, 100 °C, N$_2$	1	73	-	-	87	-
				7.5	96			93	
2	None	aniline	Heptane, 25 °C, air	1	96	-	-	83	-
3	Cu/Si	aniline	Toluene, 100 °C, H$_2$	1	>99	97	37/63	<1	1
4	CuO/Si	aniline	Toluene, 25 °C, air	1	>99	96	45/55	-	2
5	CuO/Si	p-anisidine	Toluene, 25 °C, air	1	>99	59	61/39	-	32
6	SiAl 0.6	aniline	Heptane, 25 °C, air	1	>99	71	41/59	<1	24
7	SiAl 0.6	aniline	Heptane, 0 °C, air	1	>99	79	49/51	-	17
8	SiAl 0.6	aniline	Dioxane, 25 °C, air	1	>99	82	58/42	-	14
9	SiAl 0.6	p-anisidine	Heptane, 25 °C, air	1	>99	96	51/49	<1	2
10	SiAl 0.6	toluidine	Heptane, 25 °C, N$_2$	2	>99	90	37/63	-	1
11	SiAl 13	aniline	Heptane, 25 °C, N$_2$	1	>99	64	41/59	-	23
12	SiAl 13	aniline	Heptane, 0 °C, air	1	>99	86	58/42	-	11
13	Mont K10	aniline	Heptane, 25 °C, air	1	>99	70	55/45	-	14
14	Mont KSF	aniline	Heptane, 25 °C, air	1	>99	92	71/29	-	5
15	Mont KSF	aniline	Heptane, 0 °C, air	1	>99	85	73/27	-	10

It is worth emphasizing that only Lewis acid sites could be detected on the surface of SiAl0.6, as shown by FT IR spectra of adsorbed pyridine, while the other solids used showed also Brønsted acid sites (Table 2). OHA: octahydroacridine.

Table 2. Quantitative determination of acidic sites versus products distribution in the reaction of citronellal with aniline.

Catalyst	Acidic Sites (mmol$_{py/gcat}$)		% OHA	% Cis	% Isopulegols
KSF	0.157	0.025	92	71	5
K10	0.077	0.042	70	55	14
SiAl13	0.039	0.163	64	41	23
SiAl 0.6	–	0.005	71	41	24

Figure 1 shows the comparison of the two amorphous silica–alumina catalysts and the two clays: it is evident that both Brønsted and Lewis sites were present on the surface of the catalyst containing 13% of alumina and on the two clays, whereas the Brønsted ones were absent on the low-loading one.

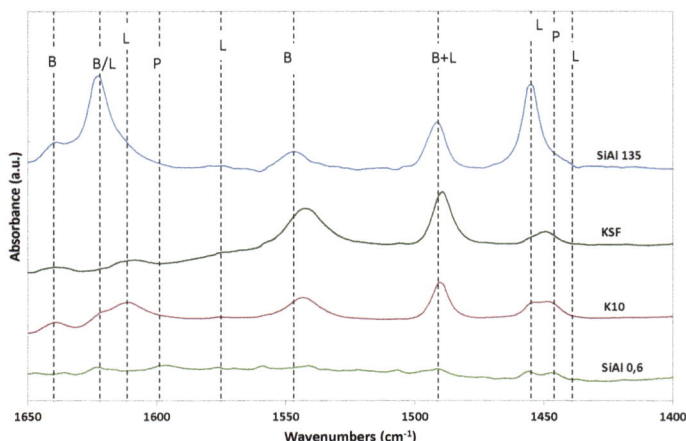

Figure 1. FTIR spectra of pyridine adsorbed on the silica–alumina catalysts: P = physisorbed, B = Brønsted sites, L = Lewis acid sites.

Thus, in SiAl 13, the bands at 1640 cm^{-1}, 1547 cm^{-1}, and 1492 cm^{-1} are the intense modes of pyridinium cations associated with a total proton transfer from the Brønsted acidic surface OH group to the basic molecule, whereas the bands at 1623 cm^{-1} and 1455 cm^{-1} are due to the pyridine molecularly coordinated on Al^{3+} cations, acting as Lewis acid sites [21].

However, the spectra of pyridine adsorbed on clays, KSF, and K10, showed the presence of both Lewis and Brønsted sites, although the former in lower concentration.

The presence of weak absorption bands at 1545 cm^{-1}, 1490 cm^{-1}, and 1450 cm^{-1}, indicating the presence of both Brønsted and Lewis acidity in low concentrations, has already been reported, although in the Diffuse Reflectance Infrared Fourier Transform (DRIFT) spectra of pyridine adsorbed on the surface of montmorillonite K10 [22], while the presence of medium-weak Brønsted sites and strong Lewis sites was evidenced by a range of complementary experimental techniques by Lenarda et al. [23]. However, the number of these sites has never been quantified. Table 2 shows that B sites in KSF are twice the number of those in SiAl 13, that in turn are twice the number of those in K10, in agreement with previous determinations through NH$_3$ chemisorption of acidic sites in KSF and K10 [24].

Comparing the results obtained with the relative number of acidic sites reported in Table 2, it is apparent that, for catalysts containing both B and L sites, the yield in OHA increases with the number of B sites, while the amount of isopulegols increases with the number of L sites (Figure 2).

The catalyst that contains only L sites showed an intermediate behavior, giving only 71% of OHA in the reaction of citronellal and aniline (Table 1, entry 6). However, by increasing the electro-donor ability of the substituent, this catalyst gave excellent results, reaching a 96% yield in the reaction with anisidine (Table 1, entry 9).

According to our results, differences in acidity can also influence products stereochemistry.

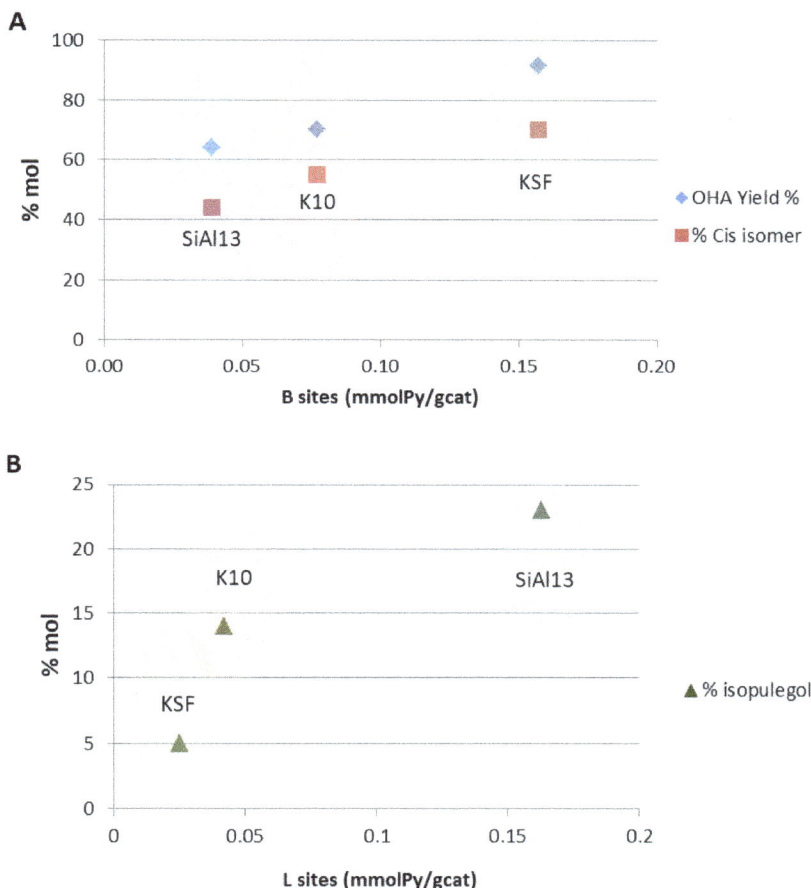

Figure 2. Trend in products distribution versus the concentration of B acidic sites (**A**) and L sites (**B**).

Laschat and Lauterwein [25] reported on the use of several homogeneous Lewis acids at −78 °C in CH$_2$Cl$_2$ and drew the conclusion that the catalyst plays only a minor role in determining the *cis/trans* ratio. In agreement with this work, Kouznetsov [26] found that the bulkiness of the aniline derivative has a pivotal role when both Lewis and Brønsted acids are used as a catalyst, allowing to reach 97% of the *trans* isomer for the N-benzyl-aniline derivative at room temperature. However, in the presence of TiCl$_3$, the *cis/trans* ratio did not change significantly by changing the amine [9].

Reports on heterogeneous catalysts in this specific reaction are rare; ZnCl$_2$ supported on silica was used in the presence of microwaves, giving a 78% yield in the case of aniline, with a *cis/trans* ratio of 1:1 [13]. The introduction of electron-donor or electron-withdrawer substituents on the aniline moiety influenced both yield and stereochemistry of the products.

In the present case, differences in stereochemistry are evident. Specifically, pure Lewis solids (both CuO/silica and SiAl 0.6) promote 40/60 mixtures with a slight excess of *trans* isomers. When introducing B acid sites, the number of *cis* isomers increases, and the solid with a more prominent B acid character, namely, KSF, favors the formation of the *cis*-isomer (Table 1, entries 14 and 15, Figure 2A). This could be because of the different adsorption of the intermediate imine on the catalyst surface, with Brønsted sites binding strongly to the N atom and favoring the transition state leading to the *cis* isomer.

The solid catalyst is also reusable, allowing to obtain the desired OHA product, although in lower yield (75%) under non-optimized conditions, such as without washing the catalyst.

This protocol can be successfully applied directly to essential oils. Thus, the reaction of kaffir lime oil with aniline in heptane and in the presence of clay KSF produced an oil containing 70% of octahydroacridine in one hour (Scheme 5).

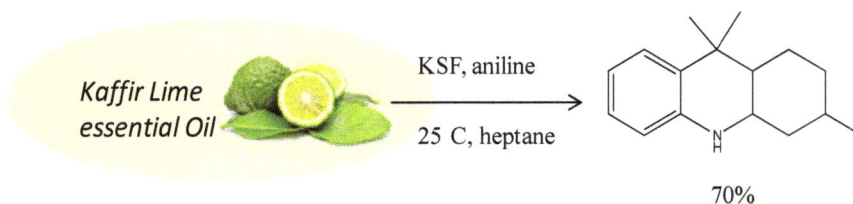

Scheme 5. Octahydroacridine formation, starting directly from kaffir lime.

3. Experimental Section

3.1. Materials and Methods

(±)-Citronellal (>95%), aniline (>99%), *p*-anisidine (>99%) and toluidine (99%) were purchased from Sigma-Aldrich (Milan, Italy) and used without pretreatment. Kaffir lime (*Citrus hystrix D.C*) oil was obtained from the Essential Oil's Institute Atsiri, Brawijaya University, Malang (RI).

SiO_2–Al_2O_3 13 (Surface Area = 485 m^2/g, Pore Volume = 0.79 mL/g), was purchased from Sigma-Aldrich; SiO_2–Al_2O_3 0.6 (Surface Area = 483 m^2/g, Pore Volume = 1.43 mL/g) was kindly provided by GRACE Davison. Montmorillonite KSF and Montmorillonite K10 were purchased from Sigma-Aldrich; SiO_2 (Surface Area = 460 m^2/g, Pore Volume = 0.74 mL/g) was purchased from Fluka (Milan, Italy).

3.2. Copper Catalyst Preparation

CuO/SiO_2 (with 16% of copper loading) catalyst was prepared by the chemisorption–hydrolysis technique as reported [16], by adding the support to an aqueous $[Cu(NH_3)_4]^{2+}$ solution prepared by dropping NH_4OH into a $Cu(NO_3)_2 \cdot 3H_2O$ solution until pH 9 had been reached. After 20 min under stirring, the slurry, held in an ice bath at 0 °C, was diluted with water. The solid was separated by filtration with a Büchner funnel, washed with water, dried overnight at 120 °C, and calcined in air at 350 °C.

3.3. Catalyst Characterization

FTIR Spectra of adsorbed pyridine. The FT-IR studies of probe molecules (pyridine) adsorption and desorption were carried out with a BioRad FTS-60 (Segrate, Italy) spectrophotometer equipped with a mid-IR MCT detector. The experiments were performed on a sample disk (15–20 mg) after a simple calcination treatment (180 °C, 20 min air + 20 min under vacuum). One spectrum was collected before probe molecule adsorption as a blank experiment. Therefore, pyridine adsorption was carried out at room temperature, and the following desorption steps were performed from room temperature to 250 °C. The spectrum of each desorption step was acquired every 50 °C after cooling the sample. For quantitative analysis, the amount of adsorbed pyridine ($\text{mmol}_{Py}/\text{g}_{cat}$) was calculated on the basis of the relationship reported by Emeis [27] from the integration of diagnostic bands evaluated in the spectra registered at 150 °C.

3.4. Catalytic Tests

Prior to the catalytic tests, the catalysts (100 mg) were pretreated as follows: SiO_2–Al_2O_3 was calcined for 20 min in air at 180 °C and 20 min under vacuum at the same temperature; Montmorillonite KSF and Montmorillonite K10 were dried in an oven at 120 °C for one night; CuO/SiO_2 was calcined for 20 min in air at 270 °C and 20 min under vacuum at the same temperature, and a following reduction step with H_2 (1 atm) was performed for Cu/SiO_2-catalyzed reactions.

In a glass reaction vessel containing the pretreated catalyst, a mixture of citronellal (0.55 mmol) and amine (1.66 mmol) in 5 mL of solvent was charged. Therefore, the reaction proceeded with the proper temperature conditions under magnetic stirring. At the desired reaction time, the mixture was separated by simple filtration from the catalysts, and the products were analyzed by GC–MS (5%-phenyl-methyl polysiloxane column) and NMR (Bruker 500 MHz).

For the recycling tests, the solution was separated from the catalyst after the first run by simple decantation, and a fresh reaction mixture was charged in the reactor without treating the catalyst.

In the experiment carried out with Kaffir lime oil, a mixture containing 100 μL of essential oil and 1.66 mmol of aniline in 5 mL of solvent was charged in a glass reaction vessel containing 100 mg of KSF dried overnight at 120 °C.

[1]*HNMR data (300 MHz, CDCl3)* (See Supplementary Materials for NMR Spectra) A) *cis* isomer: δ = 0.98 (d, J = 6.4, 3 H, CH₃), 1.07–1.89 (m, 2H, CH₂), 1.13–1.97 (m, 2H, CH₂), 1.31 (s, 3H, CH₃), 1.33 (s, 1 H, CH), 1.38 (s, 1 H, CH), 1.41 (s, 3H, CH₃), 1.83 (m, 1H, CH), 1.86 (m, 1H, CH), 3.69 (bs, NH), 3.92 (dt, 1H, CH), 6.51 (dd, 1H, CH), 6.68 (m, 1H, CH), 7.04 (m, 1H, CH), 7.22 (dd, 1H, CH).

MS (EI): m/z (%) = 229.2 (34), 214.2 (100), 158.1 (12), 144.1 (29).

B) *trans* isomer: 0.99–1.77 (m, 2H, CH₂), (d, J = 6.5, 3H, CH₃), 1.13 (m, 1H, CH), 1.19 (s, 3H, CH₃), 1.33–1.83 (m, 2H, CH₂), 1.40 (m, 1H, CH), 1.42 (s, 3H, CH₃), 1.97 (m, 1H, CH), 3.16 (dt, 1H, CH), 3.69 (bs, NH), 6.52 (dd, 1H, CH), 6.73 (s, 1H, CH), 7.04 (m, 1H, CH), 7.04 (m, 1H, CH), 7.31 (dd, 1H, CH).

MS (EI): m/z (%) = 229.2 (55), 214.2 (100), 158.1 (18), 144.1 (33).

4. Conclusions

The introduction of solid acid materials in organic synthesis is an ambitious target of green chemistry. Thus, a heterogeneous catalyst could be filtered off at the end of the reaction without the need of a neutralization step followed by washing and the time-consuming workup of the reaction mixture producing significant amounts of inorganic salts. N-containing heterocycles are the most abundant class of compounds synthesized by the pharmaceutical industry [28]. Here, we show that very simple inorganic materials such as Montmorillonite, that is a natural acidic clay, can effectively promote the synthesis of N-containing octahydroacridines with a 92% yield in one hour and at room temperature, starting from anilines and citronellal, the main component of several essential oils. In some cases, yields were almost quantitative due to high selectivity with respect to the competing *ene*-reaction to isopulegols. Both high selectivity and the use of solid acid catalyst allowed us to obtain a negligible waste process, showing high sustainability.

Supplementary Materials: The following are available online at http://www.mdpi.com/2073-4344/8/10/410/s1, Figure S1: ^{1}HNMR of octahydroacridine obtained from citronellal and aniline.

Author Contributions: Conceptualization, F.Z. and N.R.; Investigation, F.S. and E.D.I.; Writing-Original Draft Preparation, F.Z.; Writing-Review & Editing, N.R.; Supervision, N.R.

Funding:: This research received no external funding.

Acknowledgments: Institut Atsiri Universitas Brawijaya, Malang, Indonesia is kindly acknowledged for a sample of Kaffir Lime essential oil.

Conflicts of Interest: The authors declare no conflict of interest.

References

1. Monteiro, J.L.F.; Veloso, C.O. Catalytic conversion of terpenes into fine chemicals. *Top. Catal.* **2004**, *27*, 169–180. [CrossRef]
2. Ravasio, N.; Zaccheria, F.; Guidotti, M.; Psaro, R. Mono- and bifunctional heterogeneous catalytic transformation of terpenes and terpenoids. *Top. Catal.* **2004**, *27*, 157–168. [CrossRef]
3. Zhang, D.; del Rio-Chanona, E.A.; Shah, N. Screening synthesis pathways for biomass-derived sustainable polymer production. *ACS Sustain. Chem. Eng.* **2017**, *5*, 4388–4398. [CrossRef]
4. Wilbon, P.A.; Chu, F.; Tang, C. Progress in renewable polymers from natural terpenes, terpenoids, and rosin. *Macromol. Rapid Commun.* **2013**, *34*, 8–37. [CrossRef] [PubMed]
5. Kinanthi, R.; Puspitasari; Dwi, F.; Warsito, S.; Farid Rahman, M. Synthesis of schiff base from citronellal in kaffir lime oil (Citrus hystrix D.C) using acid catalyst. In Proceedings of the 1st International Conference of Essential Oils (ICEO2017), Malang, Indonesia, 11–12 October 2017.
6. Warsito; Ramadhan, S.D.; Al Karoma, D.; Zulfa, A. Comparison of synthesis of some benzimidazole derivatives of cytronellal and citral. In Proceedings of the 1st International Conference of Essential Oils (ICEO2017), Malang, Indonesia, 11–12 October 2017.
7. Kouznetsov, V.V. Recent synthetic developments in a powerful imino Diels–Alder reaction (Povarov reaction): Application to the synthesis of N-polyheterocycles and related alkaloids. *Tetrahedron* **2009**, *65*, 2721–2750. [CrossRef]
8. Canas-Rodriguez, A.; Canas, R.G.; Mateo-Bernardo, A. Tricyclic inhibitors of gastric acid secretion. Part, V. Octahydroacridines. *An. Quim. Ser. C.* **1987**, *83*, 24–27.
9. Mayekar, N.V.; Nayak, S.K.; Chattopadhyay, S. Two convenient one-pot strategies for the synthesis of octahydroacridines. *Synth. Commun.* **2004**, *34*, 3111–3119. [CrossRef]
10. Sakane, S.; Matsumura, Y.; Yamamura, Y.; Ishida, Y.; Maruoka, K.; Yamamoto, H. Olefinic cyclizations promoted by beckmann rearrangement of oxime sulfonate. *J. Am. Chem. Soc.* **1983**, *105*, 672–674. [CrossRef]
11. Sakanishi, K.; Mochida, I.; Okazaki, H.; Soeda, M. Selective hydrogenation of 9-aminoacridine over supported noble metal catalysts. *Chem. Lett.* **1990**, *19*, 319–322.
12. Kouznetsov, V.; Palma, A.; Rozo, W.; Stashenko, E.; Bahsas, A.; Amaro-Luis, J. A facile Brønsted acidic-mediated cyclisation of 2-allyl-1-arylaminocyclohexanes to octahydroacridine derivatives. *Tetrahedron Lett.* **2000**, *41*, 6985–6988. [CrossRef]
13. Jacob, R.G.; Perin, G.; Botteselle, G.V.; Lenardão, E.J. Clean and atom-economic synthesis of octahydroacridines: Application to essential oil of citronella. *Tetrahedron Lett.* **2003**, *44*, 6809–6812. [CrossRef]
14. Sabitha, G.; Reddy, E.V.; Yadav, J.S. Bismuth(III) Chloride: An efficient catalyst for the one-pot stereoselective synthesis of octahydroacridines. *Synthesis* **2002**, *3*, 409–412. [CrossRef]
15. Santoro, F.; Psaro, R.; Ravasio, N.; Zaccheria, F. Reductive amination of ketones or amination of alcohols over heterogeneous Cu catalysts: Matching the catalyst support with the N-Alkylating agent. *ChemCatChem* **2012**, *4*, 1249–1254. [CrossRef]
16. Zaccheria, F.; Scotti, N.; Marelli, M.; Psaro, R.; Ravasio, N. Unravelling the properties of supported copper oxide: Can the particle size induce acidic behaviour? *Dalton Trans.* **2013**, *42*, 1319–1328. [CrossRef] [PubMed]
17. Zaccheria, F.; Santoro, F.; Psaro, R.; Ravasio, N. CuO/SiO$_2$: A simple and effective solid acid catalyst for epoxides ring opening. *Green Chem.* **2011**, *13*, 545–548. [CrossRef]
18. Scotti, N.; Dangate, M.; Gervasini, A.; Evangelisti, C.; Ravasio, N.; Zaccheria, F. Unraveling the role of low coordination sites in a Cu metal nanoparticle: A step forwards the selective synthesis of second generation biofuels. *ACS Catal.* **2014**, *4*, 2818–2826. [CrossRef]
19. Campanati, M.; Savini, P.; Tagliani, A.; Vaccari, A.; Piccolo, O. Environmentally friendly vapour phase synthesis of alkylquinolines. *Catal. Lett.* **1997**, *47*, 247–250. [CrossRef]
20. Ravasio, N.; Antenori, M.; Babudri, F.; Gargano, M. Intramolecular ene reaction promoted by mixed cogels. *Stud. Surf. Sci. Catal.* **1997**, *108*, 625–632.

21. Trombetta, M.; Busca, G.; Rossini, S.; Piccoli, V.; Cornaro, U.; Guercio, A.; Catani, R.; Willey, R. FT-IR studies on light olefin skeletal isomerization catalysis: III. surface acidity and activity of amorphous and crystalline catalysts belonging to the SiO_2–Al_2O_3 system. *J. Catal.* **1998**, *179*, 581–596. [CrossRef]

22. Ravindra Reddy, C.; Bhat, Y.S.; Nagendrappa, G.; Jai Prakash, B.S. Brønsted and Lewis acidity of modified montmorillonite clay catalysts determined by FT-IR spectroscopy. *Catal. Today* **2009**, *141*, 157–160. [CrossRef]

23. Flessner, U.; Jones, D.J.; Rozière, J.; Zajac, J.; Storaro, L.; Lenarda, M.; Pavanc, M.; Jiménez-López, A.; Rodrìguez-Castellón, E.; Trombetta, M.; et al. A study of the surface acidity of acid-treated montmorillonite clay catalysts. *J. Mol. Catal. A Chem.* **2001**, *168*, 247–256. [CrossRef]

24. Vodnár, J.; Farkas, J.; Békássy, S. Catalytic decomposition of 1,4-diisopropylbenzene dihydroperoxide on montmorillonite-type catalysts. *Appl. Catal. A* **2001**, *208*, 329–334. [CrossRef]

25. Laschat, S.; Lauterwein, J. Intramolecular hetero-Diels-Alder reaction of N-arylimines. Applications to the synthesis of octahydroacridine derivatives. *J. Org. Chem.* **1993**, *58*, 2856–2861. [CrossRef]

26. Acelas, M.; Romero Bohórquez, A.R.; Kouznetsov, V.V. Highly diastereoselective synthesis of new trans-fused octahydroacridines via. intramolecular cationic imino diels–alder reaction of N-protected anilines and citronellal or citronella essential oil. *Synthesis* **2017**, *49*, 2153–2162.

27. Emeis, C.A. Determination of integrated molar extinction coefficients for infrared absorption bands of pyridine adsorbed on solid acid catalysts. *J. Catal.* **1993**, *141*, 347–354. [CrossRef]

28. Carey, J.S.; Laffan, D.; Thomson, C.; Williams, M.T. Analysis of the reactions used for the preparation of drug candidate molecules. *Org. Biomol. Chem.* **2006**, *4*, 2337–2347. [CrossRef] [PubMed]

catalysts

MDPI

Review

Preparation and Application of Biochar-Based Catalysts for Biofuel Production

Feng Cheng and Xiuwei Li *

School of Energy and Power Engineering, Nanjing University of Science and Technology, Nanjing 210094. China; fengcheng@njust.edu.cn
* Correspondence: good3000best@163.com; Tel.: +86-25-8431-7344

Received: 19 July 2018; Accepted: 17 August 2018; Published: 24 August 2018

Abstract: Firstly, this paper reviews two main methods for biochar synthesis, namely conventional pyrolysis and hydrothermal carbonization (HTC). The related processes are described, and the influences of biomass nature and reaction conditions, especially temperature, are discussed. Compared to pyrolysis, HTC has advantages for processing high-moisture biomass and producing spherical biochar particles. Secondly, typical features of biochar in comparison with other carbonaceous materials are summarized. They refer to the presence of inorganics, surface functional groups, and local crystalline structures made up of highly conjugated aromatic sheets. Thirdly, various strategies for biochar modification are illustrated. They include activation, surface functionalization, in situ heteroatom doping, and the formation of composites with other materials. An appropriate modification is necessary for biochar used as a catalyst. Fourthly, the applications of biochar-based catalysts in three important processes of biofuel production are reviewed. Sulfonated biochar shows good catalytic performance for biomass hydrolysis and biodiesel production. Biodiesel production can also be catalyzed by biochar-derived or -supported solid-alkali catalysts. Biochar alone and biochar-supported metals are potential catalysts for tar reduction during or after biomass gasification. Lastly, the merits of biochar-based catalysts are summarized. Biochar-based catalysts have great developmental prospects. Future work needs to focus on the study of mechanism and process design.

Keywords: heterogeneous catalysis; biorefinery; solid-acid catalyst; biochar-supported metal catalysts; surface functional groups; hydrothermal carbonization; surface functionalization; biofuel production

---

1. Introduction

The limited reserves of fossil fuels on Earth and the environmental problems associated with their use drive people to turn to renewable alternatives such as solar energy, wind energy, and biomass energy. Biomass is the most ancient fuel, providing light and heat to human beings for centuries. However, the traditional combustion of biomass is gradually being abandoned due to its low energy efficiency and air pollution. On the other hand, biomass-derived liquid or gaseous fuels are more adaptable to modern society. Hence, extensive studies are devoted to the production of biofuels. Currently, the primary conversion technologies include (1) biomass fermentation to produce bio-ethanol, (2) transesterification of vegetable oil or fat with alcohols to produce bio-diesel, (3) biomass gasification to produce syngas, and (4) fast pyrolysis of biomass to produce bio-oil, as Figure 1 displays.

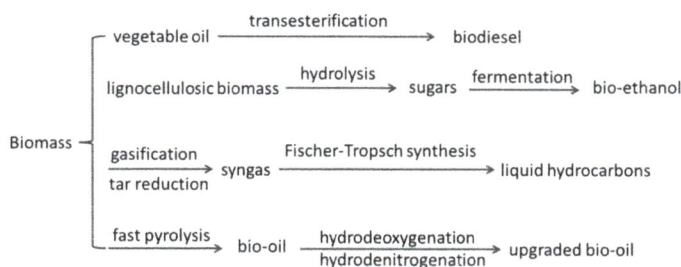

Figure 1. Primary conversion pathways from biomass to biofuel.

Catalysis, especially heterogeneous catalysis, plays an important role in biofuel production [1]. For instance, the conversion efficiency of vegetable oil to bio-diesel could be significantly enhanced using acid or alkali catalysts [2]. Tar reduction, which is crucial to the production of clean syngas via biomass gasification, also heavily relies on the use of catalysts [3]. Bio-oil cannot be directly utilized as transport fuel due to its poor fuel properties, e.g., high oxygen content and low higher heating value (HHV). The catalytic upgrading of bio-oil is considered as a promising route to overcome these shortcomings [4]. Similarly, the effective catalytic hydrolysis of cellulose to saccharides is the key to applying lignocellulosic biomass instead of cereals (e.g., corn, wheat) for bio-ethanol production [5].

Biochar is a type of carbonaceous materials obtained from the thermochemical conversion of biomass in an oxygen-limited environment [6]. Compared with conventional carbon materials such as carbon black and activated carbon from coal coke, biochar has several merits. Firstly, its feedstock is renewable, and its preparation process is simple. Biochar can be obtained as the main product of biomass carbonization, or as a by-product of biomass gasification and fast pyrolysis. Emerging techniques such as microwave pyrolysis [7,8] and hydrothermal carbonization [9] bring the production of biochar with more superiorities. Secondly, the physicochemical properties of biochar can be facilely tailored according to its specific application. A series of biochar modification strategies were developed to functionalize biochar. Because of these advantages, biochar has great potential to be used in the fields of heterogeneous catalysis, energy storage, pollution control, etc. The performances of biochar as soil amendment, carbon sequestration agents, contaminant adsorbents, electrode materials, catalysts, or catalyst supports were comprehensively reviewed in References [6,10].

The potential of biochar as a catalyst or a catalyst support was extensively studied [11]. On one hand, some features of biochar benefit its role as a catalyst material. For instance, the presence of inorganics such as K and Fe in biochar contributes to its catalytic activity for tar cracking [12]. The surface functional groups of biochar could facilitate the adsorption of metal precursors, a necessary step for the synthesis of biochar-supported metal catalysts [9,13,14]. On the other hand, some characteristics of biochar, such as a low surface area and a poor porosity, hamper its catalytic application. Thus, various modification strategies were developed to endow biochar with specific properties. These strategies include selecting proper feedstock, controlling synthesis conditions, physical or chemical activation [10,13], surface functionalization [14,15], forming composites with other materials [16], etc. Comprehensive reviews on this topic can be found in References [15–17].

The processes for which biochar-based catalysts were tested include biomass refining [13,14,18], NO$_x$ removal [19,20], ammonia ozonation [21], electrocatalytic oxygen reduction and methanol oxidation in fuel cells [22,23], photocatalytic degradation of organics [24,25], nitrate degradation in wastewater [26], etc. A good review on the application of biochar-based catalysts was provided by Lee et al. [11] and Xiong et al. [14]. Biomass refining, which mimics petroleum refining, intends to produce valuable chemicals, fuels, and functional materials from biomass. Biochar-based catalysts are used in various biomass refining processes, such as biomass hydrolysis and dehydration [13,14], biodiesel production [2,27], biomass pyrolysis [28], bio-oil upgrading [4,28], tar removal [29,30], and the

Fischer–Tropsch synthesis of liquid hydrocarbons from syngas [31], etc. Detailed information can be found in related reviews [6,13,14]. The application of biochar-based catalysts in biomass refining represents a more sustainable and more integrated process. This is because biochar is a common byproduct of biomass refinery, and the use of it can reduce the reliance of biomass refinery on metal catalysts or fossil-fuel-derived carbon catalysts.

The present review summarizes the preparation, modification, and catalytic application of biochar in three important aspects of biofuel production, namely biomass hydrolysis, biodiesel production, and tar reduction. The link between biochar properties and synthesis conditions, and the link between catalytic performance and the physicochemical properties of biochar are focused on. The objective of this review is to provide some useful information for researchers to accomplish the purpose-driven synthesis of biochar-based catalysts.

2. Synthesis of Biochar

2.1. Conventional Pyrolysis

2.1.1. Process Description

In general, biochar can be produced on various pyrolysis platforms, such as slow pyrolysis, fast pyrolysis, and pyrolytic gasification. Typical operating conditions and biochar yields of these pyrolysis platforms can be found in the literature [6,10,15,32]. The slow pyrolysis of biomass (also termed "biomass carbonization") is an ancient technology for biochar production. It is normally performed over a temperature range of 300–800 °C with a heating rate of 5–7 °C/min and a residence time of >1 h [15]. The low heating rate and the long residence time favor a secondary cracking of vapors, and thus, lead to a high yield of biochar (30–50 wt%). In comparison, a high heating rate and a short residence time in fast pyrolysis are beneficial for the production of bio-oil, with only small amounts of biochar (around 15 wt%) being generated. Biochar is also a by-product (10 wt%) of biomass gasification which is mainly for syngas generation at >750 °C in the presence of oxidizing agents (e.g., air, O_2, and steam).

In addition to biochar yields, biochar properties are also affected by pyrolysis conditions. Brewer et al. [33] observed biochar from fast pyrolysis and slow pyrolysis had a similar size of fused aromatic ring cluster (7–8 rings per cluster), while the biochar from high-temperature gasification was much more condensed (17 rings per cluster). Moreover, aromatic C–H functional groups on the slow pyrolysis char were significantly less prominent than those on the gasification char. Dutta et al. [34] found the biochar from slow pyrolysis was more porous than that from fast pyrolysis.

Slow pyrolysis has inherent drawbacks such as time consumption and low energy efficiency, although it can obtain a high biochar yield. In contrast, fast pyrolysis and gasification are more promising. For more efficient biochar production, some strategies were proposed [15]: (1) to improve the process economics and reduce pollutant emissions through recycling co-products of biochar production; (2) to tailor biochar properties by adjusting pyrolysis conditions and selecting appropriate biomass feedstock. Among the factors that influence biochar properties, the nature of the parent biomass and the pyrolysis temperature are the two important ones [35–37].

2.1.2. The Influence of Biomass Feedstock

(1) The Share of Cellulose, Hemicellulose, and Lignin in Raw Biomass

Cellulose, hemicellulose, and lignin are the main components of lignocellulosic biomass. The three components possess different structural features (e.g., constituting monomer, cross-linking, crystallinity, and branching), which lead to different pyrolysis mechanisms [13,15,38]. Figure 2 illustrates the formation pathway of biochar via cellulose pyrolysis. Consequently, the contents of these components in raw biomass affect the physicochemical properties of the resultant biochar [14,15]. For example, the pyrolysis of high-lignin biomass such as pine wood and spruce wood produces biochar with a high yield and a high fixed-carbon content [35]. It is easier for microcrystalline cellulose to produce biochar with graphitic structures at a relatively low temperature (e.g., 350 °C) than for lignocellulosic biomass [22].

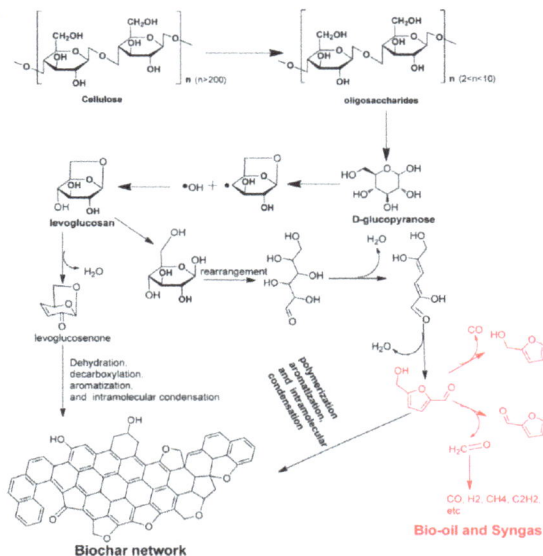

Figure 2. The mechanism of cellulose pyrolysis and biochar formation. Reprinted with permission from Reference [15]. Copyright (2015) American Chemical Society.

(2) The Biomass Matrix

For some biomasses, their microstructures are basically maintained over a pyrolysis process [14]. As Figure 3a,b display, the wood-derived biochar exhibited vertically aligned micro-channels and fibrous ridged surfaces, resembling the morphology of the raw wood [27,39,40]. The biochar from sisal leaves (Figure 3c) also inherited the natural hierarchical texture of the original biomass and showed connected porous frameworks [41]. Ordered channel arrays of hollow carbon nanofibers were prepared from the pyrolysis of crab shells [42]. These examples imply that the formation of biochar via biomass pyrolysis is highly "localized" [13].

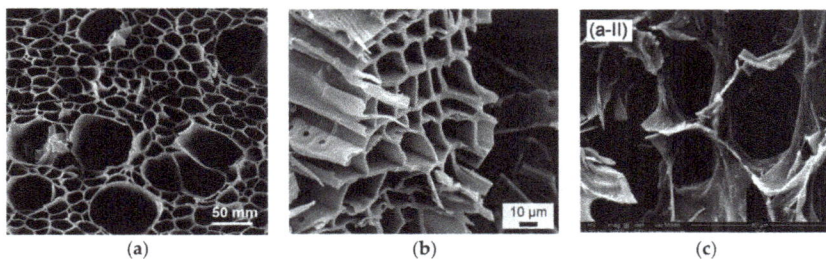

Figure 3. SEM images of biochar from the pyrolysis of (**a**) wood block reprinted with permission from [40]. Copyright (2017) American Chemical Society; (**b**) wood sawdust reprinted with permission from [39]. Copyright (2015) Royal Society of Chemistry; and (**c**) sisal leaves reprinted with permission from [41]. Copyright © 2015, Springer Nature.

(3)   The Inorganic Species

The inorganic species (i.e., ash) in biomass play an important role in the formation of biochar [14,15]. Some inorganic elements such as alkali and alkaline earth metals (AAEM) can catalyze biomass pyrolysis, thereby increasing the biochar yield remarkably [43,44]. Furthermore, the inorganic species retained in biochar endow biochar with a catalytic activity toward reactions such as methane decomposition [45] and tar cracking [12].

In some cases, inorganic species present in biomass may play a negative role. McBeath et al. [46] reported that a high ash content in feedstock retarded the formation of stable polycyclic aromatic carbon (SPAC), and hence, adversely affected the biochar stability. Yao et al. [47] observed that the presence of Si in biochar was not favorable for the dispersion of active metal Ni on the biochar support. This was because Si was easily melted at high temperatures, leading to the sintering of Ni with ash.

2.1.3. The Influence of Temperature

Biochar properties are also greatly influenced by pyrolysis temperatures [10,14,15,48]. Zhang et al. [49] and Zhao et al. [50] reported that temperature affected biochar properties more significantly than heating rate and residence time. They concluded that pyrolysis temperature had a positive correlation with pH, fixed C content, biochar stability, and ash content, whilst having a negative correlation with biochar yield, O and H mass fractions, and the number of surface functional groups.

The pyrolysis temperature also influences the surface area and pore size of biochar. As the temperature rises, more volatiles are released from the biomass surface, resulting in more pores and a larger surface area [15,37,50]. Zhao et al. [50] found that an increase in temperature led to smaller average pore sizes, while Muradov et al. [51] observed that a high pyrolysis temperature could promote the fusion of pores, and thus, expand pore sizes.

The chemical nature of biochar is also closely related to the pyrolysis temperature. Keiluweit et al. [52] revealed the evolution of biochar at a molecular level as the temperature increased. Biochar successively goes through four distinct states, as shown in Figure 4: (i) transition char, in which the constituents of the plant material are partly preserved; (ii) amorphous char, in which the heat-altered molecules and incipient aromatic polycondensates are randomly mixed; (iii) composite char, in which poorly ordered graphene stacks are embedded in amorphous phases; and (iv) turbostratic char, in which disordered graphitic crystallites dominate. McBeath et al. [46] reported that the stable polycyclic aromatic carbon (SPAC) fraction was 20% of total organic carbon (TOC) at <450 °C and >80% of TOC at >600 °C. In general, an elevated temperature could promote the formation of SPAC, and hence, enhance biochar stability. In order to achieve a balance between biochar yield and biochar stability, the optimal temperature was suggested in the range of 500–700 °C for the commonly used feedstock, such as agricultural and forestry residues [46].

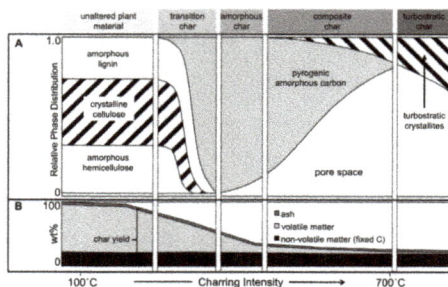

Figure 4. The evolution of biochar as the pyrolysis temperature rises: (**A**) physicochemical characteristics of organic phases; (**B**) char composition. Reprinted with permission from Reference [52]. Copyright (2010) American Chemical Society.

2.2. Hydrothermal Carbonization

2.2.1. Process Description

A hydrothermal process is normally performed in a closed reactor at temperatures above 180 °C in the presence of water. The temperature greatly affects the product distribution. Biochar [53–55], bio-oil [56,57], and gaseous products [58–60] are predominantly produced when the temperature is below 250 °C, between 250–400 °C, and above 400 °C, respectively. Accordingly, the three processes are termed as hydrothermal carbonization (HTC), hydrothermal liquefaction, and hydrothermal gasification, respectively.

The solid product from HTC is usually termed hydrochar in order to distinguish it from the char from pyrolysis (i.e., pyrochar) [13]. Depending on biomass feedstock and hydrothermal conditions, the hydrochar yield varies from 30% to 85% [9,17,61–63]. There is a complex reaction network which involves dehydration, retro-aldol condensation, isomerization, and so on [9]. Possible formation pathways of biochar during HTC were summarized and compared with conventional pyrolysis in Reference [13]. Detailed information on the hydrothermal degradation of biomass components can be found in References [9,64].

Compared to the biochar from pyrolysis, the biochar from HTC has its own characteristics [64,65]. Firstly, the hydrochar possesses more surface functional groups. Secondly, spherical micro-sized particles with a limited porosity, as shown in Figure 5, are typically produced during HTC. Due to the small pore size, the surface area and porosity analysis of hydrochar are normally performed using the CO_2 isothermal adsorption technique instead of N_2 adsorption.

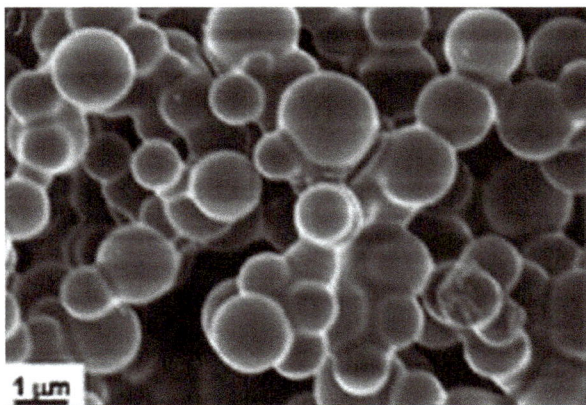

Figure 5. SEM image of biochar prepared from the hydrothermal carbonization of sucrose. Reprinted with permission from Reference [9]. Copyright (2012) Royal Society of Chemistry.

The production of biochar from HTC has several advantages over conventional pyrolysis. Firstly, biomass can be decomposed more easily in an aqueous medium than under pyrolysis conditions [64,66]. For example, the decomposition of hemicellulose and cellulose occur at 180 °C and 230 °C, respectively, under hydrothermal conditions [67,68]. In contrast, under pyrolysis conditions, hemicellulose starts decomposing between 200 and 250 °C, and totally degrades at 300 °C for 2 h. Cellulose begins decomposing at about 270–300 °C [69]. Secondly, there is no need to dry biomass feedstock for HTC, while energy intensive drying is usually required for conventional pyrolysis. Thus, HTC is very suitable for processing feedstock with high moisture contents such as aquatic plants and algae [9,10]. Lastly, HTC is a convenient technology for coating pre-formed nanostructures with carbonaceous shells.

2.2.2. Influencing Factors

(1)   Biomass Feedstock

Monosaccharides and disaccharides are the common feedstock of HTC [70]. Pentoses and hexoses have different HTC mechanisms, with furfural and hydroxymethyl furfural (HMF) as their dehydration products, respectively [71]. Consequently, they yield hydrochar with different sizes. The hydrochar from hexoses (e.g., fructose and glucose) is a spheroid with uniform micrometer size, while the hydrochar from pentoses (e.g., xylose) has a size ranging between 100 and 500 nm. Moreover, ^{13}C solid-state NMR spectroscopic analysis indicated that the pentose-derived hydrochar was more condensed than the hexose-derived hydrochar [9]. As for a disaccharide consisting of two types of monosaccharides, the hydrochar normally has a wide size distribution [72]. This is because different monosaccharides have different kinetics of char formation. Taking sucrose for instance, the size of the hydrochar particles varies from 700 nm to 2 μm [72].

The HTC mechanisms of cellulose and lignocellulosic biomass are fundamentally different from that of glucose [70]. As shown in Figure 6, the polyfuranic intermediate, which is a characteristic of D-glucose-derived hydrochar at either low temperatures or short reaction times, cannot be observed in cellulose-derived hydrochar. In contrast, the cellulose- and the biomass-derived hydrochar show a well-developed aromatic nature, even at the early stage of HTC. Lignin, a component of lignocellulosic biomass, is mildly affected by HTC, and hence, maintains the natural macrostructure of the initial biomass into the final hydrochar.

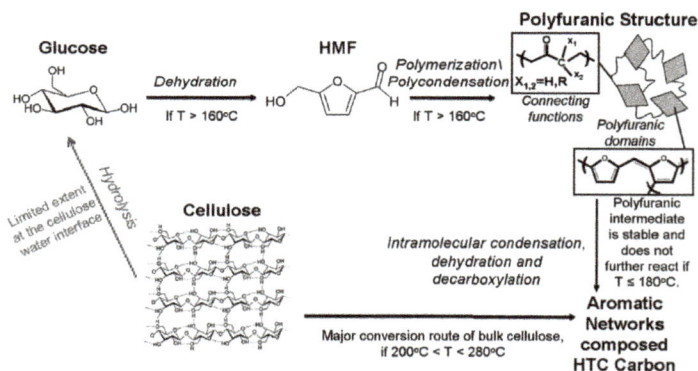

Figure 6. Different hydrothermal carbonization mechanisms for glucose and cellulose under mild processing conditions (180 °C < T < 280 °C). Reprinted with permission from Reference [70]. Copyright (2011) Royal Society of Chemistry.

HTC is also applied to various real biomass feedstocks such as paper, food waste, mixed municipal solid waste, anaerobic digestion waste, and olive mill wastewater [61,73]. An elevated operating temperature (e.g., >200 °C) is required for the HTC of real biomass compared with sugars and even pure cellulose [70]. Moreover, the conversion pathways of real biomass to biochar are difficult to construct because of the numerous intermediates.

(2)   Hydrothermal Conditions

The influencing factors of an HTC process include temperature, residence time, and biomass/water ratio [63,70]. Titirici et al. [70] demonstrated the effects of temperature on glucose HTC. A change in the temperature allowed a certain degree of control over the char diameter and the char size distribution. Furthermore, raising the temperature was able to switch the chemical structure of the char from polyfuran rich in oxygen-containing functional groups to aromatic networks. Sabio et al. [63]

investigated the influence of the processing variables on the HTC of tomato peel. A design of experiments/response surface methodology approach was implemented to identify the importance of each variable. Temperature was identified as the major variable, followed by residence time. An increase in both variables markedly decreased the solid yield, but promoted energy densification. In contrast, the biomass/water ratio had a minor effect. In general, a high HTC temperature may result in a low char yield, but a relatively low temperature would lead to the incomplete conversion of biomass and a lower degree of aromatization. Therefore, a multi-step process involving different temperature regimes is suggested for the HTC of real biomass with complex composition [9].

3. Characteristics of Biochar

The properties of biochar, which are affected by the biomass feedstock and preparation conditions, determine its potential for a specific application. For example, biochar with a high electrical conductivity and porosity is suitable for electrode materials [74]. Biochar with a high porosity and structurally bound nitrogen groups is preferred for supercapacitor electrode materials [75]. When biochar is intended as a catalyst or catalyst support, its matrix nature, surface functionality, and intrinsic inorganics are key factors.

3.1. Bulk Elements and Inorganics

In general, the carbon content of biochar is in the range of 45–60 wt%, lower than that of carbon black (>95%) and coal-derived activated carbon (80–95%) [15,76,77]. A considerable amount of hydrogen and oxygen elements also exist in biochar. Another feature of biochar is the presence of small amounts of inorganic species such as K, Na, Ca, Mg, Si, Al, Fe, etc. The quantity and the composition of inorganics are highly dependent on the nature of the raw biomass. Woody biomass-derived biochar usually shows a significantly lower content of inorganics than biochar from either herbaceous biomass or hydrophyte biomass [37,77]. Some inorganics present in biochar play a crucial role in the catalytic applications of biochar [6], such as for tar cracking [12], bio-oil upgrading [28], and methane decomposition [45].

3.2. Chemistry of Biochar Matrix

The biochar matrix is mainly amorphous with some local crystalline structures of highly conjugated aromatic sheets. These aromatic sheets are cross-linked in a random manner as shown in Figure 7. With the processing temperature rising, the size of biochar crystallites increases, and the entire structure becomes more ordered [78]. Additionally, heteroatoms such as N, P, and S from feedstock may be incorporated into the biochar aromatic structure. The difference in the electronegativity between these heteroatoms and the aromatic C causes the chemical heterogeneity of biochar, which plays an important role in catalytic applications [79,80].

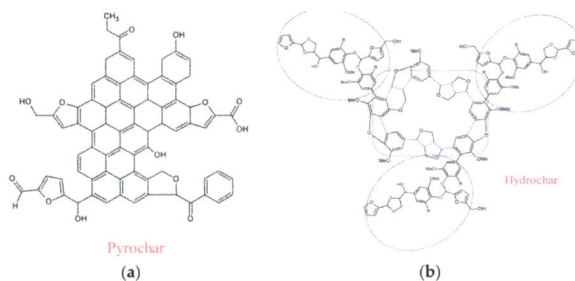

Figure 7. Chemical structures of pyrochar and hydrochar: (**a**) chemical structure of pyrochar; (**b**) chemical structure of hydrochar. Reprinted from Reference [13] published by The Royal Society of Chemistry.

3.3. Surface Functional Groups

In contrast with other carbon materials (e.g., activated carbon and carbon black), biochar usually possesses abundant surface functional groups, as Figure 8 depicts. These surface functional groups are very helpful for biochar functionalization [15,81]. For example, the loading of metal precursors into biochar, a critical step for the synthesis of biochar-supported metal catalysts, can be facilitated by surface functional groups clinching metal cations [9,13,14]. In addition, some surface functional groups can improve the performance of biochar-based catalysts for certain reactions. A typical example is shown in biochar-based solid-acid catalysts. Kitano et al. [82] reported that sulfonated carbon material could catalyze the hydrolysis of cellohexaose more effectively than SO_3H-bearing resins. This was because phenolic OH and COOH groups in the carbon material functioned as adsorption sites. They concluded that the synergetic combination of the functional groups in biochar could lead to an efficient hydrolysis of β-1,4-glucan, including cellulose, on biochar-based solid acids.

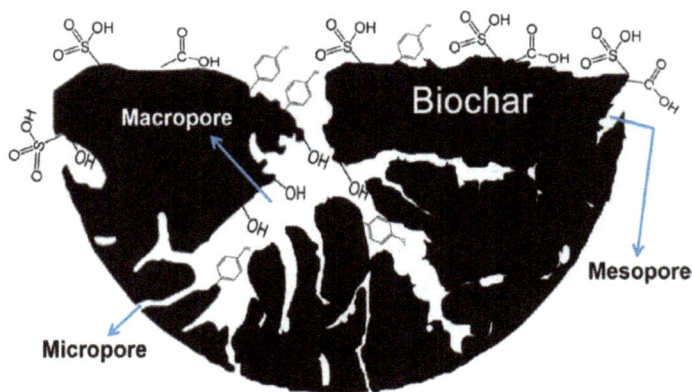

Figure 8. Schematic of porous biochar containing various functional groups. Reprinted with permission from Reference [11]. Copyright (2017) Elsevier Ltd.

4. Modifications of Biochar

4.1. To Produce Porous Structures

As depicted in Figure 9, heterogeneous catalysis on a porous catalyst usually undergoes several steps [1,11,83]: (1) reactants diffuse from the bulk to the external surface of the catalyst; (2) reactants diffuse to the internal surface through pores; (3) reactants are adsorbed on the catalyst; (4) reactions occur at the active sites; (5) products are desorbed from the catalyst surface; (6) products diffuse to the external surface of the catalyst through catalyst pores; and (7) products diffuse from the external catalyst surface to the bulk. Therefore, the performance of a catalyst highly relies on the accessibility to active sites and the quantity of active sites [83]. An appropriate pore size distribution can improve the catalytic selectivity. A large specific surface area can enhance the dispersity of active substances. A three-dimensional (3D) interconnected porous structure will facilitate the diffusion of reactant and product molecules. In general, the porosity and the surface area are two important properties of catalyst materials. However, originally formed biochar has limited surface area and porosity [77], which hinder its applications in catalysis. This section reviews approaches to preparing porous biochar.

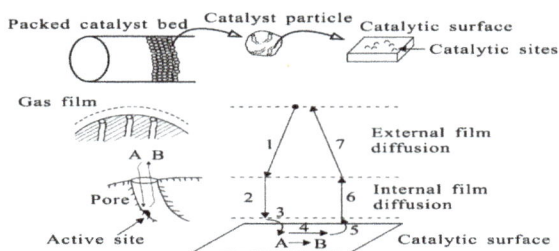

Figure 9. Schematic of the heterogeneous catalysis process on porous catalysts. Reprinted with permission from Reference [1]. Copyright (2009) Royal Society of Chemistry.

4.1.1. Physical Activation

Physical activation is usually carried out by exposing the as-prepared biochar to a flow of gasifying agents (steam, CO_2, or their mixture) at temperatures above 700 °C. During the process, carbon atoms with a higher reactivity are removed via Reaction (1) or (2) [10,15]. As a result, the porosity and the surface area of biochar are increased. Steam activation is more efficient than CO_2 activation, probably due to the faster reaction kinetics between carbon and steam. Duan et al. [84] investigated the influence of different heating methods on the yield and the pore nature of physically activated biochar. Compared to conventional heating, microwave heating could significantly increase the biochar yield for CO_2 activation, but showed a negligible effect for steam activation. In addition, the pore volume and the surface area of steam-activated biochar were doubled when using microwave heating instead of conventional heating, while the two morphology parameters of CO_2-activated biochar were not changed remarkably.

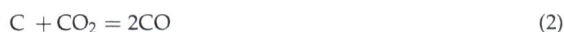

$$C + H_2O = CO + H_2 \tag{1}$$

$$C + CO_2 = 2CO \tag{2}$$

Table 1 shows the effects of physical activation on the structural characteristics of biochar. For instance, Alvarez et al. [85] demonstrated the Brunauer–Emmett–Teller (BET) surface area of rice husk char was enhanced from 227 to 1365 m^2/g by steam activation at 800 °C for only 15 min. They found that long activation times would lead to a decrease in the surface area due to the growth and eventual destruction of micropores. Yang et al. [86] increased the surface area of coconut shell char from 702 to above 2000 m^2/g using physical activation (steam, CO_2, or their mixture) under microwave heating. Moreover, they found that an increase in the activation time would lead to a larger BET surface area and a larger pore volume irrespective of the activation agent.

Table 1. The changes in surface area and pore volume of biochar caused by physical activation. BET: Brunauer–Emmett–Teller.

Raw materials	Carbonization (C) or Activation (A) Conditions	BET Surface Area (m^2/g)	Pore Volume (cm^3/g)
Rice husk [85]	C: flash pyrolysis at 500 °C followed by silica removal	227	0.17
	A: 800 °C 15 min steam	1365	1.2
Coconut shell [86]	C: 1000 °C 2 h N_2	702	0.53
	A: 900 °C by microwave heating, 75 min, steam	2079	1.212
	A: 900 °C by microwave heating, 210 min, CO_2	2288	1.299
	A: 900 °C by microwave heating, 75 min, steam + CO_2	2194	1.293

Table 1. *Cont.*

Raw materials	Carbonization (C) or Activation (A) Conditions	BET Surface Area (m^2/g)	Pore Volume (cm^3/g)
Rice straw [87]	C: 500 °C 1 h N$_2$	139.5	0.092
	A: 700 °C 1 h steam	363	0.164
Sewage sludge [87]	C: 500 °C 1 h N$_2$	18	0.018
	A: 700 °C 1 h steam	64	0.039
Sewage sludge with an acid washing treatment [88]	C: 700 °C 30 min N$_2$ followed by acid washing	188	0.09
	A: 800 °C 2–4 h CO$_2$	269	0.11
Jatropha hull [84]	C: 600 °C 1 h N$_2$	480	0.42
	A: 900 °C 22 min steam	748	0.53
	A: 950 °C 40 min CO$_2$	1207	0.86
Walnut shell [89]	C: 600 °C 1 h N$_2$	280	0.16
	A: 850 °C 30 min steam	792	0.524
Almond tree pruning [89]	C: 600 °C 1 h N$_2$	204	0.118
	A: 850 °C 30 min steam	1080	0.95
Almond shell [89]	C: 600 °C 1 h N$_2$	42	0.094
	A: 850 °C 30 min steam	601	0.375
Olive stone [89]	C: 600 °C 1 h N$_2$	53	0.036
	A: 850 °C 30 min steam	813	0.555

4.1.2. Chemical Activation

Chemical activation is usually performed by adding activators to biomass precursors followed by a routine carbonization. In this case, chemical activation and carbonization occur simultaneously (in situ activation). Sometimes, chemical activators are added to pre-formed biochar followed by a thermal treatment [15,90–92]. The post-activation method is especially useful for biochar from HTC [64]. Compared with physical activation, chemical activation can create porous structures at a relatively lower temperature [87,90]. However, several issues, such as equipment corrosion, chemical recycling, product purification, etc., need to be considered when using chemical activation. Common activators include KOH, K$_2$CO$_3$, H$_3$PO$_4$, ZnCl$_2$, etc. Different activators may have different activation mechanisms.

(1) Alkali Metal Activators

Alkali metal activators include alkali hydroxides (e.g., NaOH, KOH, and LiOH), alkali carbonates (e.g., Na$_2$CO$_3$ and Li$_2$CO$_3$), alkali hydro-carbonates (e.g., KHCO$_3$), etc.

When KOH is used as an activator, Reactions (3)–(6) may occur [8,15,93]. Pores are generated due to a synergistic effect of several factors. Firstly, KOH, as well as its intermediates K$_2$CO$_3$ and K$_2$O, etches biochar through chemical Reactions (3), (5), and (6). Secondly, the escape of gaseous products (e.g., CO and H$_2$) from the biochar interior leads to the formation of macropores, similar to a leaving process [41]. Furthermore, H$_2$O and CO$_2$ in situ formed during carbonization may cause a physical activation. Lastly, the metallic K produced intercalates into the biochar matrix during activation, resulting in an expansion of the biochar lattice [94].

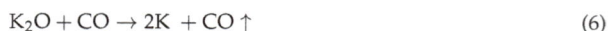

$$6KOH + 2C \; \rightarrow 2K + 3H_2 \uparrow + 2K_2CO_3 \tag{3}$$

$$6KOH + CO_2 \; \rightarrow K_2CO_3 + H_2O \uparrow \tag{4}$$

$$K_2CO_3 + 2C \; \rightarrow K_2O \; + 2CO \uparrow \tag{5}$$

$$K_2O + CO \rightarrow 2K \; + CO \uparrow \tag{6}$$

$$2KHCO_3 \rightarrow K_2CO_3 + CO_2 + H_2O \qquad (7)$$

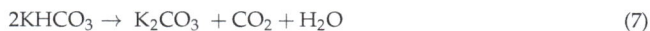

Cha et al. [87] reported that both rice-straw char and sewage char were effectively activated by KOH, and increases in their surface areas from 140 to 772 m^2/g and from 18 to 783 m^2/g were achieved, respectively. Bhandari et al. [95] enhanced the effect of KOH activation by sonicating a mixture of biochar and KOH solution. Sevilla et al. [96] employed KOH to activate an HTC biochar and achieved a high surface area (up to 2700 m^2/g) and a narrow pore size distribution (PSD; 0.7–2 nm). Li et al. [41] created both macropores and mesopores on the biochar of sisal leaves using LiOH and Li$_2$CO$_3$ activators, as Figure 10 shows. The formation of pores was attributed to the chemical corrosion of LiOH and Li$_2$CO$_3$ on biochar, similar to Reactions (3) and (5).

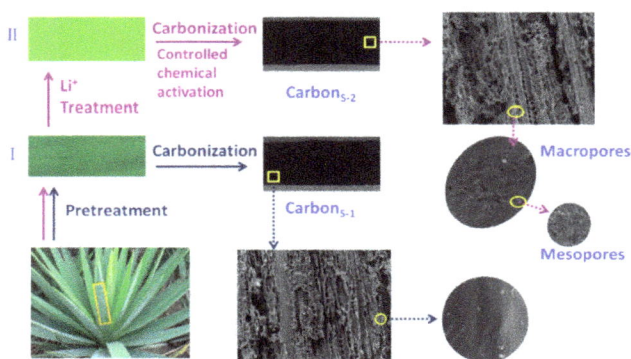

Figure 10. The preparation of porous biochar via simultaneous carbonization and controlled chemical activation. Reprinted from Reference [41].

KHCO$_3$ was also used as an activator for the synthesis of hierarchically porous biochar through a simple one-pot approach [97]. Using this method, biochar with 3D hierarchical pores (macro-, meso-, and micropores) and a large surface area (up to 1893 m^2/g) could be obtained. Macropores were formed due to the gas evolution from KHCO$_3$ decomposition (Reaction (7)), resembling bread fermentation. Mesopores and micropores were generated through reactions (5) and (6).

(2) H$_3$PO$_4$ Activator

H$_3$PO$_4$ is another commonly used activator for the production of porous biochar. Lobos et al. [98] reported that a pre-treatment of cellulose with 5% H$_3$PO$_4$ prior to pyrolysis led to an increase in the surface area of biochar from 199 to 557 m^2/g, and an increase in the pore volume from 0.026 to 0.22 cm^3/g. Moreover, the concentration of structural defects (i.e., terraces, steps, and kinks) in the biochar was also increased, which promoted the dispersion of metal precursors onto the biochar support. Shen et al. [20] also demonstrated that the surface area of biochar from cotton stalks could be significantly enhanced from 516 to 1955 m^2/g using H$_3$PO$_4$ activation.

The addition of H$_3$PO$_4$ into biomass precursors affects biomass pyrolysis [15]. At temperatures below 200 °C, H$_3$PO$_4$ promotes the hydrolysis of hemicellulose and cellulose, and cleaves the aryl ether bonds of lignin. Additionally, H$_3$PO$_4$ helps cross-link the polymer chains by forming ester linkages with −OH groups [99]. As the temperature goes up, the P−O−C bonds are broken, and cyclization and condensation reactions occur, leading to an increase in the aromaticity of biochar. Furthermore, H$_3$PO$_4$ can improve the separation of cellulose microfibers by swelling the biomass structure, resulting in the formation of open pores.

(3)   ZnCl$_2$ Activator

The activation mechanism of ZnCl$_2$ was summarized by Reference [15]. ZnCl$_2$ can intrude into the biomass interior at low temperatures, and it maintains in a liquid state throughout the entire pyrolysis process (<700 °C). As a result, ZnCl$_2$ is evenly distributed in the matrix of the resulting biochar. By removing ZnCl$_2$, a well-developed microporous biochar could be obtained. In addition, ZnCl$_2$ could decrease the carbonization temperature of biomass components due to its strong dehydration capacity at elevated temperatures. Furthermore, the presence of ZnCl$_2$ changes the decomposition pathways of biomass, and suppresses the tar formation.

Ucar et al. [100] employed ZnCl$_2$ to prepare activated biochar from pomegranate seeds. The surface area was strongly influenced by the ZnCl$_2$ impregnation ratio and the carbonization temperature. When using an impregnation ratio of 2.0 and a carbonization temperature of 600 °C, the specific surface area of the resultant biochar was as high as 978.8 m^2/g. Sun et al. [101] prepared sheet-like graphitic biochar with a surface area of 1874 m^2/g and a total pore volume of 1.21 cm^3/g by simultaneously applying a ZnCl$_2$ activator and a graphitic catalyst precursor, FeCl$_3$, on coconut shell biomass waste.

4.1.3. Templating Methods

A hard-templating approach (also termed nanocasting) is normally carried out as follows: firstly, a pre-formed porous solid template is impregnated with a biomass precursor and subjected to a carbonization procedure. Then, the original scaffold is removed, and a carbonaceous porous replica is obtained. This approach is especially useful for the production of porous biochar via HTC. Titirici's research group does a lot of work in this field [102–105]. When using silica template Si-100 in the HTC of glucose or furfural [102], biochar with different morphologies (e.g., mesoporous microspheres, mesoporous hollow spheres, carbon nanoparticles, and macroporous casts) was obtained by simply altering the polarity of the silica surface. When adding silica template SBA-15 into the HTC of furfural, ordered mesoporous biochar with functional groups residing at the pore surface was fabricated [103]. Uniform, open-ended carbonaceous tubular nanostructures were also prepared using a macroporous alumina membrane template in the HTC of furfural [104]. In order to remove templates, harsh and non-environmentally friendly solvents are usually used in the hard-templating method, whereas a soft-templating approach does not have this problem. Kubo et al. [106] fabricated both ordered microporous and mesoporous biochar materials via a combined HTC/soft-templating approach with the block copolymer, Pluronic, as a structure-directing agent, and D-fructose as a carbon source. White et al. [105] prepared hollow carbonaceous nanospheres by adding latex into the HTC of glucose. Figure 11 displays three typical porous structures synthesized via the templating method.

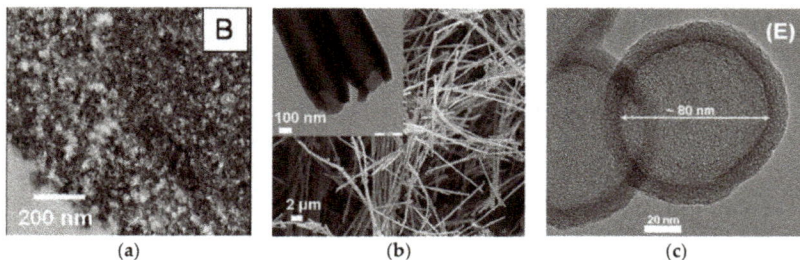

(a)   (b)   (c)

Figure 11. TEM images of porous biochar prepared via templating methods: (a) mesoporous microspheres, reprinted with permission from Reference [102]. Copyright (2007) John Wiley and Sons; (b) tubular structures, reprinted with permission from Reference [104]. Copyright (2010) American Chemical Society; (c) hollow nanospheres, reprinted with permission from Reference [105]. Copyright (2010) American Chemical Society.

In addition to artificial templates, inorganic components naturally present in biomass precursors are also employed to aid the formation of pores. For instance, prawn shells, a high-volume food waste, were successfully converted to mesoporous biochar via an HTC process followed by the removal of its inherent mineral $CaCO_3$ with acid washing [107].

4.2. Surface Functionalization

4.2.1. Surface Acidification

The surface acidification of biochar is usually used to synthesize biochar-based solid acid, a catalyst for reactions such as polysaccharide hydrolysis [5,108–113], esterification, and transesterification [2,114–116].

Sulfonation is a common method for surface acidification. Its product, biochar bearing SO_3H, represents a typical class of solid Brønsted acids. The sulfonation of biochar can be implemented via two approaches. The first approach is to immerse biochar in sulfuric acid solution [2,5,112–114,116,117]. The mixture is stirred at temperatures between 90 and 200 °C followed by washing and drying. Using this method, the concentration of acid sites on the biochar surface can be up to 2.5 mmol/g [118]. The second approach is to expose biochar to SO_3 gas (> 20%) at temperatures between 25 and 150 °C [2,114]. Kastner et al. [114] compared the two sulfonation methods. Their results indicated that SO_3 sulfonation could generate higher –SO_3H densities, whereas H_2SO_4 sulfonation could clearly increase the surface area and pore volume of biochar. Other studies [116,119] indicated that both sulfonation methods would lead to a reduction in the surface area of the biochar.

In addition, copolymerization under hydrothermal conditions can also introduce acid groups into the biochar. Liang et al. [120] prepared SO_3H-bearing biochar through the one-step HTC of furfural and hydroxyethyl-sulfonic acid at 180 °C for 4 h. Qi et al. [108] synthesized biochar-based solid acids containing –SO_3H and –COOH groups via the in situ hydrothermal copolymerization of glucose with sulfosalicylic acid and acrylic acid, respectively.

4.2.2. Surface Amination

The surface amination of biochar can be performed via three methods. The common method is to treat biochar with NH_3 at elevated temperatures [121–124]. In order to increase the surface reactivity of biochar toward NH_3, surface oxidation is usually conducted prior to surface amination. This method is simple, but attention needs to be paid to the leakage of NH_3 during handling. Another method is to anchor amino-containing reagents such as tris(2-aminoethyl)amine [125] and polyethylenimine [126] on the biochar surface through condensation reactions between amino or imino groups and carboxyl groups. Surface amination can also be achieved through nitration followed by reduction. Using this method, nitro groups are first introduced into the biochar surface, before being reduced to amino groups [15].

4.2.3. Surface Oxidation

Surface oxidation is to produce extra oxygenated functional groups (e.g., carboxyl, hydroxyl, phenolic groups, etc.) on the biochar surface. It is usually carried out by thermally treating biochar under aerobic conditions. Chen et al. [127] markedly increased the concentration of oxygen-containing functional groups, especially carboxyl groups (from 0.53 to 3.70 mmol/g) on the surface of hydrochar by simply heating the hydrochar at 300 °C in air. In addition to thermal treatment in air, the utilization of oxidizing reagents such as H_2O_2, O_3, HNO_3, and $KMnO_4$ can also achieve surface oxidation [128–131]. Li et al. [131] modified the surface of bamboo-derived biochar via the chemical method. A large number of acidic functional groups were created on the biochar surface, with HNO_3 being more effective than $KMnO_4$. They also found that the chemical treatment increased the hydrophilicity of biochar, while the thermal treatment caused an opposite effect.

4.3. In Situ Heteroatom Doping

Heteroatom doping (e.g., N [107,132–134], S [135], and P [136]) can alter the physicochemical properties of carbonaceous materials and improve their performance in catalysis, adsorption, or energy storage. For example, Nagy et al. [134] reported that the electrocatalytic activity of mesoporous carbon aerogels for oxygen reduction reactions increased with the content of N doping. Various strategies were proposed to incorporate N into biochar. Compared to annealing biomass precursors under NH_3 [132] or with NH_4NO_3 [137], the direct carbonization of N-containing biomass is more favorable [138,139] because of the facile operation and sustainable feedstock.

4.3.1. Utilize Amino-Containing Biomass Feedstock

Amino-containing carbohydrates are common feedstock for the production of N-doped biochar. Zhao et al. [138] successfully synthesized N-doped biochar (>8.9% N and 6.6% N) through the HTC of chitosan and D-glucosamine, respectively, at 180 °C. Under hydrothermal conditions, D-glucosamine firstly decomposed into HMF and ammonia, before forming an aromatic network containing pyrrole and pyridine-type N functionalities. Prawn shells and king-crab shells, natural sources of chitin (poly-β(1-4)-N-acetyl-D-glucosamine), were also successfully utilized to fabricate N-doped porous biochar via HTC [107] or pyrolysis [140].

4.3.2. Add N-Containing Molecules to Biomass Feedstock

Another method for in situ N-doping is to add N-containing molecules such as protein and amino acids into biomass feedstock. White et al. [141] reported the hydrothermal synthesis of N-doped monolithic carbon aerogels with D-glucose and ovalbumin as feedstock. The protein acted as not only an N donor, but also a surface stabilizing agent during the HTC of glucose, resulting in the formation of high-surface-area aerogels with continuous mesopores. Baccile et al. [142] prepared N-doped biochar via the HTC of glucose and glycine, and studied the chemical structure of the resulting biochar. They found N-containing aromatic domains and polyfuran networks coexisted in this material. In contrast to biochar obtained from pure carbohydrates, the N-containing biochar displayed a higher degree of aromatization. Algae, a class of N-containing biomass abundant on Earth, can also be used as a secondary biomass precursor for the production of N-doped biochar [9].

4.4. To Form Composites

For some reactions, biochar itself has little catalytic activity, and thus, it is necessary to load the biochar with active substances. This section summarizes the synthesis method of three common biochar-based composites.

4.4.1. Biochar/Metal Composites

(1) One-Step Method

Biochar/metal composites can be synthesized by subjecting the mixture of biomass and metal precursor to a carbonization condition (pyrolysis [143–147] or HTC [148–150]). During the process, the biomass is converted to biochar and the high-valence metal precursor is reduced in situ [15]. The one-step approach has several advantages. Firstly, the operation is simple. Secondly, the gaseous or solid products of biomass decomposition could act as reducing agents, thus avoiding the use of additional reagents (atom economy). Lastly, the metal nanoparticles formed can catalyze the biomass carbonization and improve the quality of the biochar [143–145,151].

Various biochar/base-metal composites were prepared using this method. Richardson et al. [143] synthesized biochar-supported Ni metal nanoparticles (NPs) by pyrolyzing Ni^{2+}-impregnated wood between 400 and 500 °C. The formation mechanism of Ni NPs was deeply investigated. Firstly, Ni ions were adsorbed onto the biochar with surface functional groups as adsorption sites. Then, an amorphous $Ni_xO_yH_z$ phase was formed during the pyrolysis process. Lastly, the $Ni_xO_yH_z$ phase was reduced to metallic Ni (Ni^0) by gaseous or solid products of biomass pyrolysis. Their group [144] later found that the in situ generated Ni^0 NPs exhibited a higher catalytic activity for tar conversion than pre-formed Ni^0 NPs that were inserted into the biomass prior to pyrolysis. Liu et al. [147] utilized fir sawdust to absorb metal ions (i.e., Cu^{2+} and Fe^{3+}) from synthetic wastewater, and then converted the metal-polluted biomass to metallic Cu-anchored magnetic biochar (Cu&Fe$_3$O$_4$-mC) via fast pyrolysis.

Similarly, the one-step method was also employed to prepare biochar/noble-metal composites. Yu et al. [150] synthesized noble metals supported on various carbon nanoarchitectures (e.g., nanocables, hollow tubes, and hollow spheres) via the mild HTC of starch in the presence of noble-metal salts. Here, starch acted as the carbon source and reducing agent. Noble-metal salts functioned as the metal source and the catalyst for biomass carbonization. Makowski et al. [148] prepared a Pd@hydrophilic-C nanocomposite via the HTC of furfural and palladium acetylacetonate at 190 °C for 14 h. Yu et al. from the Chinese University of Hong Kong [149] demonstrated the one-pot synthesis of coaxial Ag/C nanocables via the microwave-assisted HTC of sucrose and AgNO$_3$. Fang et al. [152] prepared Ag/C core/shell nanoparticles and nanocables via the HTC of ascorbic acid in the presence of Ag$^+$ and cetrimonium bromide (CTAB). The concentration of CTAB determined the final morphology of the product.

(2) Two-Step Method

Biochar/metal composites can also be prepared using a two-step procedure. The first step is biomass carbonization (pyrolysis or HTC) to produce biochar. The second step is to impregnate the as-formed biochar with metal salts followed by the transformation of metal ions to metals through carbothermal reduction [153] or surface redox reactions [154].

Yao et al. [47] prepared a biochar-supported Ni catalyst by impregnating biochar with an aqueous solution of Ni(NO$_3$)$_2$·6H$_2$O, then drying at 105 °C overnight, followed by calcination under N$_2$ atmosphere at 800 °C. Their results suggested the Ni catalyst was efficient for hydrogen production from biomass steam gasification. Yan et al. [31] prepared a carbon-encapsulated Fe catalyst by thermally treating Fe(NO$_3$)$_3$-impregnated pine-wood char at 1000 °C for 1 h. Shen et al. [153] prepared biochar-supported Ni/Fe bimetallic catalysts by using the two-step approach with Fe(NO$_3$)$_3$·9H$_2$O and Ni(NO$_3$)$_2$·6H$_2$O as metal precursors. Figure 12 illustrates the preparation process.

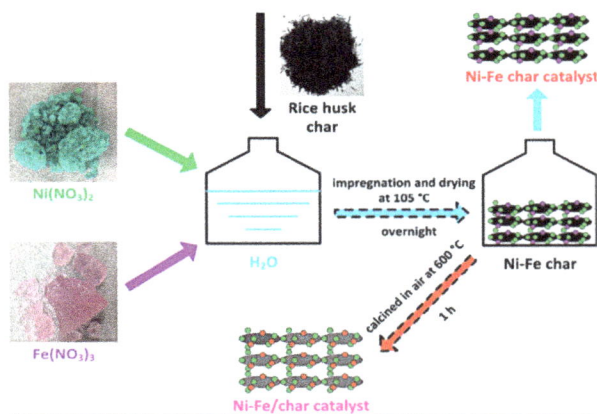

Figure 12. Schematic of a two-step method for the synthesis of an Ni/Fe–char catalyst, reprinted with permission from Reference [153]. Copyright (2014) Elsevier B.V.

Biochar derived from HTC has abundant surface functionalities. These surface functional groups could adsorb and in situ reduce noble-metal ions at a relatively low temperature, resulting in the formation of noble-metal/biochar hybrids [154–157]. Using this method, Sun and Li [154] successfully loaded Ag and Pt NPs onto the surface of colloidal hydrochar spheres at room temperature or in a reflux setup. After the metal loading, the number of reducing functional groups (e.g., –OH) on the hydrochar surface was reduced due to the occurrence of a surface redox reaction. Qian et al. [156] decorated carbon nanofibers prepared via HTC with fine noble-metal NPs such as Pd, Pt, and Au, through a redox reaction between the carbon fibers and noble-metal salts. These carbon-nanofiber-supported noble metals possessed stability, a large surface area, and high chemical reactivity. Therefore, they were considered as ideal candidates for heterogeneous catalysis.

(3) Coating Pre-Formed Metals with Biomass Followed by Carbonization

Biochar/metal composites can also be prepared by coating the pre-formed metal nanostructures with biomass before carbonization. This method is very useful for cases where (1) the carbonization condition (e.g., HTC) is unable to reduce metal ions; and (2) metals with special nanostructures are desired, but cannot be achieved by in situ growth. Sun and Li [154] prepared Ag-cored biochar spheres using Ag NPs as nuclei for the growth of biochar spheres under HTC conditions. Qian et al. [158] successfully synthesized uniform Te@C nanocables by adding pre-formed Te nanowires into an HTC container of D-glucose. The Te nanowires not only restrained the homogeneous nucleation of carbonaceous spheres, but also promoted the deposition of a carbonaceous shell. Moreover, the diameter of the final nanocables could be easily adjusted by changing the HTC duration or the ratio of Te to D-glucose.

4.4.2. Biochar/Carbide Composites

When metal-salt-impregnated biomass or biochar is subjected to a thermal treatment, the metal element may transform in the following order with the temperature rising: metal salt $\rightarrow$ metal oxide $\rightarrow$ metal $\rightarrow$ metal carbide. Biomass or biochar acts not only as a reducing agent, but also as a C-provider for the production of metal carbides.

Yan et al. [159] synthesized tungsten carbide (WC) NPs that were embedded in a biochar matrix by thermally processing ammonium-tungstate-impregnated biochar at 1000 °C under N_2. The WC NPs were produced through the following process: $WO_3 \rightarrow WO_2 \rightarrow W \rightarrow W_2C \rightarrow WC$. Zhang et al. [160] prepared a mesoporous biochar-supported WC_x catalyst by impregnating the biochar with ammonium metatungstate followed by carbothermal hydrogen reduction at 900 °C. In addition to WC_x, other carbides such as Fe_3C [39], SiC [161,162], B_4C [163,164], TiC [165,166], and MoC [167] were also prepared in a biochar matrix using this method.

4.4.3. Biochar/Nanostructured Carbon Material Composites

The hybridization of biochar with other nanostructured carbon materials can achieve unique physiochemical properties. Surface functional groups and inorganic species present in biochar are beneficial to this combination to some extent. Inyang et al. [168] coated biochar with multi-walled carbon nanotubes (CNTs) by impregnating biomass with carboxyl-functionalized CNT solutions followed by slow pyrolysis. The physiochemical properties (e.g., surface area, porosity, and thermal stability) of the resultant composite were much better than that of the biochar itself. Chen et al. [169] reported the direct synthesis of carbon nanofibers on activated biochar using the chemical vapor deposition of ethylene. The Fe present in the ash content of the biochar in situ catalyzed the growth of carbon nanofibers according to a "tip-growing" mechanism [170]. Consequently, Fe particles were on the tip of the carbon nanofibers, as displayed in Figure 13.

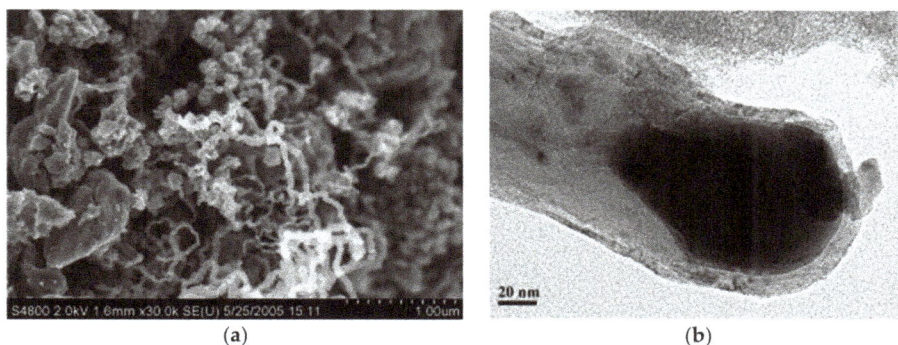

Figure 13. (a) SEM image of carbon nanofibers supported on biochar; (b) TEM image of a carbon nanofiber with an Fe particle on the tip. Reprinted with permission from Reference [169]. Copyright (2008) Elsevier Ltd.

5. Applications of Biochar-Based Catalysts for Biofuel Production

The use of biochar as a catalyst or catalyst support has several advantages [16]. Firstly, the production process of biochar is simple and low-cost owing to sustainable biomass resources and well-developed synthesis techniques. Secondly, the physicochemical properties of biochar can be easily tuned using various strategies. Thirdly, some intrinsic features of biochar, such as surface functional groups, the presence of inorganic species, a hierarchical structure inherited from biomass matrix, etc., may play an important role in its catalytic applications [14]. Finally, a synergic effect between active metals and the biochar support for its catalytic performance may exist in some cases [31,171]. This section discusses the application of biochar-based catalysts in biodiesel production, biomass hydrolysis, and tar reduction, which represent three main branches of biomass upgrading.

5.1. Biodiesel Production

Biodiesel is an ideal substitute for fossil diesel as it is biodegradable, nontoxic, renewable, and shows a similar fuel property to fossil diesel. Biodiesel can be produced from the transesterification of vegetable oils (Figure 14a) or the esterification of free fatty acids (FFAs) with alcohols (Figure 14b). Ethanol and methanol are commonly used alcohols.

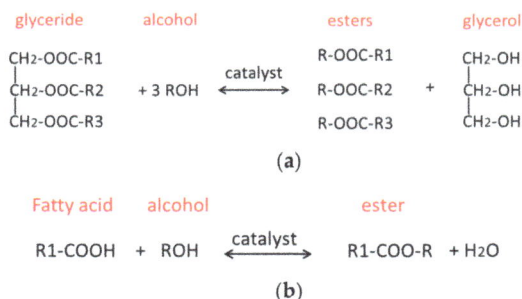

Figure 14. Reaction equations for biodiesel production: (a) transesterification of glyceride with alcohol; (b) esterification of fatty acid and alcohol.

Various catalysts were explored for biodiesel production. Common homogeneous catalysts include KOH, NaOH, H_2SO_4, and NaOMe. The recycling of catalysts and the purification of products are two problems involved in the use of homogeneous catalysts [172–175]. In contrast, heterogeneous catalysts such as CaO, MgO, ZnO, and CaZrO₃ can be easily separated from products and reused for several times. However, expensive metal precursors are required for the synthesis of these metal-oxide-based catalysts. Activated biochar is used to prepare catalysts for biodiesel production [10,11,27,172,176,177]. These biochar-based catalysts can be classified into two types: (1) solid-acid catalysts and (2) solid-alkali catalysts.

5.1.1. Solid-Acid Catalysts

Acid-functionalized biochar catalysts are usually prepared by sulfonating biochar with liquid H_2SO_4 or gaseous SO_3 (see Section 4.2.1). Examples of their catalytic activity and reusability in biodiesel production are shown in Table 2.

Zeng et al. [178] synthesized a solid-acid catalyst via the sulfonation of partially carbonized peanut shell. The acid strength of the catalyst was stronger than that of HZSM-5 (Si/Al = 75). When using the catalyst in the transesterification of cottonseed oil with methanol, a conversion of 90.2% and a good reusability was obtained. Joyleene et al. [179] also prepared a biochar-based solid-acid catalyst via KOH activation and fuming H_2SO_4 sulfonation. They found that the transesterification yield of canola oil was 44.2% at 150 °C and 1.52 MPa.

Low-grade or waste oils usually contain a large amount of FFAs, which are likely to lower the reaction rate and biodiesel yield. Therefore, the development of catalysts capable of simultaneously catalyzing esterification and transesterification is desirable [180]. Dehkhoda et al. [116] developed a biochar-based solid-acid catalyst for the simultaneous transesterification and esterification of a canola oil and oleic acid mixture at 150 °C under 1.52 MPa. The yield of alkyl ester could be up to 48% in 3 h if the proportion of the three reactants was appropriate. The reaction yield decreased by ~8% upon reusing the catalyst. Shu et al. [181] applied a solid acid derived from the sulfonation of carbonized vegetable oil asphalt for the conversion of a waste vegetable oil rich in free fatty acids to biodiesel. The maximum conversions of triglyceride and FFA reached 80.5 wt% and 94.8 wt%, respectively, after 4.5 h at 220 °C, when using a 16.8 molar ratio of methanol to oil and 0.2 wt% of catalyst to oil. Dawodu et al. [182] reported the conversion of non-edible seed oil, *Calophyllum inophyllum*, with 15% FFAs over sulfonated biochar. Under the optimized conditions, the biochar-based catalyst could convert 99.0% of the oil into biodiesel, which was comparable to conventional acid catalysts. González et al. [119] introduced –SO_3H groups to biochar using a microwave reactor. The microwave-assisted transesterification and esterification of waste cooking oil over this catalyst achieved a methyl ester yield of 75% in 15 min.

(1) The Number of Acid Sites

The number of acid sites on a biochar surface is one of the important parameters that determines its catalytic activity. González et al. [180] observed that an increase in the number of –SO_3H groups was responsible for a higher biodiesel yield. Shu et al. [181] also believed that a high density of acid sites was one of the factors that led to a high catalytic activity.

In order to load more –SO_3H groups, biochar with a large surface area and a higher porosity is desirable. Dehkhoda et al. [2] reported that the catalyst with a higher surface area showed a higher transesterification activity among the catalysts with similar acid densities. Kastner et al. [114] also observed the solid-acid catalysts derived from activated biochar had significantly higher activity than that of biochar without activation. In addition to surface area, intrinsic acidic functional groups of biochar also contribute to the total number of acid sites [82]. Both the surface area and the surface functional groups of biochar are affected by its carbonization condition. A high carbonization temperature could increase the surface area of the resultant biochar, but decrease the total acid density [179]. A high carbonization temperature also increases the rigidity of the biochar's carbon sheets, which makes the incorporation of –SO_3H groups more difficult. That was why Yu et al. [179] observed that the catalyst carbonized at 675 °C exhibited the highest catalytic performance compared to that carbonized at 450 °C or 875 °C. Therefore, the biochar obtained from incomplete carbonization is usually used for the preparation of biochar-based solid-acid catalysts [178,182].

Apart from the factors of biochar synthesis, the sulfonation methods also influence the number of acid sites. Dehkhoda et al. [2] tested the catalytic performances of two biochar-based solid acids in the transesterification of canola oil. The catalyst treated with fuming H_2SO_4 exhibited a much higher activity than that sulfonated with concentrated H_2SO_4. Kastner et al. [114] demonstrated that sulfonation of biochar with gaseous SO_3 at 23 °C yielded a higher –SO_3H density than sulfonation with concentrated H_2SO_4 at 100 °C.

Table 2. Biochar-based solid-acid catalysts for biodiesel production.

Catalysts (Feedstock, Synthesis Conditions)	Biodiesel Production (Feedstock, Reaction Conditions)	Catalytic Activity	Reusability	Reference
Peanut shell, carbonized at 450 °C for 15 h, sulfonated with H_2SO_4 at 200 °C for 10 h	Cottonseed oil and methanol (1:9), 85 °C, 2 h	conversion of 90.2%	50.3% in five consecutive cycles	[178]
Biochar, activated with KOH at 675 °C; sulfonated with SO_3 at 150 °C for 15 h	Canola oil and methanol (1:15), 150 °C, 1.52 Mpa, 3 h	a reaction yield of 44.2 %	drop to 0.9%, poor reusability	[179]
Biochar, activated with KOH at 675 °C; sulfonated with SO_3 at 150 °C for 15 h	Canola oil, oleic acid, and ethanol (3:1:30), 150 °C, 1.52 MPa, 3 h	a reaction yield of 48.1%	29% in the first recycling	[116]
Carbonized vegetable oil asphalt, sulfonated with concentrated H_2SO_4 at 210 °C for 10 h	A mixture of cottonseed oil and oleic acid (1:1 in weight), a molar ratio of methanol/oil of 16.8, 220 °C, 4.5 h	The conversion of triglyceride and free fatty acid were 80.5 wt% and 94.8 wt%, respectively.	75.5% and 97%, respectively, in 5 cycles	[181]
Glucose, carbonized at 400 °C for 5 h, sulfonated with H_2SO_4 at 150 °C for 10 h	*Calophyllum inophyllum* oil with free fatty acid of 15%, a molar ratio of methanol/oil of 30:1, 180 °C for 5 h	Conversion of 99%	50.3% in 5 cycles	[182]
Oat hull, carbonized at 600 °C for 3 h, sulfonated with H_2SO_4 at 140 °C for 30 min assisted by microwave	Waste cooking oil and methanol (1:10), 140 °C, 30 min, a microwave reactor	A biodiesel yield of 75% in 15 min	33% in 6 cycles	[119]
Coconut shell, carbonized at 422 °C for 4 h, sulfonated with concentrated H_2SO_4 at 100 °C for 15 h	Palm oil and methanol (1:30), 60 °C, 6h	A biodiesel yield of 88.15 %	Not applicable	[183]

The catalytic performance of biochar-based acid catalysts is affected by several factors.

(2) The Accessibility of Active Sites

Shu et al. [181] and Juan et al. [115] suggested that hydrophilic functional groups present on the biochar surface could facilitate the adsorption of hydrophilic reactants such as ethanol. In addition to the hydrophilic surface, a large pore size allows reactants to have easy access to the active sites. The superiority in the pore morphology of biochar may compensate for the shortcomings of low –SO_3H density in some cases [112].

(3) The Leaching of –SO_3H

The leaching of –SO_3H is a main reason for the loss of activity of biochar-based solid acids. A hydrophobic biochar matrix, which prevents the hydration of sulfonic groups, is crucial to the stability of the catalyst. Once –SO_3H groups are hydrated, they are easily leached as H_2SO_4 in the aqueous system [184,185]. In addition, the electron-withdrawing COOH groups present on biochar could also reduce the leaching rate of the –SO_3H groups by increasing the electron density between the carbon and sulfur atoms (C–SO_3H) [82]. On the contrary, water and FFAs which usually exist in vegetable oils favor the leaching of the active centers [172].

(4)   Impurities in Crude Oil or Biochar

Impurities present in crude oil, such as chlorophyll and phospholipids, were reported to inhibit the catalytic activity of biochar-based solid acids in the transesterification reaction [27,182]. In order to remove these impurities, a refinement of crude oil prior to the production of biodiesel was suggested [27]. Some inorganic species such as Cl and P present in the biochar derived from microalgae also have the potential to impair the catalytic performance of biochar catalysts and lower the quality of the biodiesel produced [27].

5.1.2. Solid-Alkali Catalysts

Another type of biochar-based catalyst for biodiesel production is biochar-supported alkali catalysts or biomass-derived solid-alkali catalysts. Their catalytic activity and reusability are listed in Table 3 and are discussed below.

McKay and co-workers [186,187] utilized palm-kernel-shell biochar rich in $CaCO_3$ to prepare a CaO/biochar catalyst for sunflower-oil metholysis. CaO was believed to be responsible for the catalytic activity. The conversion of sunflower oil was as high as 99% at 65 °C, with a methanol-to-oil ratio of 9:1 and a catalyst loading of 3 wt%. The catalyst could be reused for three consecutive cycles with no significant drop in activity. Chakraborty et al. [188] prepared a fly-ash-supported CaO catalyst with egg shells as the source of CaO. Its catalytic activity for the transesterification of soybean oil was examined. A fatty acid methyl ester (FAME) yield of 96.97% was obtained under its optimal conditions. Other Ca-rich biomass materials such as bovine bone waste [189] and crab shell [190] were also used to prepare CaO catalysts through a simple calcination process. These low-cost catalysts exhibited a high biodiesel yield (97% and 94%, respectively) and a good reusability, making them an attractive alternative to existing transesterification catalyst systems.

Similarly, K_2CO_3- or KOH-functionalized biochar were also used as catalysts for biodiesel production. Wang et al. [173] synthesized a peat-biochar-supported K_2CO_3 catalyst using a wet impregnation method. When the K_2CO_3 loading was 30% and the activation temperature was 600 °C, the catalyst achieved the maximum biodiesel yield, which was 98.6% for the transesterification of palm oil. The catalytic activity was mainly attributed to its total basicity. Dhawane et al. [191] firstly prepared activated biochar from flamboyant pods, and then loaded KOH onto the biochar via an impregnation method. When the biochar-supported KOH catalyst was used for the transesterification of *Hevea brasiliensis* oil, a biodiesel yield of 89.3% was achieved.

Table 3. Biochar-based solid-alkali catalysts for biodiesel production. FAME: fatty acid methyl ester.

Catalysts (Feedstock, Synthesis Conditions)	Biodiesel Production (Feedstock, Reaction Conditions)	Catalytic Activity	Reusability	References
Palm-kernel-shell biochar, calcined at 800 °C for 2 h to form CaO/biochar catalyst	Sunflower oil and methanol (1:9), 65 °C, 300 min	Conversion of 99%	No significant activity drop in 3 cycles	[186,187]
Waste egg shells as CaO source, fly ash as support, wet-impregnation method	Soybean oil and methanol (1:6.9), 70 °C, 5 h	A fame yield of 96.97%	Without major loss of activity for 16 cycles	[188]
Bovine bone waste, calcined at 750 °C for 6 h	Soybean oil and methanol (1:6), 65 °C, 3 h	A fame yield of 97%	>90% in four consecutive runs	[189]
Crab shell, calcined at 800 °C	Karanja oil and methanol (1:8), 65 °C, 120 min	A biodiesel yield of 94%	84% in the fifth run	[190]
Chicken manure containing $CaCO_3$, carbonized at 350 °C	Waste cooking oil and methanol (1:20 in volume), 350 °C	A fame yield of 95%	Not applicable	[192]
Pig meat and bone meal, carbonized at 650 °C, activated by KOH, alkalization with K_2CO_3	Palm oil and methanol (1:7), 65 °C, 150 min	A biodiesel yield of 98.2%	84% in the 10th cycles	[193]

5.2. Biomass Hydrolysis

The hydrolysis of lignocellulosic biomass to produce monosaccharides or oligosaccharides is a crucial step for the conversion of biomass to various platform chemicals (e.g., HMF and furfural) and biofuel (e.g., ethanol) [194,195]. Hydrolysis can be catalyzed by Brønsted acids at the temperature range of 90–260 °C [196]. During this process, protons provided by Brønsted acids attack the oxygen atom of glycosidic linkages, leading to the formation of a cyclic carbonium ion. The cyclic carbonium ion subsequently accepts a hydroxide ion and produces a monosaccharide or an oligosaccharide [196,197].

Common catalysts for biomass hydrolysis include enzymes, liquid mineral acids, and solid acids. The enzymatic hydrolysis is slow and costly. The mineral acids used for biomass hydrolysis are difficult to recycle. In contrast, sulfonated biochar is a promising solid-acid catalyst due to its low cost, high catalytic activity, and good recyclability [5,112,198]. The catalytic performances of various biochar-based acid catalysts were summarized in the literature [13]. Some typical examples are shown below.

Jiang et al. [199] prepared a biochar-based solid acid using hydrolyzed corncob residues, and then applied the catalyst to corncob hydrolysis under microwave irradiation. At the temperature range of 110–140 °C, corncobs were effectively hydrolyzed with yields of 34.6% glucose, 77.3% xylose, and 100% arabinose. No substantial decrease in the catalytic activity was observed when the catalyst was reused three times. Li et al. [5] investigated the catalytic performance of a corn-stover-char-based solid acid for the hydrolysis of corn stover, switch grass, and prairie cord grass in comparison with sulfuric acid that had a similar acid concentration. The biochar solid acid exhibited a higher selectivity to glucose, xylose, and total reducing sugar (TSR), as well as a higher glucan conversion, than sulfuric acid. Qi et al. [108] reported that the biochar solid acid derived from the HTC of glucose and sulfosalicylic acid exhibited good catalytic activity for the hydrolysis of cellulose in ionic liquid. A TRS yield of 59.4% was achieved at 130 °C for three hours. Moreover, after five recycle runs, the TRS yield remained at 57.4%, indicating the good stability of the solid-acid catalyst. Bai et al. [109] also reported the microwave-assisted conversion of bamboo hemicelluloses into xylo-oligosaccharides (XOS) using sulfonated bamboo char. The maximum XOS yield of 54.7 wt% was obtained at 150 °C for 45 min with a solid-acid-to-water-solvent mass ratio of 1:200. Liu et al. [111] derived sulfonated biochar from lignocellulose residues and concentrated saccharide solution. The catalytic effect of the solid acid on the hydrothermal degradation of corncob was investigated. A furfural yield of 37.75% and overall corncob conversion rate of 62.00% was achieved.

Surface functional groups of biochar play an important role in the catalytic performance for biomass hydrolysis. Functional groups that have a strong affinity to beta-1,4-glycosidic bonds, such as –COOH and phenolic –OH, could promote the accessibility of beta-1,4-glucans, including cellulose, to the –SO$_3$H sites, leading to a more efficient hydrolysis [82,113,200]. In addition, electrophilic –COOH groups present on the biochar surface could increase the electron density between the carbon and sulfur atoms, and stabilize the aromatic carbon–SO$_3$H bonds so that the leaching of SO$_3$H groups is inhibited [82,110].

Pore volumes of biochar also affect the catalytic performance. Ormsby et al. [112] observed that a biochar-based catalyst with a smaller SO$_3$H density showed a much higher turnover frequency (TOF) for xylan hydrolysis than sulfonated commercial resin (Amberlyst 15). This was due to the large surface area and appropriate pore volume of the biochar-based catalyst. There was a threshold for pore volume, below which the accessibility to active sites, as well as absorption capacity, was limited.

5.3. Tar Reduction

Biomass gasification is a promising technology for the utilization of biomass energy. Tar formation during biomass gasification is a main bottleneck of this technology for several reasons. Firstly, the generation of tar consumes carbon atoms in the feedstock, and hence, reduces the yield of syngas. Secondly, the condensable tar may block downstream pipes or equipment. Lastly, the tar contaminates syngas and adversely affects its application. For example, syngas containing tar would produce more aerosols and soot during combustion [201]. Therefore, effective tar reduction is critical to the commercialization of biomass gasification.

The primary constituents of tar are aromatic hydrocarbons including toluene, naphthalene, styrene, phenol, and polycyclic aromatic hydrocarbons (PAHs) [11]. Common methods for tar removal from syngas involve water washing, activated-carbon adsorption, biomass or ceramic material filtration, and catalytic tar cracking. Catalytic tar cracking is considered as the most effective method. Dolomite, olivine, nickel, alkali metals, and noble metals are commonly used catalysts for tar cracking. Recently, the catalytic activities of both biochar itself and biochar-supported active metals were tested [3,91,95,201–203].

5.3.1. Biochar without Active Metal Loading

Biochar without active metal loading showed good catalytic activity for tar removal [3,30,204,205]. El-Rub et al. [3] experimentally compared biochar with other common catalysts in terms of their catalytic performances for tar reduction. When naphthalene was used as a model tar compound, the ranking of catalytic activity at 900 °C was nickel > pine-wood char > dolomite > olivine > silica sand. In other words, biochar exhibited the highest activity for naphthalene conversion among the low-cost catalysts. Mani et al. [204] utilized pine-bark char to catalyze the decomposition of toluene (a model tar compound) in the presence of steam. The toluene conversion could be as high as 94% at 900 °C. The activation energy of toluene decomposition was 91 kJ/mol for the biochar catalyst, comparable to those for synthetic catalysts (e.g., 80 kJ/mol for Ni/mayenite). However, the biochar catalyst showed a lower tar removal rate than olivine and nickel catalysts. Luo et al. [30] reported that a rice-straw-derived biochar catalyst could achieve a tar removal efficiency of 94.03% after an 8-min reaction at 600 °C under microwave heating. They believed that the catalytic performance of the biochar catalyst was comparable to conventional technologies at 700–900 °C.

The catalytic activity of biochar for tar removal is affected by its mineral content, surface area, and surface functional groups. The inorganic species present in biochar play an important role in the catalytic behavior, despite their quantity being small. Klinghoffer et al. [45] demonstrated that >95% removal of Ca, K, P, and Mg from the char led to an 18% decrease in the decomposition of methane. An elevated temperature would cause the combination of inorganics with oxygen to form metal oxides on the surface of the biochar, which also caused a 40% loss in the catalytic activity. The surface area of biochar also affects its catalytic activity. Bhandari et al. [95] reported that switchgrass biochar activated with KOH exhibited a higher toluene removal efficiency than that without activation (92% vs. 69%). They attributed the better catalytic performance to the higher surface area, larger pore diameter, and larger pore volume caused by activation. Alkaline oxygenated groups (e.g., carbonylic, quinonyl, and pyrone structures) that remain attached on biochar at the tar-cracking temperature may also contribute to the catalytic activity [45]. In contrast, acidic groups (e.g., carboxylic and lactones), which disappear at an elevated temperature, are not likely to participate in the cracking reactions [45].

5.3.2. Biochar-Supported Active Metals

Loading biochar with active metals (e.g., Ni and Fe) can achieve a higher catalytic activity for tar decomposition than using biochar alone. Kastner et al. [204,206] reported the loading of Fe on biochar lowered the activation energy of toluene reformation from 91 kJ/mol to 48 kJ/mol. At 800 °C for the catalyst loaded with 13% Fe, the conversion of toluene approached 100%. If maintaining the reaction stream for four days, a mean toluene conversion of 91% was obtained. Wang et al. [202] prepared an Ni/char catalyst by mechanically mixing NiO and biochar particles followed by in situ reduction. This catalyst could remove more than 96% of the tar in syngas under optimal conditions, while the biochar without Ni could only remove 90% of the tar. Shen et al. [153] prepared a biochar-supported Ni/Fe catalyst by impregnating rich husk char in metal salts followed by a pyrolysis process. Under the optimized conditions, the conversion efficiency of condensable tar via in situ dry reforming over the biochar-supported Ni/Fe catalyst could reach about 92.3%, much higher than that over the biochar itself (42%). Shen et al. [201] also prepared a rice-husk char supported (HCS) Ni catalyst via a facile one-step method (see Section 4.4.1). The HCS Ni catalyst exhibited considerable catalytic activity on tar reformation, even at a relatively lower temperature (500 °C). When the reaction temperature was at 700 °C, the efficiency of tar reformation could be up to 99.8%, and kept almost unchanged after five cycles. Richardson et al. [144] also demonstrated that the Ni NPs in situ generated on biochar via the one-step method exhibited a higher catalytic activity for aromatic tar conversion than pre-formed Ni NPs that were inserted into the biomass prior to pyrolysis.

For metal/biochar catalysts, biochar has multiple functions [6,153]. Firstly, biochar acts as a support to disperse and stabilize the active metals. Secondly, biochar works as an intermediate reductant to convert metal oxides (normally from the decomposition of metal salts) to a metallic state, which enhances the tar conversion. Thirdly, the biochar itself has some catalytic activity owing to the presence of inorganics [45]. Lastly, biochar also plays the role of absorbent for tar components if its surface is tuned to be hydrophobic.

6. Conclusions and Perspectives

6.1. Conclusions

This paper reviewed the synthesis, characteristics, and modification of biochar, as well as the application of biochar-based catalysts in three important processes for biofuel production.

Conventionally, biochar is synthesized on various pyrolysis platforms. The properties of biochar are affected by the composition and morphology of biomass feedstock, as well as pyrolysis conditions, especially temperature. In recent times, hydrothermal carbonization (HTC), which is performed at 180–250 °C in the presence of water, was developed for biochar production. Compared to pyrolysis, HTC has advantages when processing high-moisture biomass and producing spherical biochar particles, but the feedstock for HTC is presently limited by simple carbohydrates. The resultant biochar from either pyrolysis or HTC exhibits typical features, which are the presence of inorganics, surface functional groups, and local crystalline structures made up of highly conjugated aromatic sheets. Various strategies are adopted to modify biochar so that it is more capable of being a catalyst or catalyst support. The surface area and porosity of biochar can be increased via physical activation or chemical activation. The biochar surface can be functionalized via surface sulfonation, surface amination, or surface oxidation. The physiochemical properties can also be finely tuned by in situ heteroatom doping. In addition, forming composites with other materials, such as metals, metal carbides, and nanostructured carbon materials, is also a common method to make use of biochar as a catalyst material. Through these modification strategies, a series of biochar-based catalysts were prepared and tested in biorefinery processes. Sulfonated biochar showed good catalytic performance for biomass hydrolysis and biodiesel production. Biodiesel production can also be catalyzed by biochar-derived or -supported solid-alkali catalysts. Biochar alone and biochar-supported metal catalysts are potential catalysts for tar reduction, an important step to obtaining clean syngas from biomass gasification.

In summary, biochar-based catalysts have several advantages. Firstly, the production of biochar is cheap and convenient owing to its renewable feedstock and well-developed synthesis techniques. Secondly, various strategies were developed to tune the physicochemical properties of biochar according to its specific purpose. Lastly, some intrinsic features of biochar, such as surface functional groups and the presence of inorganic species, may be favorable for its role as a catalyst or catalyst support. Therefore, biochar-based catalysts are a promising alternative to expensive or non-renewable traditional catalysts. The use of biochar-based catalysts in biofuel production represents a more sustainable and more integrated biorefinery scheme.

6.2. Perspectives

Biochar-based catalysts are still at the stage of laboratory research. For an industrial application in the future, purpose-driven synthesis and modification are required. This objective could be realized by mechanism studies. Firstly, the relationship between the catalytic activity and physicochemical properties of biochar needs to be deeply understood. This can be achieved by combining advanced characterization techniques for materials and theoretical modeling for catalytic mechanisms. Secondly, it is also important to reveal the influences of synthesis conditions and feedstock on the properties of resultant biochar. The complex composition of biomass and complicated formation mechanism of biochar make this work quite difficult. Advanced characterization techniques, especially those for online process monitoring from several aspects (e.g., thermogravimetric analysis/Fourier-transform infrared spectroscopy/mass spectrometry (TGA/FTIR/MS) and pyrolysis/gas chromatography/mass spectrometry (Py/GC/MS), are likely to play a crucial role.

As for the aspect of process optimization, much more attention needs to be paid to the use of catalysts in biochar synthesis. Although some inorganic species present in biomass feedstock show catalytic activity for pyrolysis, their autocatalysis alone is insufficient. It is necessary to develop catalysts that can achieve one of the following goals at least: (1) to produce biochar more efficiently by reducing reaction temperature or residence time; (2) to be able to prepare biochar with desired properties in one pot instead of synthesis and modification as two steps. Hopefully, in the future, a catalyst can convert biomass into biochar with the desired functional groups and porous structure in one step, and later combine with the biochar to directly produce biochar-supported catalysts. Furthermore, the industrial production of biochar-based catalysts could be closely combined with the production of biofuels and biochemicals in a biomass refinery, so that the process of biomass utilization is more integrated and more sustainable.

Author Contributions: F.C. and X.L. collected the related literature; X.L. constructed the outline of the review and provided comments; F.C. wrote the paper.

Funding:: This research was funded by the National Natural Science Foundation of China [grant number 51676098], the Natural Science Foundation of Jiangsu Province [grant numbers BK20160822 and BK20170095], and the Fundamental Research Funds for the Central Universities [grant number 30917011327]. The APC was funded by the Natural Science Foundation of Jiangsu Province [grant number BK20160822].

Acknowledgments: The authors gratefully acknowledge the financial supports from the National Natural Science Foundation of China under contract No. 51676098, the Natural Science Foundation of Jiangsu Province under contract No. BK20160822 and BK20170095, and the Fundamental Research Funds for the Central Universities under contract No. 30917011327.

Conflicts of Interest: The authors declare no conflict of interest. The founding sponsors had no role in the design of the study; in the collection, analyses, or interpretation of data; in the writing of the manuscript, and in the decision to publish the results. This manuscript has not been published or presented elsewhere in part or entirety and it is not under consideration by another journal. All the authors approved the manuscript and agreed to its submission to your esteemed journal.

References

1. Lin, Y.C.; Huber, G.W. The critical role of heterogeneous catalysis in lignocellulosic biomass conversion. *Energy Environ. Sci.* **2009**, *2*, 68–80. [CrossRef]
2. Dehkhoda, A.M.; West, A.H.; Ellis, N. Biochar based solid acid catalyst for biodiesel production. *Appl. Catal. A* **2010**, *382*, 197–204. [CrossRef]
3. Abu El-Rub, Z.; Bramer, E.A.; Brem, G. Experimental comparison of biomass chars with other catalysts for tar reduction. *Fuel* **2008**, *87*, 2243–2252. [CrossRef]
4. Nguyen, H.K.D.; Pham, V.V.; Do, H.T. Preparation of Ni/biochar catalyst for hydrotreating of bio-oil from microalgae biomass. *Catal. Lett.* **2016**, *146*, 2381–2391. [CrossRef]
5. Li, S.; Gu, Z.; Bjornson, B.E.; Muthukumarappan, A. Biochar based solid acid catalyst hydrolyze biomass. *Chem. Eng. J.* **2013**, *1*, 1174–1181. [CrossRef]

6. Qian, K.; Kumar, A.; Zhang, H.; Bellmer, D.; Huhnke, R. Recent advances in utilization of biochar. *Renew. Sustain. Energy Rev.* **2015**, *42*, 1055–1064. [CrossRef]
7. Li, J.; Dai, J.; Liu, G.; Zhang, H.; Gao, Z.; Fu, J.; He, Y.; Huang, Y. Biochar from microwave pyrolysis of biomass: A review. *Biomass Bioenergy* **2016**, *94*, 228–244. [CrossRef]
8. Lam, S.S.; Liew, R.K.; Wong, Y.M.; Yek, P.N.Y.; Ma, N.L.; Lee, C.L.; Chase, H.A. Microwave-assisted pyrolysis with chemical activation, an innovative method to convert orange peel into activated carbon with improved properties as dye adsorbent. *J. Clean. Prod.* **2017**, *162*, 1376–1387. [CrossRef]
9. Titirici, M.M.; White, R.J.; Falco, C.; Sevilla, M. Black perspectives for a green future: Hydrothermal carbons for environment protection and energy storage. *Energy Environ. Sci.* **2012**, *5*, 6796–6822. [CrossRef]
10. Cha, J.S.; Park, S.H.; Jung, S.C.; Ryu, C.; Jeon, J.K.; Shin, M.C.; Park, Y.K. Production and utilization of biochar: A review. *J. Ind. Eng. Chem.* **2016**, *40*, 1–15. [CrossRef]
11. Lee, J.; Kim, K.-H.; Kwon, E.E. Biochar as a catalyst. *Renew. Sustain. Energy Rev.* **2017**, *77*, 70–79. [CrossRef]
12. Hervy, M.; Berhanu, S.; Weiss-Hortala, E.; Chesnaud, A.; Gerente, C.; Villot, A.; Minh, D.P.; Thorel, A.; Le Coq, L.; Nzihou, A. Multi-scale characterisation of chars mineral species for tar cracking. *Fuel* **2017**, *189*, 88–97. [CrossRef]
13. Cao, X.; Sun, S.; Sun, R. Application of biochar-based catalysts in biomass upgrading: A review. *RSC Adv.* **2017**, *7*, 48793–48805. [CrossRef]
14. Xiong, X.; Yu, I.K.M.; Cao, L.; Tsang, D.C.W.; Zhang, S.; Ok, Y.S. A review of biochar-based catalysts for chemical synthesis, biofuel production, and pollution control. *Bioresour. Technol.* **2017**, *246*, 254–270. [CrossRef] [PubMed]
15. Liu, W.J.; Jiang, H.; Yu, H.Q. Development of biochar-based functional materials: Toward a sustainable platform carbon material. *Chem. Rev.* **2015**, *115*, 12251–12285. [CrossRef] [PubMed]
16. Tan, X.F.; Liu, Y.G.; Gu, Y.L.; Xu, Y.; Zeng, G.M.; Hu, X.J.; Liu, S.B.; Wang, X.; Liu, S.M.; Li, J. Biochar-based nano-composites for the decontamination of wastewater: A review. *Bioresour. Technol.* **2016**, *212*, 318–333. [CrossRef] [PubMed]
17. Cheng, B.H.; Zeng, R.J.; Jiang, H. Recent developments of post-modification of biochar for electrochemical energy storage. *Bioresour. Technol.* **2017**, *246*, 224–233. [CrossRef] [PubMed]
18. Dong, Q.; Li, H.; Niu, M.; Luo, C.; Zhang, J.; Qi, B.; Li, X.; Zhong, W. Microwave pyrolysis of moso bamboo for syngas production and bio-oil upgrading over bamboo-based biochar catalyst. *Bioresour. Technol.* **2018**, *266*, 284–290. [CrossRef] [PubMed]
19. Singh, S.; Nahil, M.A.; Sun, X.; Wu, C.; Chen, J.; Shen, B.; Williams, P.T. Novel application of cotton stalk as a waste derived catalyst in the low temperature SCR-deNO$_x$ process. *Fuel* **2013**, *105*, 585–594. [CrossRef]
20. Shen, B.; Chen, J.; Yue, S.; Li, G. A comparative study of modified cotton biochar and activated carbon based catalysts in low temperature SCR. *Fuel* **2015**, *156*, 47–53. [CrossRef]
21. Kastner, J.R.; Miller, J.; Kolar, P.; Das, K.C. Catalytic ozonation of ammonia using biomass char and wood fly ash. *Chemosphere* **2009**, *75*, 739–744. [CrossRef] [PubMed]
22. Da, Z.L.; Niu, X.H.; Li, X.; Zhang, W.C.; He, Y.F.; Pan, J.M.; Qiu, F.X.; Yan, Y.S. From moldy orange waste to natural reductant and catalyst support: Active palladium/biomass-derived carbonaceous hybrids for promoted methanol electro-oxidation. *ChemElectroChem* **2017**, *4*, 1372–1377. [CrossRef]
23. Jia, Y.; Feng, H.; Shen, D.; Zhou, Y.; Chen, T.; Wang, M.; Chen, W.; Ge, Z.; Huang, L.; Zheng, S. High-performance microbial fuel cell anodes obtained from sewage sludge mixed with fly ash. *J. Hazard. Mater.* **2018**, *354*, 27–32. [CrossRef] [PubMed]
24. Kim, J.R.; Kan, E. Heterogeneous photocatalytic degradation of sulfamethoxazole in water using a biochar-supported TiO$_2$ photocatalyst. *J. Environ. Manag.* **2016**, *180*, 94–101. [CrossRef] [PubMed]
25. Liu, X.Q.; Chen, W.J.; Jiang, H. Facile synthesis of Ag/Ag$_3$PO$_4$/AMB composite with improved photocatalytic performance. *Chem. Eng. J.* **2017**, *308*, 889–896. [CrossRef]
26. Li, P.J.; Lin, K.R.; Fang, Z.Q.; Wang, K.M. Enhanced nitrate removal by novel bimetallic Fe/Ni nanoparticles supported on biochar. *J. Clean. Prod.* **2017**, *151*, 21–33. [CrossRef]
27. Dong, T.; Gao, D.; Miao, C.; Yu, X.; Degan, C.; Garcia-Pérez, M.; Rasco, B.; Sablani, S.S.; Chen, S. Two-step microalgal biodiesel production using acidic catalyst generated from pyrolysis-derived bio-char. *Energy Convers. Manag.* **2015**, *105*, 1389–1396. [CrossRef]

28. Ren, S.; Lei, H.; Wang, L.; Bu, Q.; Chen, S.; Wu, J. Hydrocarbon and hydrogen-rich syngas production by biomass catalytic pyrolysis and bio-oil upgrading over biochar catalysts. *RSC Adv.* **2014**, *4*, 10731–10737. [CrossRef]

29. Shen, Y. Chars as carbonaceous adsorbents/catalysts for tar elimination during biomass pyrolysis or gasification. *Renew. Sustain. Energy Rev.* **2015**, *43*, 281–295. [CrossRef]

30. Luo, H.; Bao, L.; Wang, H.; Kong, L.; Sun, Y. Microwave-assisted in-situ elimination of primary tars over biochar: Low temperature behaviours and mechanistic insights. *Bioresour. Technol.* **2018**, *267*, 333–340. [CrossRef] [PubMed]

31. Yan, Q.; Wan, C.; Liu, J.; Gao, J.; Yu, F.; Zhang, J.; Cai, Z. Iron nanoparticles in situ encapsulated in biochar-based carbon as an effective catalyst for the conversion of biomass-derived syngas to liquid hydrocarbons. *Green Chem.* **2013**, *15*, 1631–1640. [CrossRef]

32. DeSisto, W.J.; Hill, N.; Beis, S.H.; Mukkamala, S.; Joseph, J.; Baker, C.; Ong, T.H.; Stemmler, E.A.; Wheeler, M.C.; Frederick, B.G. Fast pyrolysis of pine sawdust in a fluidized-bed reactor. *Energy Fuels* **2010**, *24*, 2642–2651. [CrossRef]

33. Brewer, C.E.; Schmidt-Rohr, K.; Satrio, J.A.; Brown, R.C. Characterization of biochar from fast pyrolysis and gasification systems. *Environ. Prog. Sustain. Energy* **2009**, *28*, 386–396. [CrossRef]

34. Dutta, B.; Raghavan, G.S.V.; Ngadi, M. Surface characterization and classification of slow and fast pyrolyzed biochar using novel methods of pycnometry and hyperspectral imaging. *J. Wood Chem. Technol.* **2012**, *32*, 105–120. [CrossRef]

35. Antal, M.J.; Gronli, M. The art, science, and technology of charcoal production. *Ind. Eng. Chem. Res.* **2003**, *42*, 1619–1640. [CrossRef]

36. Brendova, K.; Szakova, J.; Lhotka, M.; Krulikovska, T.; Puncochar, M.; Tlustos, P. Biochar physicochemical parameters as a result of feedstock material and pyrolysis temperature: Predictable for the fate of biochar in soil? *Environ. Geochem. Health* **2017**, *39*, 1381–1395. [CrossRef] [PubMed]

37. Zhao, L.; Cao, X.; Masek, O.; Zimmerman, A. Heterogeneity of biochar properties as a function of feedstock sources and production temperatures. *J. Hazard. Mater.* **2013**, *256*, 1–9. [CrossRef] [PubMed]

38. Yang, H.; Yan, R.; Chen, H.; Lee, D.H.; Zheng, C. Characteristics of hemicellulose, cellulose and lignin pyrolysis. *Fuel* **2007**, *86*, 1781–1788. [CrossRef]

39. Thompson, E.; Danks, A.E.; Bourgeois, L.; Schnepp, Z. Iron-catalyzed graphitization of biomass. *Green Chem.* **2015**, *17*, 551–556. [CrossRef]

40. Xue, G.; Liu, K.; Chen, Q.; Yang, P.; Li, J.; Ding, T.; Duan, J.; Qi, B.; Zhou, J. Robust and low-cost flame-treated wood for high-performance solar steam generation. *ACS Appl. Mater. Interfaces* **2017**, *9*, 15052–15057. [CrossRef] [PubMed]

41. Li, Y.; Zhang, Q.; Zhang, J.; Jin, L.; Zhao, X.; Xu, T. A top-down approach for fabricating free-standing bio-carbon supercapacitor electrodes with a hierarchical structure. *Sci. Rep.* **2015**, *5*, 14155. [CrossRef] [PubMed]

42. Yao, H.; Zheng, G.; Li, W.; McDowell, M.T.; Seh, Z.; Liu, N.; Lu, Z.; Cui, Y. Crab shells as sustainable templates from nature for nanostructured battery electrodes. *Nano Lett.* **2013**, *13*, 3385–3390. [CrossRef] [PubMed]

43. Abu Bakar, M.S.; Titiloye, J.O. Catalytic pyrolysis of rice husk for bio-oil production. *J. Anal. Appl. Pyrolysis* **2013**, *103*, 362–368. [CrossRef]

44. Nowakowski, D.J.; Jones, J.M.; Brydson, R.M.D.; Ross, A.B. Potassium catalysis in the pyrolysis behaviour of short rotation willow coppice. *Fuel* **2007**, *86*, 2389–2402. [CrossRef]

45. Klinghoffer, N.B.; Castaldi, M.J.; Nzihou, A. Influence of char composition and inorganics on catalytic activity of char from biomass gasification. *Fuel* **2015**, *157*, 37–47. [CrossRef]

46. McBeath, A.V.; Wurster, C.M.; Bird, M.I. Influence of feedstock properties and pyrolysis conditions on biochar carbon stability as determined by hydrogen pyrolysis. *Biomass Bioenergy* **2015**, *73*, 155–173. [CrossRef]

47. Yao, D.; Hu, Q.; Wang, D.; Yang, H.; Wu, C.; Wang, X.; Chen, H. Hydrogen production from biomass gasification using biochar as a catalyst/support. *Bioresour. Technol.* **2016**, *216*, 159–164. [CrossRef] [PubMed]

48. Uchimiya, M.; Wartelle, L.H.; Klasson, K.T.; Fortier, C.A.; Lima, I.M. Influence of pyrolysis temperature on biochar property and function as a heavy metal sorbent in soil. *J. Agric. Food. Chem.* **2011**, *59*, 2501–2510. [CrossRef] [PubMed]

49. Zhang, J.; Liu, J.; Liu, R. Effects of pyrolysis temperature and heating time on biochar obtained from the pyrolysis of straw and lignosulfonate. *Bioresour. Technol.* **2015**, *176*, 288–291. [CrossRef] [PubMed]

50. Zhao, B.; O'Connor, D.; Zhang, J.; Peng, T.; Shen, Z.; Tsang, D.C.W.; Hou, D. Effect of pyrolysis temperature, heating rate, and residence time on rapeseed stem derived biochar. *J. Clean. Prod.* **2018**, *174*, 977–987. [CrossRef]

51. Muradov, N.; Fidalgo, B.; Gujar, A.C.; Garceau, N.; T-Raissi, A. Production and characterization of lemna minor bio-char and its catalytic application for biogas reforming. *Biomass Bioenergy* **2012**, *42*, 123–131. [CrossRef]

52. Keiluweit, M.; Nico, P.S.; Johnson, M.G.; Kleber, M. Dynamic molecular structure of plant biomass-derived black carbon (biochar). *Environ. Sci. Technol.* **2010**, *44*, 1247–1253. [CrossRef] [PubMed]

53. Román, S.; Nabais, J.M.V.; Laginhas, C.; Ledesma, B.; González, J.F. Hydrothermal carbonization as an effective way of densifying the energy content of biomass. *Fuel Process. Technol.* **2012**, *103*, 78–83. [CrossRef]

54. Xiao, L.P.; Shi, Z.J.; Xu, F.; Sun, R.C. Hydrothermal carbonization of lignocellulosic biomass. *Bioresour. Technol.* **2012**, *118*, 619–623. [CrossRef] [PubMed]

55. Hu, B.; Wang, K.; Wu, L.; Yu, S.H.; Antonietti, M.; Titirici, M.M. Engineering carbon materials from the hydrothermal carbonization process of biomass. *Adv. Mater.* **2010**, *22*, 813–828. [CrossRef] [PubMed]

56. De Caprariis, B.; De Filippis, P.; Petrullo, A.; Scarsella, M. Hydrothermal liquefaction of biomass: Influence of temperature and biomass composition on the bio-oil production. *Fuel* **2017**, *208*, 618–625. [CrossRef]

57. Tekin, K.; Karagoz, S.; Bektas, S. Hydrothermal liquefaction of beech wood using a natural calcium borate mineral. *J. Supercrit. Fluids* **2012**, *72*, 134–139. [CrossRef]

58. He, C.; Chen, C.L.; Giannis, A.; Yang, Y.; Wang, J.Y. Hydrothermal gasification of sewage sludge and model compounds for renewable hydrogen production: A review. *Renew. Sustain. Energy Rev.* **2014**, *39*, 1127–1142. [CrossRef]

59. Kong, L.; Li, G.; Zhang, B.; He, W.; Wang, H. Hydrogen production from biomass wastes by hydrothermal gasification. *Energy Source Part A* **2008**, *30*, 1166–1178. [CrossRef]

60. Kruse, A. Hydrothermal biomass gasification. *J. Supercrit. Fluids* **2009**, *47*, 391–399. [CrossRef]

61. Poerschmann, J.; Baskyr, I.; Weiner, B.; Koehler, R.; Wedwitschka, H.; Kopinke, F.D. Hydrothermal carbonization of olive mill wastewater. *Bioresour. Technol.* **2013**, *133*, 581–588. [CrossRef] [PubMed]

62. Kruse, A.; Funke, A.; Titirici, M.-M. Hydrothermal conversion of biomass to fuels and energetic materials. *Curr. Opin. Chem. Biol.* **2013**, *17*, 515–521. [CrossRef] [PubMed]

63. Sabio, E.; Álvarez-Murillo, A.; Román, S.; Ledesma, B. Conversion of tomato-peel waste into solid fuel by hydrothermal carbonization: Influence of the processing variables. *Waste Manag.* **2016**, *47*, 122–132. [CrossRef] [PubMed]

64. Kambo, H.S.; Dutta, A. A comparative review of biochar and hydrochar in terms of production, physico-chemical properties and applications. *Renew. Sustain. Energy Rev.* **2015**, *45*, 359–378. [CrossRef]

65. Licursi, D.; Antonetti, C.; Fulignati, S.; Vitolo, S.; Puccini, M.; Ribechini, E.; Bernazzani, L.; Galletti, A.M.R. In-depth characterization of valuable char obtained from hydrothermal conversion of hazelnut shells to levulinic acid. *Bioresour. Technol.* **2017**, *244*, 880–888. [CrossRef] [PubMed]

66. Kim, D.; Lee, K.; Park, K.Y. Upgrading the characteristics of biochar from cellulose, lignin, and xylan for solid biofuel production from biomass by hydrothermal carbonization. *J. Ind. Eng. Chem.* **2016**, *42*, 95–100. [CrossRef]

67. Cao, X.; Peng, X.; Sun, S.; Zhong, L.; Sun, R. Hydrothermal conversion of bamboo: Identification and distribution of the components in solid residue, water-soluble and acetone-soluble fractions. *J. Agric. Food. Chem.* **2014**, *62*, 12360–12365. [CrossRef] [PubMed]

68. Sasaki, M.; Adschiri, T.; Arai, K. Fractionation of sugarcane bagasse by hydrothermal treatment. *Bioresour. Technol.* **2003**, *86*, 301–304. [CrossRef]

69. Bilgic, E.; Yaman, S.; Haykiri-Acma, H.; Kucukbayrak, S. Is torrefaction of polysaccharides-rich biomass equivalent to carbonization of lignin-rich biomass? *Bioresour. Technol.* **2016**, *200*, 201–207. [CrossRef] [PubMed]

70. Falco, C.; Baccile, N.; Titirici, M.-M. Morphological and structural differences between glucose, cellulose and lignocellulosic biomass derived hydrothermal carbons. *Green Chem.* **2011**, *13*, 3273–3281. [CrossRef]

71. Chheda, J.N.; Roman-Leshkov, Y.; Dumesic, J.A. Production of 5-hydroxymethylfurfural and furfural by dehydration of biomass-derived mono- and poly-saccharides. *Green Chem.* **2007**, *9*, 342–350. [CrossRef]

72. Titirici, M.-M.; Antonietti, M.; Baccile, N. Hydrothermal carbon from biomass: A comparison of the local structure from poly- to monosaccharides and pentoses/hexoses. *Green Chem.* **2008**, *10*, 1204–1212. [CrossRef]

73. Berge, N.D.; Ro, K.S.; Mao, J.; Flora, J.R.V.; Chappell, M.A.; Bae, S. Hydrothermal carbonization of municipal waste streams. *Environ. Sci. Technol.* **2011**, *45*, 5696–5703. [CrossRef] [PubMed]

74. Huggins, T.; Wang, H.; Kearns, J.; Jenkins, P.; Ren, Z.J. Biochar as a sustainable electrode material for electricity production in microbial fuel cells. *Bioresour. Technol.* **2014**, *157*, 114–119. [CrossRef] [PubMed]

75. Zhao, L.; Fan, L.Z.; Zhou, M.Q.; Guan, H.; Qiao, S.; Antonietti, M.; Titirici, M.M. Nitrogen-containing hydrothermal carbons with superior performance in supercapacitors. *Adv. Mater.* **2010**, *22*, 5202–5206. [CrossRef] [PubMed]

76. Brown, T.R.; Wright, M.M.; Brown, R.C. Estimating profitability of two biochar production scenarios: Slow pyrolysis vs fast pyrolysis. *Biofuels Bioprod. Biorefining* **2011**, *5*, 54–68. [CrossRef]

77. Mullen, C.A.; Boateng, A.A.; Goldberg, N.M.; Lima, I.M.; Laird, D.A.; Hicks, K.B. Bio-oil and bio-char production from corn cobs and stover by fast pyrolysis. *Biomass Bioenergy* **2010**, *34*, 67–74. [CrossRef]

78. Lua, A.C.; Yang, T.; Guo, J. Effects of pyrolysis conditions on the properties of activated carbons prepared from pistachio-nut shells. *J. Anal. Appl. Pyrolysis* **2004**, *72*, 279–287. [CrossRef]

79. Liu, S.S.; Wang, M.F.; Sun, X.Y.; Xu, N.; Liu, J.; Wang, Y.Z.; Qian, T.; Yan, C.L. Facilitated oxygen chemisorption in heteroatom-doped carbon for improved oxygen reaction activity in all-solid-state zinc-air batteries. *Adv. Mater.* **2018**, *30*, 1704898. [CrossRef] [PubMed]

80. Ren, Q.; Wang, H.; Lu, X.F.; Tong, Y.X.; Li, G.R. Recent progress on MOF-derived heteroatom-doped carbon-based electrocatalysts for oxygen reduction reaction. *Adv. Sci.* **2018**, *5*, 1700515. [CrossRef] [PubMed]

81. Son, E.B.; Poo, K.M.; Mohamed, H.O.; Choi, Y.J.; Cho, W.C.; Chae, K.J. A novel approach to developing a reusable marine macro-algae adsorbent with chitosan and ferric oxide for simultaneous efficient heavy metal removal and easy magnetic separation. *Bioresour. Technol.* **2018**, *259*, 381–387. [CrossRef] [PubMed]

82. Kitano, M.; Yamaguchi, D.; Suganuma, S.; Nakajima, K.; Kato, H.; Hayashi, S.; Hara, M. Adsorption-enhanced hydrolysis of beta-1,4-glucan on graphene-based amorphous carbon bearing SO_3H, COOH, and OH groups. *Langmuir* **2009**, *25*, 5068–5075. [CrossRef] [PubMed]

83. Hagen, J. Heterogeneous catalysis: Fundamentals. In *Industrial Catalysis*; John Wiley and Sons, Inc.: Hoboken, NJ, USA, 2015; pp. 99–100, ISBN 9783527684625.

84. Duan, X.H.; Srinivasakannan, C.; Peng, J.H.; Zhang, L.B.; Zhang, Z.Y. Comparison of activated carbon prepared from jatropha hull by conventional heating and microwave heating. *Biomass Bioenergy* **2011**, *35*, 3920–3926.

85. Alvarez, J.; Lopez, G.; Amutio, M.; Bilbao, J.; Olazar, M. Upgrading the rice husk char obtained by flash pyrolysis for the production of amorphous silica and high quality activated carbon. *Bioresour. Technol.* **2014**, *170*, 132–137. [CrossRef] [PubMed]

86. Yang, K.; Peng, J.; Srinivasakannan, C.; Zhang, L.; Xia, H.; Duan, X. Preparation of high surface area activated carbon from coconut shells using microwave heating. *Bioresour. Technol.* **2010**, *101*, 6163–6169. [CrossRef] [PubMed]

87. Cha, J.S.; Choi, J.C.; Ko, J.H.; Park, Y.K.; Park, S.H.; Jeong, K.E.; Kim, S.S.; Jeon, J.K. The low-temperature scr of no over rice straw and sewage sludge derived char. *Chem. Eng. J.* **2010**, *156*, 321–327. [CrossRef]

88. Ros, A.; Lillo-Ródenas, M.A.; Fuente, E.; Montes-Morán, M.A.; Martín, M.J.; Linares-Solano, A. High surface area materials prepared from sewage sludge-based precursors. *Chemosphere* **2006**, *65*, 132–140. [CrossRef] [PubMed]

89. González, J.F.; Román, S.; Encinar, J.M.; Martínez, G. Pyrolysis of various biomass residues and char utilization for the production of activated carbons. *J. Anal. Appl. Pyrolysis* **2009**, *85*, 134–141. [CrossRef]

90. Angın, D.; Altintig, E.; Köse, T.E. Influence of process parameters on the surface and chemical properties of activated carbon obtained from biochar by chemical activation. *Bioresour. Technol.* **2013**, *148*, 542–549. [CrossRef] [PubMed]

91. Jin, H.; Capareda, S.; Chang, Z.; Gao, J.; Xu, Y.; Zhang, J. Biochar pyrolytically produced from municipal solid wastes for aqueous as(v) removal: Adsorption property and its improvement with koh activation. *Bioresour. Technol.* **2014**, *169*, 622–629. [CrossRef] [PubMed]

92. Dehkhoda, A.M.; Ellis, N.; Gyenge, E. Electrosorption on activated biochar: Effect of thermo-chemical activation treatment on the electric double layer capacitance. *J. Appl. Electrochem.* **2014**, *44*, 141–157. [CrossRef]

93. Tay, T.; Ucar, S.; Karagöz, S. Preparation and characterization of activated carbon from waste biomass. *J. Hazard. Mater.* **2009**, *165*, 481–485. [CrossRef] [PubMed]

94. Gratuito, M.K.B.; Panyathanmaporn, T.; Chumnanklang, R.A.; Sirinuntawittaya, N.; Dutta, A. Production of activated carbon from coconut shell: Optimization using response surface methodology. *Bioresour. Technol.* **2008**, *99*, 4887–4895. [CrossRef] [PubMed]

95. Bhandari, P.N.; Kumar, A.; Bellmer, D.D.; Huhnke, R.L. Synthesis and evaluation of biochar-derived catalysts for removal of toluene (model tar) from biomass-generated producer gas. *Renew. Energy* **2014**, *66*, 346–353. [CrossRef]

96. Sevilla, M.; Fuertes, A.B.; Mokaya, R. High density hydrogen storage in superactivated carbons from hydrothermally carbonized renewable organic materials. *Energy Environ. Sci.* **2011**, *4*, 1400–1410. [CrossRef]

97. Deng, J.; Xiong, T.; Xu, F.; Li, M.; Han, C.; Gong, Y.; Wang, H.; Wang, Y. Inspired by bread leavening: One-pot synthesis of hierarchically porous carbon for supercapacitors. *Green Chem.* **2015**, *17*, 4053–4060. [CrossRef]

98. Nieva Lobos, M.L.; Manuel Sieben, J.; Comignani, V.; Duarte, M.; Alicia Volpe, M.; Laura Moyano, E. Biochar from pyrolysis of cellulose: An alternative catalyst support for the electro-oxidation of methanol. *Int. J. Hydrogen Energy* **2016**, *41*, 10695–10706. [CrossRef]

99. Jagtoyen, M.; Derbyshire, F. Activated carbons from yellow poplar and white oak by H3PO4 activation. *Carbon* **1998**, *36*, 1085–1097. [CrossRef]

100. Ucar, S.; Erdem, M.; Tay, T.; Karagoz, S. Preparation and characterization of activated carbon produced from pomegranate seeds by ZnCl$_2$ activation. *Appl. Surf. Sci.* **2009**, *255*, 8890–8896. [CrossRef]

101. Sun, L.; Tian, C.; Li, M.; Meng, X.; Wang, L.; Wang, R.; Yin, J.; Fu, H. From coconut shell to porous graphene-like nanosheets for high-power supercapacitors. *J. Mater. Chem. A* **2013**, *1*, 6462–6470. [CrossRef]

102. Titirici, M.M.; Thomas, A.; Antonietti, M. Replication and coating of silica templates by hydrothermal carbonization. *Adv. Funct. Mater.* **2007**, *17*, 1010–1018. [CrossRef]

103. Titirici, M.M.; Thomas, A.; Antonietti, M. Aminated hydrophilic ordered mesoporous carbons. *J. Mater. Chem.* **2007**, *17*, 3412–3418. [CrossRef]

104. Kubo, S.; Tan, I.; White, R.J.; Antonietti, M.; Titirici, M.M. Template synthesis of carbonaceous tubular nanostructures with tunable surface properties. *Chem. Mater.* **2010**, *22*, 6590–6597. [CrossRef]

105. White, R.J.; Tauer, K.; Antonietti, M.; Titirici, M.M. Functional hollow carbon nanospheres by latex templating. *J. Am. Chem. Soc.* **2010**, *132*, 17360–17363. [CrossRef] [PubMed]

106. Kubo, S.; White, R.J.; Yoshizawa, N.; Antonietti, M.; Titirici, M.M. Ordered carbohydrate-derived porous carbons. *Chem. Mater.* **2011**, *23*, 4882–4885. [CrossRef]

107. White, R.J.; Antonietti, M.; Titirici, M.M. Naturally inspired nitrogen doped porous carbon. *J. Mater. Chem.* **2009**, *19*, 8645–8650. [CrossRef]

108. Qi, X.; Lian, Y.; Yan, L.; Smith, R.L. One-step preparation of carbonaceous solid acid catalysts by hydrothermal carbonization of glucose for cellulose hydrolysis. *Catal. Commun.* **2014**, *57*, 50–54. [CrossRef]

109. Bai, Y.Y.; Xiao, L.P.; Sun, R.C. Microwave-assisted conversion of biomass derived hemicelluloses into xylo-oligosaccharides by novel sulfonated bamboo-based catalysts. *Biomass Bioenergy* **2015**, *75*, 245–253. [CrossRef]

110. Guo, H.; Lian, Y.; Yan, L.; Qi, X.; Smith, R.L., Jr. Cellulose-derived superparamagnetic carbonaceous solid acid catalyst for cellulose hydrolysis in an ionic liquid or aqueous reaction system. *Green Chem.* **2013**, *15*, 2167–2174. [CrossRef]

111. Liu, Q.Y.; Yang, F.; Sun, X.F.; Liu, Z.H.; Li, G. Preparation of biochar catalyst with saccharide and lignocellulose residues of corncob degradation for corncob hydrolysis into furfural. *J. Mater. Cycles Waste Manag.* **2017**, *19*, 134–143. [CrossRef]

112. Ormsby, R.; Kastner, J.R.; Miller, J. Hemicellulose hydrolysis using solid acid catalysts generated from biochar. *Catal. Today* **2012**, *190*, 89–97. [CrossRef]

113. Wu, Y.; Fu, Z.; Yin, D.; Xu, Q.; Liu, F.; Lu, C.; Mao, L. Microwave-assisted hydrolysis of crystalline cellulose catalyzed by biomass char sulfonic acids. *Green Chem.* **2010**, *12*, 696–700. [CrossRef]

114. Kastner, J.R.; Miller, J.; Geller, D.P.; Locklin, J.; Keith, L.H.; Johnson, T. Catalytic esterification of fatty acids using solid acid catalysts generated from biochar and activated carbon. *Catal. Today* **2012**, *190*, 122–132. [CrossRef]

115. Maciá-Agulló, J.A.; Sevilla, M.; Diez, M.A.; Fuertes, A.B. Synthesis of carbon-based solid acid microspheres and their application to the production of biodiesel. *ChemSusChem* **2010**, *3*, 1352–1354. [CrossRef] [PubMed]

116. Dehkhoda, A.M.; Ellis, N. Biochar-based catalyst for simultaneous reactions of esterification and transesterification. *Catal. Today* **2013**, *207*, 86–92. [CrossRef]

117. Arneil Arancon, R.; Barros, H.R., Jr.; Balu, A.M.; Vargas, C.; Luque, R. Valorisation of corncob residues to functionalised porous carbonaceous materials for the simultaneous esterification/transesterification of waste oils. *Green Chem.* **2011**, *13*, 3162–3167. [CrossRef]

118. Toda, M.; Takagaki, A.; Okamura, M.; Kondo, J.N.; Hayashi, S.; Domen, K.; Hara, M. Green chemistry biodiesel made with sugar catalyst. *Nature* **2005**, *438*, 178. [CrossRef] [PubMed]

119. González, M.E.; Cea, M.; Reyes, D.; Romero-Hermoso, L.; Hidalgo, P.; Meier, S.; Benito, N.; Navia, R. Functionalization of biochar derived from lignocellulosic biomass using microwave technology for catalytic application in biodiesel production. *Energy Convers. Manag.* **2017**, *137*, 165–173. [CrossRef]

120. Liang, X.; Zeng, M.; Qi, C. One-step synthesis of carbon functionalized with sulfonic acid groups using hydrothermal carbonization. *Carbon* **2010**, *48*, 1844–1848. [CrossRef]

121. Shafeeyan, M.S.; Daud, W.M.A.W.; Houshmand, A.; Arami-Niya, A. The application of response surface methodology to optimize the amination of activated carbon for the preparation of carbon dioxide adsorbents. *Fuel* **2012**, *94*, 465–472. [CrossRef]

122. Jansen, R.J.J.; van Bekkum, H. Amination and ammoxidation of activated carbons. *Carbon* **1994**, *32*, 1507–1516. [CrossRef]

123. Stöhr, B.; Boehm, H.P.; Schlögl, R. Enhancement of the catalytic activity of activated carbons in oxidation reactions by thermal treatment with ammonia or hydrogen cyanide and observation of a superoxide species as a possible intermediate. *Carbon* **1991**, *29*, 707–720. [CrossRef]

124. Xu, L.; Yao, O.; Zhang, Y.; Fu, Y. Integrated production of aromatic amines and N-doped carbon from lignin via ex situ catalytic fast pyrolysis in the presence of ammonia over zeolites. *Energy Convers. Manag.* **2017**, *5*, 2960–2969. [CrossRef]

125. Zhao, L.; Bacsik, Z.; Hedin, N.; Wei, W.; Sun, Y.; Antonietti, M.; Titirici, M.M. Carbon dioxide capture on amine-rich carbonaceous materials derived from glucose. *Chemsuschem* **2010**, *3*, 840–845. [CrossRef] [PubMed]

126. Ma, Y.; Liu, W.J.; Zhang, N.; Li, Y.S.; Jiang, H.; Sheng, G.P. Polyethylenimine modified biochar adsorbent for hexavalent chromium removal from the aqueous solution. *Bioresource Technol.* **2014**, *169*, 403–408. [CrossRef] [PubMed]

127. Chen, Z.; Ma, L.; Li, S.; Geng, J.; Song, Q.; Liu, J.; Wang, C.; Wang, H.; Li, J.; Qin, Z. Simple approach to carboxyl-rich materials through low-temperature heat treatment of hydrothermal carbon in air. *Appl. Surf. Sci.* **2011**, *257*, 8686–8691. [CrossRef]

128. Anfruns, A.; Garcia-Suarez, E.J.; Montes-Moran, M.A.; Gonzalez-Olmos, R.; Martin, M.J. New insights into the influence of activated carbon surface oxygen groups on H_2O_2 decomposition and oxidation of pre-adsorbed volatile organic compounds. *Carbon* **2014**, *77*, 89–98. [CrossRef]

129. Wu, L.; Sitamraju, S.; Xiao, J.; Liu, B.; Li, Z.; Janik, M.J.; Song, C. Effect of liquid-phase O_3 oxidation of activated carbon on the adsorption of thiophene. *Chem. Eng. J.* **2014**, *242*, 211–219. [CrossRef]

130. Gokce, Y.; Aktas, Z. Nitric acid modification of activated carbon produced from waste tea and adsorption of methylene blue and phenol. *Appl. Surf. Sci.* **2014**, *313*, 352–359. [CrossRef]

131. Li, Y.; Shao, J.; Wang, X.; Deng, Y.; Yang, H.; Chen, H. Characterization of modified biochars derived from bamboo pyrolysis and their utilization for target component (furfural) adsorption. *Energy Fuels* **2014**, *28*, 5119–5127. [CrossRef]

132. Luo, W.; Wang, B.; Heron, C.G.; Allen, M.J.; Morre, J.; Maier, C.S.; Stickle, W.F.; Ji, X. Pyrolysis of cellulose under ammonia leads to nitrogen-doped nanoporous carbon generated through methane formation. *Nano Lett.* **2014**, *14*, 2225–2229. [CrossRef] [PubMed]

133. Wang, L.; Yan, W.; He, C.; Wen, H.; Cai, Z.; Wanga, Z.X.; Chen, Z.Z.; Liu, W.F. Microwave-assisted preparation of nitrogen-doped biochars by ammonium acetate activation for adsorption of acid red 18. *Appl. Surf. Sci.* **2018**, *433*, 222–231. [CrossRef]

134. Nagy, B.; Villar-Rodil, S.; Tascón, J.M.D.; Bakos, I.; László, K. Nitrogen doped mesoporous carbon aerogels and implications for electrocatalytic oxygen reduction reactions. *Microporous Mesoporous Mater.* **2016**, *230*, 135–144. [CrossRef]

135. Gao, Y.; Xu, S.P.; Yue, Q.Y.; Ortaboy, S.; Gao, B.Y.; Sun, Y.Y. Synthesis and characterization of heteroatom-enriched biochar from keratin-based and algous-based wastes. *Adv. Powder Technol.* **2016**, *27*, 1280–1286. [CrossRef]

136. Chen, Z.; Li, K.; Pu, L. The performance of phosphorus (p)-doped activated carbon as a catalyst in air-cathode microbial fuel cells. *Bioresour. Technol.* **2014**, *170*, 379–384. [CrossRef] [PubMed]

137. Zhu, S.; Huang, X.; Ma, F.; Wang, L.; Duan, X.; Wang, S. Catalytic removal of aqueous contaminants on n-doped graphitic biochars: Inherent roles of adsorption and nonradical mechanisms. *Environ. Sci. Technol.* **2018**. [CrossRef] [PubMed]

138. Zhao, L.; Baccile, N.; Gross, S.; Zhang, Y.; Wei, W.; Sun, Y.; Antonietti, M.; Titirici, M.M. Sustainable nitrogen-doped carbonaceous materials from biomass derivatives. *Carbon* **2010**, *48*, 3778–3787. [CrossRef]

139. Guo, Z.; Zhou, Q.; Wu, Z.; Zhang, Z.; Zhang, W.; Zhang, Y.; Li, L.; Cao, Z.; Wang, H.; Gao, Y. Nitrogen-doped carbon based on peptides of hair as electrode materials for surpercapacitors. *Electrochim. Acta* **2013**, *113*, 620–627. [CrossRef]

140. Thanh-Dinh, N.; Shopsowitz, K.E.; MacLachlan, M.J. Mesoporous nitrogen-doped carbon from nanocrystalline chitin assemblies. *J. Mater. Chem. A* **2014**, *2*, 5915–5921.

141. White, R.J.; Yoshizawa, N.; Antonietti, M.; Titirici, M.-M. A sustainable synthesis of nitrogen-doped carbon aerogels. *Green Chem.* **2011**, *13*, 2428–2434. [CrossRef]

142. Baccile, N.; Laurent, G.; Coelho, C.; Babonneau, F.; Zhao, L.; Titirici, M.M. Structural insights on nitrogen-containing hydrothermal carbon using solid-state magic angle spinning 13C and 15N nuclear magnetic resonance. *J. Phys. Chem. C* **2011**, *115*, 8976–8982. [CrossRef]

143. Richardson, Y.; Blin, J.; Volle, G.; Motuzas, J.; Julbe, A. In situ generation of ni metal nanoparticles as catalyst for H$_2$-rich syngas production from biomass gasification. *Appl. Catal. A* **2010**, *382*, 220–230. [CrossRef]

144. Richardson, Y.; Motuzas, J.; Julbe, A.; Volle, G.; Blin, J. Catalytic investigation of in situ generated ni metal nanoparticles for tar conversion during biomass pyrolysis. *J. Phys. Chem. C* **2013**, *117*, 23812–23831. [CrossRef]

145. Shen, Y.; Yoshikawa, K. Tar conversion and vapor upgrading via in situ catalysis using silica-based nickel nanoparticles embedded in rice husk char for biomass pyrolysis/gasification. *Ind. Eng. Chem. Res.* **2014**, *53*, 10929–10942. [CrossRef]

146. Shen, Y.; Areeprasert, C.; Prabowo, B.; Takahashi, F.; Yoshikawa, K. Metal nickel nanoparticles in situ generated in rice husk char for catalytic reformation of tar and syngas from biomass pyrolytic gasification. *Rsc Adv.* **2014**, *4*, 40651–40664. [CrossRef]

147. Liu, W.J.; Tian, K.; Jiang, H.; Yu, H.Q. Harvest of cu np anchored magnetic carbon materials from fe/cu preloaded biomass: Their pyrolysis, characterization, and catalytic activity on aqueous reduction of 4-nitrophenol. *Green Chem.* **2014**, *16*, 4198–4205. [CrossRef]

148. Makowski, P.; Demir Cakan, R.; Antonietti, M.; Goettmann, F.; Titirici, M.M. Selective partial hydrogenation of hydroxy aromatic derivatives with palladium nanoparticles supported on hydrophilic carbon. *Chem. Commun.* **2008**, 999–1001. [CrossRef] [PubMed]

149. Yu, J.C.; Hu, X.; Li, Q.; Zhang, L. Microwave-assisted synthesis and in-situ self-assembly of coaxial Ag/C nanocables. *Chem. Commun.* **2005**, 2704–2706. [CrossRef] [PubMed]

150. Yu, S.H.; Cui, X.J.; Li, L.L.; Li, K.; Yu, B.; Antonietti, M.; Cölfen, H. From starch to metal/carbon hybrid nanostructures: Hydrothermal metal-catalyzed carbonization. *Adv. Mater.* **2004**, *16*, 1636–1640. [CrossRef]

151. Liu, W.J.; Tian, K.; Jiang, H.; Zhahg, X.S.; Ding, H.S.; Yu, H.Q. Selectively improving the bio-oil quality by catalytic fast pyrolysis of heavy-metal-polluted biomass: Take copper (Cu) as an example. *Environ. Sci. Technol.* **2012**, *46*, 7849–7856. [CrossRef] [PubMed]

152. Zhen, F.; Kaibin, T.; Shuijin, L.; Tanwei, L. CTAB-assisted hydrothermal synthesis of Ag/C nanostructures. *Nanotechnology* **2006**, *17*, 3008. [CrossRef]

153. Shen, Y.; Zhao, P.; Shao, Q.; Ma, D.; Takahashi, F.; Yoshikawa, K. In-situ catalytic conversion of tar using rice husk char-supported nickel-iron catalysts for biomass pyrolysis/gasification. *Appl. Catal. B-Environ.* **2014**, *152*, 140–151. [CrossRef]

154. Xiaoming, S.; Yadong, L. Colloidal carbon spheres and their core/shell structures with noble-metal nanoparticles. *Angew. Chem. Int. Ed.* **2004**, *43*, 597–601.

155. Hu, B.; Zhao, Y.; Zhu, H.Z.; Yu, S.H. Selective chromogenic detection of thiol-containing biomolecules using carbonaceous nanospheres loaded with silver nanoparticles as carrier. *ACS Nano* **2011**, *5*, 3166–3171. [CrossRef] [PubMed]

156. Qian, H.-S.; Antonietti, M.; Yu, S.-H. Hybrid "golden fleece": Synthesis and catalytic performance of uniform carbon nanofibers and silica nanotubes embedded with a high population of noble-metal nanoparticles. *Adv. Funct. Mater.* **2007**, *17*, 637–643. [CrossRef]

157. Tang, S.; Vongehr, S.; Meng, X. Carbon spheres with controllable silver nanoparticle doping. *J. Phys. Chem. C* **2010**, *114*, 977–982. [CrossRef]

158. Qian, H.S.; Yu, S.H.; Luo, L.B.; Gong, J.Y.; Fei, L.F.; Liu, X.M. Synthesis of uniform te@carbon-rich composite nanocables with photoluminescence properties and carbonaceous nanofibers by the hydrothermal carbonization of glucose. *Chem. Mater.* **2006**, *18*, 2102–2108. [CrossRef]

159. Yan, Q.; Lu, Y.; To, F.; Li, Y.; Yu, F. Synthesis of tungsten carbide nanoparticles in biochar matrix as a catalyst for dry reforming of methane to syngas. *Catal. Sci. Technol.* **2015**, *5*, 3270–3280. [CrossRef]

160. Zhang, Y.; Wang, A.; Zhang, T. A new 3D mesoporous carbon replicated from commercial silica as a catalyst support for direct conversion of cellulose into ethylene glycol. *Chem. Commun.* **2010**, *46*, 862–864. [CrossRef] [PubMed]

161. Wen, L.; Ma, Y.; Dai, B.; Zhou, Y.; Liu, J.; Pei, C. Preparation and dielectric properties of sic nanowires self-sacrificially templated by carbonated bacterial cellulose. *Mater. Res. Bull.* **2013**, *48*, 687–690. [CrossRef]

162. Church, T.L.; Fallani, S.; Liu, J.; Zhao, M.; Harris, A.T. Novel biomorphic ni/sic catalysts that enhance cellulose conversion to hydrogen. *Catal. Today* **2012**, *190*, 98–106. [CrossRef]

163. Xinyong, T.; Lixin, D.; Xinnan, W.; Wenkui, Z.; Bradley, J.N.; Xiaodong, L.Z. B$_4$C-nanowires/carbon-microfiber hybrid structures and composites from cotton t-shirts. *Adv. Mater.* **2010**, *22*, 2055–2059.

164. Du, J.; Li, Q.; Xia, Y.; Cheng, X.; Gan, Y.; Huang, H.; Zhang, W.; Tao, X. Synthesis of boron carbide nanoflakes via a bamboo-based carbon thermal reduction method. *J. Alloys Compd.* **2013**, *581*, 128–132. [CrossRef]

165. Shin, Y.; Li, X.S.; Wang, C.; Coleman, J.R.; Exarhos, G.J. Synthesis of hierarchical titanium carbide from titania-coated cellulose paper. *Adv. Mater.* **2004**, *16*, 1212–1215. [CrossRef]

166. Gotoh, Y.; Fujimura, K.; Koike, M.; Ohkoshi, Y.; Nagura, M.; Akamatsu, K.; Deki, S. Synthesis of titanium carbide from a composite of TiO$_2$ nanoparticles/methyl cellulose by carbothermal reduction. *Mater. Res. Bull.* **2001**, *36*, 2263–2275. [CrossRef]

167. Li, R.; Shahbazi, A.; Wang, L.; Zhang, B.; Chung, C.C.; Dayton, D.; Yan, Q. Nanostructured molybdenum carbide on biochar for CO$_2$ reforming of CH$_4$. *Fuel* **2018**, *225*, 403–410. [CrossRef]

168. Inyang, M.; Gao, B.; Zimmerman, A.; Zhang, M.; Chen, H. Synthesis, characterization, and dye sorption ability of carbon nanotube-biochar nanocomposites. *Chem. Eng. J.* **2014**, *236*, 39–46. [CrossRef]

169. Chen, X.W.; Timpe, O.; Hamid, S.B.A.; Schloegl, R.; Su, D.S. Direct synthesis of carbon nanofibers on modified biomass-derived activated carbon. *Carbon* **2009**, *47*, 340–343. [CrossRef]

170. Baker, R.T.K. Catalytic growth of carbon filaments. *Carbon* **1989**, *27*, 315–323. [CrossRef]

171. Chen, S.; Rotaru, A.E.; Shrestha, P.M.; Malvankar, N.S.; Liu, F.; Fan, W.; Nevin, K.P.; Lovley, D.R. Promoting interspecies electron transfer with biochar. *Sci. Rep.* **2014**, *4*. [CrossRef] [PubMed]

172. Konwar, L.J.; Boro, J.; Deka, D. Review on latest developments in biodiesel production using carbon-based catalysts. *Renew. Sustain. Energy Rev.* **2014**, *29*, 546–564. [CrossRef]

173. Wang, S.; Zhao, C.; Shan, R.; Wang, Y.; Yuan, H. A novel peat biochar supported catalyst for the transesterification reaction. *Energy Convers. Manag.* **2017**, *139*, 89–96. [CrossRef]

174. Atadashi, I.M.; Aroua, M.K.; Abdul Aziz, A.R.; Sulaiman, N.M.N. The effects of catalysts in biodiesel production: A review. *J. Ind. Eng. Chem.* **2013**, *19*, 14–26. [CrossRef]

175. Melero, J.A.; Iglesias, J.; Morales, G. Heterogeneous acid catalysts for biodiesel production: Current status and future challenges. *Green Chem.* **2009**, *11*, 1285–1308. [CrossRef]

176. Li, M.; Zheng, Y.; Chen, Y.; Zhu, X. Biodiesel production from waste cooking oil using a heterogeneous catalyst from pyrolyzed rice husk. *Bioresour. Technol.* **2014**, *154*, 345–348. [CrossRef] [PubMed]

177. Lee, J.; Jung, J.M.; Oh, J.I.; Ok, Y.S.; Lee, S.R.; Kwon, E.E. Evaluating the effectiveness of various biochars as porous media for biodiesel synthesis via pseudo-catalytic transesterification. *Bioresour. Technol.* **2017**, *231*, 59–64. [CrossRef] [PubMed]

178. Zeng, D.; Liu, S.; Gong, W.; Wang, G.; Qiu, J.; Chen, H. Synthesis, characterization and acid catalysis of solid acid from peanut shell. *Appl. Catal. A* **2014**, *469*, 284–289. [CrossRef]

179. Yu, J.T.; Dehkhoda, A.M.; Ellis, N. Development of biochar-based catalyst for transesterification of canola oil. *Energy Fuels* **2011**, *25*, 337–344. [CrossRef]

180. Chen, S.S.; Maneerung, T.; Tsang, D.C.W.; Ok, Y.S.; Wang, C.H. Valorization of biomass to hydroxymethylfurfural, levulinic acid, and fatty acid methyl ester by heterogeneous catalysts. *Chem. Eng. J.* **2017**, *328*, 246–273. [CrossRef]

181. Shu, Q.; Gao, J.; Nawaz, Z.; Liao, Y.; Wang, D.; Wang, J. Synthesis of biodiesel from waste vegetable oil with large amounts of free fatty acids using a carbon-based solid acid catalyst. *Appl. Energy* **2010**, *87*, 2589–2596. [CrossRef]

182. Dawodu, F.A.; Ayodele, O.; Xin, J.; Zhang, S.; Yan, D. Effective conversion of non-edible oil with high free fatty acid into biodiesel by sulphonated carbon catalyst. *Appl. Energy* **2014**, *114*, 819–826. [CrossRef]

183. Endut, A.; Abdullah, S.H.Y.S.; Hanapi, N.H.M.; Hamid, S.H.A.; Lananan, F.; Kamarudin, M.K.A.; Umar, R.; Juahir, H.; Khatoon, H. Optimization of biodiesel production by solid acid catalyst derived from coconut shell via response surface methodology. *Int. Biodeterior. Biodegrad.* **2017**, *124*, 250–257. [CrossRef]

184. Xiangcheng, L.; Kaihao, P.; Xiaohui, L.; Qineng, X.; Yanqin, W. Comprehensive understanding of the role of brønsted and lewis acid sites in glucose conversion into 5–hydromethylfurfural. *ChemCatChem* **2017**, *9*, 2739–2746.

185. Gallo, J.M.R.; Alamillo, R.; Dumesic, J.A. Acid-functionalized mesoporous carbons for the continuous production of 5-hydroxymethylfurfural. *J. Mol. Catal. A: Chem.* **2016**, *422*, 13–17. [CrossRef]

186. Bazargan, A.; Kostić, M.D.; Stamenković, O.S.; Veljković, V.B.; McKay, G. A calcium oxide-based catalyst derived from palm kernel shell gasification residues for biodiesel production. *Fuel* **2015**, *150*, 519–525. [CrossRef]

187. Kostić, M.D.; Bazargan, A.; Stamenković, O.S.; Veljković, V.B.; McKay, G. Optimization and kinetics of sunflower oil methanolysis catalyzed by calcium oxide-based catalyst derived from palm kernel shell biochar. *Fuel* **2016**, *163*, 304–313. [CrossRef]

188. Chakraborty, R.; Bepari, S.; Banerjee, A. Transesterification of soybean oil catalyzed by fly ash and egg shell derived solid catalysts. *Chem. Eng. J.* **2010**, *165*, 798–805. [CrossRef]

189. Smith, S.M.; Oopathum, C.; Weeramongkholert, V.; Smith, C.B.; Chaveanghong, S.; Ketwong, P.; Boonyuen, S. Transesterification of soybean oil using bovine bone waste as new catalyst. *Bioresour. Technol.* **2013**, *143*, 686–690. [CrossRef] [PubMed]

190. Madhu, D.; Chavan, S.B.; Singh, V.; Singh, B.; Sharma, Y.C. An economically viable synthesis of biodiesel from a crude millettia pinnata oil of jharkhand, india as feedstock and crab shell derived catalyst. *Bioresour. Technol.* **2016**, *214*, 210–217. [CrossRef] [PubMed]

191. Dhawane, S.H.; Kumar, T.; Halder, G. Central composite design approach towards optimization of flamboyant pods derived steam activated carbon for its use as heterogeneous catalyst in transesterification of hevea brasiliensis oil. *Energy Convers. Manag.* **2015**, *100*, 277–287. [CrossRef]

192. Jung, J.M.; Oh, J.I.; Baek, K.; Lee, J.; Kwon, E.E. Biodiesel production from waste cooking oil using biochar derived from chicken manure as a porous media and catalyst. *Energy Convers. Manag.* **2018**, *165*, 628–633. [CrossRef]

193. Wang, S.; Yuan, H.; Wang, Y.; Shan, R. Transesterification of vegetable oil on low cost and efficient meat and bone meal biochar catalysts. *Energy Convers. Manag.* **2017**, *150*, 214–221. [CrossRef]

194. Tuck, C.O.; Perez, E.; Horvath, I.T.; Sheldon, R.A.; Poliakoff, M. Valorization of biomass: Deriving more value from waste. *Science* **2012**, *337*, 695–699. [CrossRef] [PubMed]

195. Wataniyakul, P.; Boonnoun, P.; Quitain, A.T.; Sasaki, M.; Kida, T.; Laosiripojana, N.; Shotipruk, A. Preparation of hydrothermal carbon as catalyst support for conversion of biomass to 5-hydroxymethylfurfural. *Catal. Commun.* **2018**, *104*, 41–47. [CrossRef]

196. Zhou, C.H.; Xia, X.; Lin, C.X.; Tong, D.S.; Beltramini, J. Catalytic conversion of lignocellulosic biomass to fine chemicals and fuels. *Chem. Soc. Rev.* **2011**, *40*, 5588–5617. [CrossRef] [PubMed]

197. Zhang, X.; Wilson, K.; Lee, A.F. Heterogeneously catalyzed hydrothermal processing of C5–C6 sugars. *Chem. Rev.* **2016**, *116*, 12328–12368. [CrossRef] [PubMed]

198. Deng, A.; Lin, Q.; Yan, Y.; Li, H.; Ren, J.; Liu, C.; Sun, R. A feasible process for furfural production from the pre-hydrolysis liquor of corncob via biochar catalysts in a new biphasic system. *Bioresour. Technol.* **2016**, *216*, 754–760. [CrossRef] [PubMed]

199. Jiang, Y.; Li, X.; Wang, X.; Meng, L.; Wang, H.; Peng, G.; Wang, X.; Mu, X. Effective saccharification of lignocellulosic biomass over hydrolysis residue derived solid acid under microwave irradiation. *Green Chem.* **2012**, *14*, 2162–2167. [CrossRef]

200. Okamura, M.; Takagaki, A.; Toda, M.; Kondo, J.N.; Domen, K.; Tatsumi, T.; Hara, M.; Hayashi, S. Acid-catalyzed reactions on flexible polycyclic aromatic carbon in amorphous carbon. *Chem. Mater.* **2006**, *18*, 3039–3045. [CrossRef]

201. Shen, Y.; Chen, M.; Sun, T.; Jia, J. Catalytic reforming of pyrolysis tar over metallic nickel nanoparticles embedded in pyrochar. *Fuel* **2015**, *159*, 570–579. [CrossRef]

202. Wang, D.; Yuan, W.; Ji, W. Char and char-supported nickel catalysts for secondary syngas cleanup and conditioning. *Appl. Energy* **2011**, *88*, 1656–1663. [CrossRef]

203. Bhandari, P.N.; Kumar, A.; Huhnke, R.L. Simultaneous removal of toluene (model tar), NH_3, and H_2S, from biomass-generated producer gas using biochar-based and mixed-metal oxide catalysts. *Energy Fuels* **2014**, *28*, 1918–1925. [CrossRef]

204. Mani, S.; Kastner, J.R.; Juneja, A. Catalytic decomposition of toluene using a biomass derived catalyst. *Fuel Process. Technol.* **2013**, *114*, 118–125. [CrossRef]

205. Feng, D.; Zhao, Y.; Zhang, Y.; Sun, S. Effects of H_2O and CO_2 on the homogeneous conversion and heterogeneous reforming of biomass tar over biochar. *Int. J. Hydrogen Energy* **2017**, *42*, 13070–13084. [CrossRef]

206. Kastner, J.R.; Mani, S.; Juneja, A. Catalytic decomposition of tar using iron supported biochar. *Fuel Process. Technol.* **2015**, *130*, 31–37. [CrossRef]

catalysts

MDPI

Article

Pyrolyzing Renewable Sugar and Taurine on the Surface of Multi-Walled Carbon Nanotubes as Heterogeneous Catalysts for Hydroxymethylfurfural Production

Huiping Ji [1], Jie Fu [2] and Tianfu Wang [3,*]

[1] University of Chinese Academy of Sciences, Beijing 100049, China; jihuiping15@mails.ucas.ac.cn
[2] Key Laboratory of Biomass Chemical Engineering of Ministry of Education, College of Chemical and Biological Engineering, Zhejiang University, Hangzhou 310027, China; jiefu@zju.edu.cn
[3] Xinjiang Technical Institute of Physics & Chemistry, Chinese Academy of Sciences, Urumqi 830011, China
* Correspondence: tianfuwang@ms.xjb.ac.cn

Received: 12 August 2018; Accepted: 21 September 2018; Published: 5 November 2018

Abstract: Conversion of biorenewable feedstocks into transportation fuels or chemicals likely necessitates the development of novel heterogeneous catalysts with good hydrothermal stability, due to the nature of highly oxygenated biomass compounds and the prevalence of water as a processing solvent. The use of carbon-based materials, derived from sugars as catalyst precursors, can achieve hydrothermal stability while simultaneously realizing the goal of sustainability. In this work, the simultaneous pyrolysis of glucose and taurine in the presence of multi-walled carbon nanotubes (MWCNTs), to obtain versatile solid acids, has been demonstrated. Structural and textural properties of the catalysts have been characterized by TEM, TGA, and XPS. Additionally, solid state nuclear magnetic resonance (ssNMR) spectroscopy has been exploited to elucidate the chemical nature of carbon species deposited on the surface of MWCNTs. $Al(OTf)_3$, a model Lewis acidic metal salt, has been successfully supported on sulfonic groups tethered to MWCNTs. This catalyst has been tested for C_6 sugar dehydration for the production of HMF in a tetrahydrofuran (THF)/water solvent system with good recyclability.

Keywords: carbon nanotubes; carbohydrates; HMF; Lewis acids; NMR

---

1. Introduction

Due to the rapidly growing world energy and materials consumption, and diminishing petroleum reserves, economical and environmental concerns have risen, urging the utilization of renewable sources of carbon for liquid transportation fuels and chemicals. Biomass is uniquely situated to provide an alternative to petroleum-sourced carbon because of its abundance and sustainability [1]. However, in contrast to traditional feedstocks, molecules derived from the carbohydrate fractions of biomass consist of highly oxygenated compounds. This necessitates the removal of functionality before biomass can replace the highly reduced carbon platform molecules typical of the modern fuel and chemical industry [2]. Therefore, processing of these materials will greatly differ from the functionality-adding processes that have been well-researched and employed in the petrochemical industry. The need to adapt to these new feedstocks and processing conditions is particularly challenging for the field of heterogeneous catalysis [3]. It is important that the heterogeneous catalysis community performs research with common biorenewable compounds, in order to gain understanding and subsequent ability to control the active site environment, as well as develop catalysts with hydrothermal stability [3].

Common catalysts employed in the petroleum industry, such as zeolites, have been shown to be unstable with exposure to hot water. Stability must be considered for biomass upgrading [4]. Consequently, hydrothermal stability concerns have promoted recent research into carbon-based supports for biorenewable applications [5]. One particular carbon catalyst support receiving a significant amount of attention in recent years is that of carbon nanotubes. Multi-walled carbon nanotubes (MWCNTs) are a hydrothermally stable material, and they offer other reasons that make them an interesting material as well. For instance, they offer uniform and tunable textural properties, such as porosity, that can be easily manipulated. This allows for greater control over the catalyst active site than is available when using conventional carbon-based materials, such as activated carbon. MWCNTs also have a relatively high surface area and relatively less microporosity, making them less likely than conventional carbon-based materials to encounter mass transport limitations when they are used as catalyst supports [6].

Previously, a simple method to produced functionalized carbon rich materials, from glucose and a wide number of molecules containing a primary amine, was demonstrated, including $(CH_2)_2$–SO_3H groups, which would be ideal to coordinate metal cations, and are strong Brønsted acids [7]. The hydrothermal stability of sulfonated carbon materials has been called into question in the past, and it has been shown that SO_3H groups directly bonded to aromatic carbons are substantially less stable than SO_3H groups bonded to alkyl CH_2 [8]. In an earlier study, it was found that glucose and taurine can be pyrolyzed together to form alkyl-linked sulfonic acids for various acid-catalyzed reactions, such as dehydration and esterification [9], and this class of materials has been thoroughly characterized by NMR [10]. However, solid acids produced from this strategy suffer from their intrinsically low surface area, typically ~1 m^2/g, limiting their potential applications in catalysis. In order to solve this problem, we reasoned that, in the presence of a porous support such as MWCNT, co-pyrolysis of glucose and taurine could form, ideally, a thin layer of carbonaceous material containing alkyl-linked sulfonic acid groups. The resulting high surface area solid acids could be applicable in a range of acid-catalyzed reactions.

Recently, increasing interest has been attracted for the development of biorenewable plastics derived from the carbohydrate portion of lignocellulosic biomass [11]. Among all the proposed platform chemicals that can be converted to monomers for subsequent production of polymers, 5-hydroxymethylfurfural (HMF) has been identified as a promising platform chemical candidate for the production of polyethylene furanics (PEF) to replace petroleum-based polyethylene terephthalate (PET). In the past decade, various catalysts, such as homogeneous sulfonic acid [11], heteropolyacids (HPAs) [12], heterogeneous metal-oxide-based catalysts, like tin-beta zeolites [13] and Nb_2O_5 [14], have all been reported to catalyze the dehydration reaction of C_6 monosaccharide to HMF. In addition, numerous green solvent and catalytic systems have been demonstrated to show high glucose-HMF conversion activity and good yields of HMF have been reported, inspiring the development of new class of catalysts [15–17]. Very recently, a class of metal salts, such as $AlCl_3$ [18], $SnCl_4$ [19], $GeCl_4$ [20], and $YbCl_3$ [21], dissolved in either aqueous or ionic liquid media, have been investigated as extremely effective catalysts for HMF production. The unique catalytic performance of these metal salt catalytic systems has been attributed to the combination of the Lewis and Brønsted nature possessed by metal cations and, hence, facile interaction with sugar molecules to promote HMF synthesis [18]. Although promising, recycling and separation of the homogeneous metal salts may raise concerns on the practicality of such systems. In order to solve this dilemma, it could be advantageous to immobilize these metal salts on a solid support.

In this study, we co-pyrolyzed glucose and taurine to disperse a carbonaceous layer onto MWCNTs, and tested the reaction kinetics, hydrothermal stability, and SO_3H bonding environment of this catalyst. Co-pyrolysis of glucose and taurine with MWCNT, under controlled experimental conditions (200 °C and 250 °C), resulted in a hybrid material in which sulfonic groups were present on the surface of MWCNT. Solid-state NMR was used to determine the chemical structure of the hybrid material and the bonding nature of the sulfonic group using ^{13}C-labeled glucose and ^{13}C- or

^{15}N-labeled taurine. The electrostatic interaction between sulfonate and the aluminum cation was exploited for immobilization of the model Lewis acid catalyst, Al(OTf)$_3$. Lewis acid-immobilized MWCNT catalysts were subsequently tested for their activities in glucose and fructose dehydration reactions, as shown in Figure 1. In addition, catalyst recyclability studies using fructose were carried out, and Al leaching tests with ICP analysis were performed after each run, to probe catalyst stability under reaction conditions.

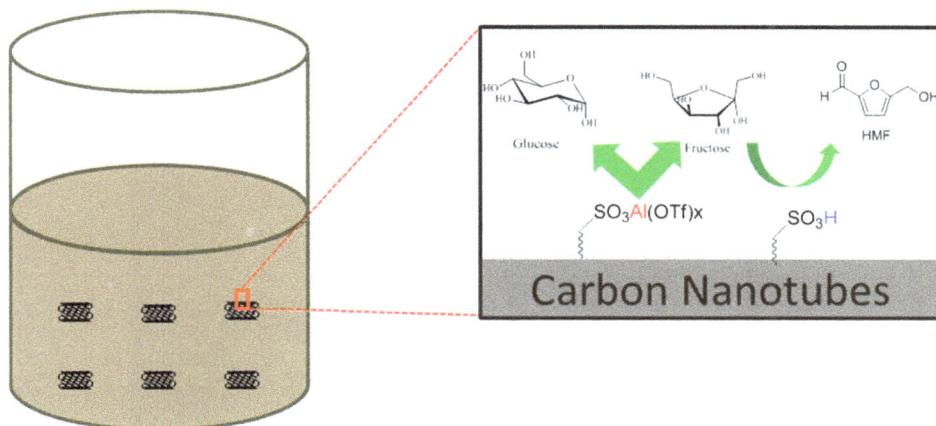

Figure 1. Solid Lewis/Brønsted acid catalysts for hydroxymethylfurfural (HMF) production from C$_6$ sugars.

2. Results and Discussion

2.1. Catalyst Characterizations

Sulfonic acid groups were introduced to the surface of MWCNT by co-pyrolyzing glucose and taurine. Before pyrolysis, the parent MWCNT has a BET surface area of 304 m^2/g and pore volume of 1.13 cm^3/g, which were reduced to 118 m^2/g and 0.33 cm^3/g, respectively. This change in porosity properties provides indirect evidence of the successful deposition of functional groups on the surface of MWCNT. TEM and EDS imaging studies have provided further confirmation of the successful synthesis of sulfonate-tethered MWCNT catalysts. As shown in Figure 2, the representative solid Lewis acid catalysts (after the immobilization of Al(OTf)$_3$) revealed the typical morphology of a multi-walled carbon nanotube (see Figure 2a). EDS mapping revealed a homogeneous distribution of sulfur, indicating that the sulfonic groups are impregnated throughout the MWCNT surface. Further EDS data is summarized in Figure 3 for sulfur, carbon, oxygen, and aluminum. By comparing the EDS data, it can be concluded that both Al and S are well-dispersed throughout the MWCNT surface, proving the effectiveness of the pyrolysis and immobilization as catalyst preparation methods. Additionally, shown in Figure 2c,d are the images of the Lewis acid catalyst after being subjected to the reaction condition of 160 °C in THF/water for 3 h, referred to as the "spent" catalyst. Although ICP analysis of the supernatant would be a more definitive method to quantify the stability of the immobilized Al, the visual TEM/EDS images demonstrate a similar morphology and S distribution for both the intact and spent Lewis acid catalysts. This issue is discussed further below.

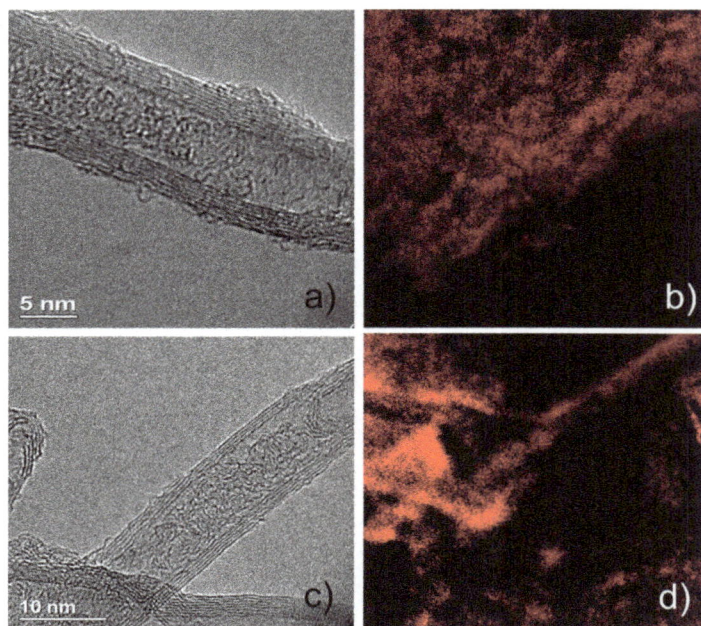

Figure 2. TEM and EDS images of (**a**) LA-MWCNT (Lewis acid-multi-wall carbon nanotube); (**b**) S mapping of LA-MWCNT; (**c**) spent LA-MWCNT; and (**d**) S mapping of spent LA-MWCNT.

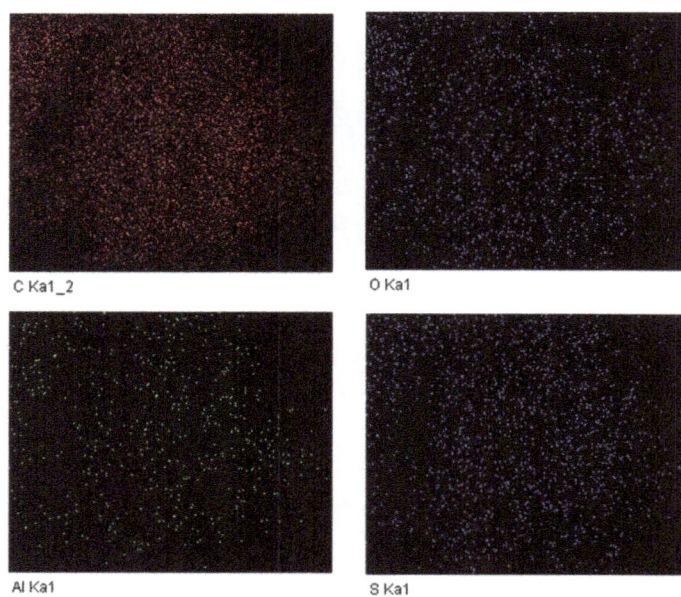

Figure 3. EDS elemental mapping of LA-MWCNT.

2.2. Solid-State NMR

The aromaticity of each material is quantified by measuring the area under the aromatic peak in the ^{13}C spectrum as a percentage of the total area under the spectrum. Since glucose makes up 30 mol % of the carbon in the MWCNT-containing samples, and only 1% of the remaining carbon is ^{13}C, the intensity in the DP/MAS spectra shown in Figure 4 may be considered to be entirely from the ^{13}C nuclei that originated in the glucose. By this method, the fate of 69% of the glucose carbon in GT250/1-GC is to become aromatic, while 72% of the glucose carbon in GT250/10-CNT-GC becomes aromatic. This indicates that the presence of the MWCNT during the pyrolysis reaction at 250 °C does not significantly change the aromatic fraction of the pyrolysis product.

Figure 4 also shows DP/MAS experiments with dipolar dephasing, a method which suppresses signals from ^{13}C species with strong dipolar couplings to ^{1}H. Therefore, the spectra with dipolar dephasing selectively show signal from nonprotonated and very mobile ^{13}C species. Sample GT250/10-CNT-GC loses less signal than GT250/1-GC in the dipolar dephasing experiments (Figure 4), indicating that fewer of the ^{13}C nuclei contributed to the material by glucose end up protonated when the reaction is carried out with MWCNT. To further probe this reduced protonation, we applied DP-based ^{13}C [22] REDOR experiments to the glucose-labeled samples. The aromatic ^{13}C nuclei in GT250/10-CNT-GC are dephased more slowly than those in GT250/1-GC. Dephasing in a REDOR experiment occurs slower when ^{1}H nuclei are farther away from the ^{13}C. The REDOR results indicate that the aromatic domains are larger when MWCNTs are present, than in the 250 °C pyrolysis product alone, even though the aromatic fraction is similar.

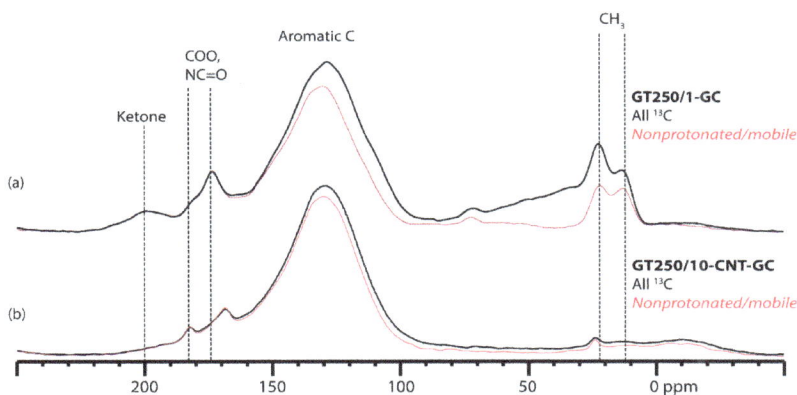

Figure 4. ^{13}C NMR spectra of ^{13}C-glucose-labeled glucose–taurine materials. Synthesis details in Table 1. (a) Pyrolyzed product prepared without MWCNTs. Thick black trace: DP spectrum of all ^{13}C species, 69% of which are aromatic. Thin red trace: nonprotonated or highly mobile ^{13}C species, obtained by dipolar dephasing; (b) Pyrolyzed product prepared with MWCNTs. Thick black trace: multiCP spectrum of all ^{13}C species, 72% of which are aromatic. Thin red trace: nonprotonated or highly mobile ^{13}C species, obtained by dipolar dephasing.

In the 4 mm probehead experiments on samples containing ^{13}C-labeled taurine, no indication was found of methylene groups, which would be characteristic of an alkyl link between the taurine sulfonate and the bulk of the pyrolyzed product. However, as described in the experimental section, incomplete ^{1}H decoupling, due to conductivity-induced power loss, can cause alkyl ^{13}C to be underrepresented. For that reason, NMR experiments on the ^{13}C-taurine samples were repeated using the 2.5 mm probehead to ensure that any existing methylene groups were observed (Figure 5). In these experiments, fewer scans were needed to compile a spectrum with a similar level of signal-to-noise (also due to the lower loss of power in the 2.5 mm probehead). The ^{1}H decoupling achieved in the 2.5 mm probehead was over 71 kHz, as determined by the 90° ^{1}H pulse time of 3.5 µs. For the 2.5 mm probehead spectra, the quantitative multiCP experiment was used instead of DP, since that experiment is much faster. The multiCP experiment was not used in the 4 mm probehead because it relies heavily on efficient ^{1}H pulses, which were not achieved in the 4 mm probehead, due to power loss.

In the GT250/10-CNT-TC quantitative spectrum from the 2.5 mm probehead (Figure 5a), as in the 4 mm probehead, no characteristic peaks from the alkyl moiety contributed by taurine are seen. Since a failure to sufficiently decouple the methylene ^{13}C nuclei from their ^{1}H bonding partners is ruled out in the 2.5-mm probehead, we can say that the alkyl moiety is, in fact, missing from that material, and no alkyl-linked sulfonate is expected to exist.

Two variations of the ^{13}C-taurine-labeled material with MWCNT were prepared to examine the effect on reaction temperature and the rate of temperature increase on the retention of the taurine methylene moiety in the post-pyrolysis structure. Both of these samples were analyzed using multiCP with and without recoupled dipolar dephasing in the 2.5 mm probehead. The spectra are shown for comparison with that of GT250/10-CNT-TC (Figure 5). The sample that was pyrolyzed at 250 °C, but was brought up to that temperature more slowly (GT250/1-CNT-TC), shows similar relative intensities in the aromatic and aliphatic regions as GT250/10-CNT-TC (Figure 5). The ^{13}C from taurine is distributed between the aromatic framework, and any remaining alkyl groups in about the same way as in the faster-ramped sample. However, the slower-ramped sample shows a larger loss from dipolar dephasing in the aromatic regions. This indicates that more of the aromatic ^{13}Cs in the slower-ramped sample are protonated. A larger fraction of protonated carbon indicates that the taurine ^{13}C nuclei become incorporated into smaller aromatic domains in the slower-ramped sample.

Compared to the products of pyrolysis at 250 °C, the ^{13}C-taurine sample pyrolyzed at 200 °C (GT200/1-CNT-TC) shows less intensity in the aromatic region and more intensity in the region associated with methylene groups. This demonstrates that a lower reaction temperature can increase the retention of the alkyl moiety of taurine in the final product. The intensity in the alkyl region is in the range expected for N–CH$_2$ and S–CH$_2$ bonding, which is evidence for the retention of the entire N(CH$_2$)$_2$SO$_3$ sequence. Since the intensity in this region is composed of overlapping peaks, the determination of the exact bonding environment of these methylene groups would require some modeling of potential structures, probably in combination with 2D NMR experiments. The predicted ^{13}C and ^{15}N NMR spectra based on those structures could then be compared to the experimental results for the ^{13}C-labeled and ^{15}N-labeled samples.

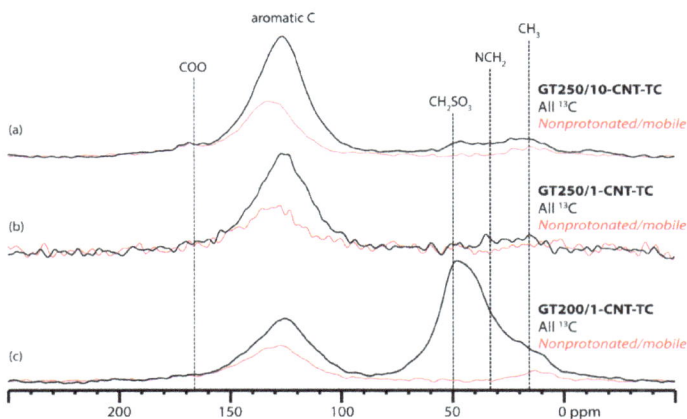

Figure 5. MultiCP ^{13}C NMR spectra of ^{13}C-taurine-labeled glucose-taurine pyrolysis products, all prepared with MWCNTs. Synthesis details in Table 1. Thick black traces: multiCP spectrum of all ^{13}C species. Thin red traces: nonprotonated or highly mobile ^{13}C species, obtained by dipolar dephasing. Fraction of aromatic species (a) 80%; (b) 76%; and (c) 24%.

2.3. Reaction Testing

Initial reaction testing of each catalyst was performed in a binary solvent consisting of THF and DMSO. This is a well-known reaction solvent to minimize side reactions. As shown in Supplemental Figure 2, the glucose/fructose conversion kinetics are highly dependent on the catalysts used. Very little glucose conversion was observed in the absence of catalysts added. All of the added catalyst combinations can promote glucose conversion to HMF, as can be seen in the plot. The greatest improvement was shown by the solid Lewis and Brønsted acid catalyst, and the solid Lewis acid catalyst, both of which were more effective than the solid Brønsted acid catalyst. This is in agreement with our previous work. However, for all of the glucose reactions, low selectivity, ranging from 0 to 20 mol %, was obtained. This can be ascribed to undesired side reactions on the surface of MWCNTs, and likely be associated with metal cation catalysis to induce the formation of humins and other side products. Since solid Lewis acid alone was found to be quite active for the catalysis, a combination of Lewis and Brønsted acid was chosen to be used for later conversions. Unlike for glucose, catalytic performance of the solid catalysts for fructose was improved significantly by adding Brønsted acid functionality. The conversion kinetics showed that at a reaction temperature lower than 120 °C, fructose conversion was completed within the first 3 h with selectivity greater than 85 mol %.

Based on the initial reactivity/selectivity trends, we chose to use the THF/water binary solvent system and a combination of solid Lewis and Brønsted acid catalysts for further reaction testing. Although DMSO can generally enhance sugar dehydration selectivity by minimizing side reactions, its practical implementation can be problematic, due to its high boiling point and cost. Figure 6 summarizes the reaction result of fructose conversion using reaction media of THF/water and reaction temperature of 160 °C. Conversion rate using fructose as feedstock was faster than glucose, as was observed in THF/DMSO solvent. In addition, selectivity of HMF from fructose was about 60 mol %, while less than 10 mol % selectivity from glucose was achieved.

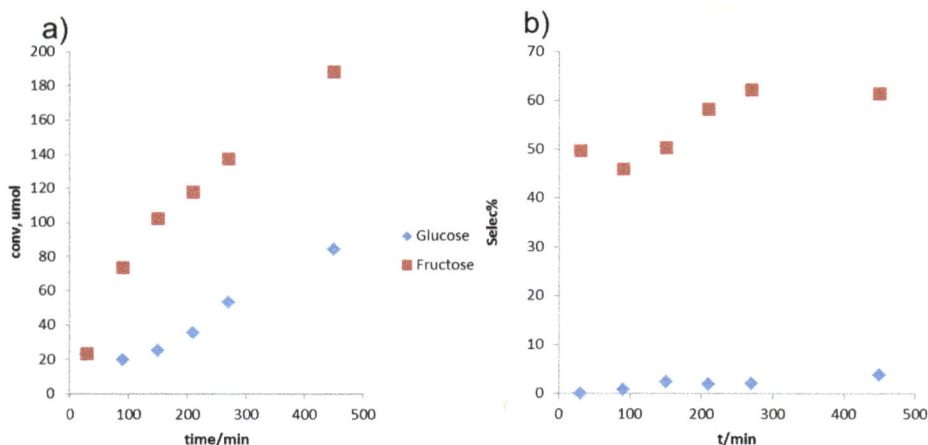

Figure 6. Reaction results of glucose/fructose conversion in THF/water solvent. (**a**) Conversions; (**b**) selectivities. Reaction conditions: solvent of THF/water 9:1 (*v/v*), reaction temperature of 160 °C, 2 wt % glucose/fructose, 5 mg each of Lewis and Brønsted catalyst/g of reaction media.

Catalyst stability and recyclability under reaction conditions are of great importance in industrial applications. Regarding this aspect, the recyclability of the solid catalysts for the fructose reaction was tested. As shown in Figure 7 and described above, three consecutive reactions were run with fructose as feedstock following the same procedure. The reaction kinetic profile showed a relatively similar reaction rate between the three runs, with little loss of catalyst activity. Selectivity for HMF was also kept around 60 mol %, which was slightly lower than THF/DMSO solvent system. The supernatant after each reaction was also analyzed for possible Al leaching. The calculated Al leaching was found to be 3.3%, 2.2%, and 1.8%, respectively.

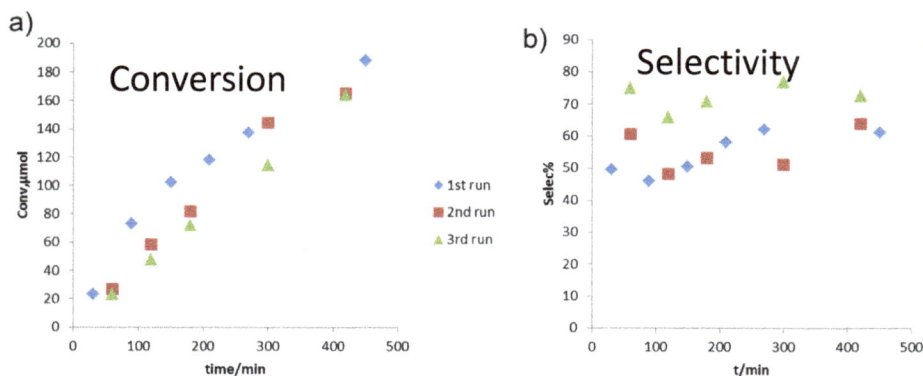

Figure 7. Recyclability study with fructose as feedstock. Reaction conditions: solvent of THF/water 9:1 (*v/v*), reaction temperature of 160 °C, 2 wt % fructose, 5 mg each of Lewis and Brønsted catalyst/g of reaction media.

3. Experimental

3.1. Catalyst Synthesis

Before the sulfonate-tethering reaction, the MWCNTs (Timenano, Chengdu, China) were ultrasonicated using a Branson Digital Sonifier® (Emerson, St. Louis, MO, USA) for 15 min with 90% power output, to remove amorphous carbon and introduce oxygenated groups to the surface. To synthesize sulfonate-tethered MWCNTs, an aqueous solution of 0.71 M glucose (Fisher, Hampton, NH, USA), 0.355 M taurine (Sigma-Aldrich, St. Louis, MO, USA), and 0.355 M NaOH (Fisher) was mixed with MWCNT (combined weight glucose and taurine accounts for 15 wt % carbon content relative to the weight of MWCNT). The mixture was then sonicated for 5 min, followed by heating to 120 °C with ramp rate of 1 °C /min or 10 °C/min, and held for 3 h until dry. The pyrolysis was done at 200 °C or 250 °C under 200 mL/min flowing N_2 protection, with a temperature ramping rate of either 1 °C/min or 10 °C/min, and held at the final temperature for additional 10 h. The same protocol was repeated again to increase the number of functional groups introduced. Table 1 summarizes the synthesis conditions for different pyrolysis products.

To make the solid Lewis acid (LA)-MWCNT, 0.8 g of the sulfonate-tethered MWCNTs, synthesized by the method described above, were first immersed in 1 M NaCl solution and stirred for 24 h, followed by contacting with $Al(OTf)_3$ (Sigma-Aldrich, 1.0 g) in ethanol at 80 °C for 24 h, to immobilize Al catalyst via cation exchange. Solid Brønsted acid (BA-MWCNT) was synthesized by contacting excess amount of 1.6 M HCl aqueous solution with sulfonate-tethered MWCNT after pyrolysis to fully protonate the sulfonate group.

Table 1. Synthesis conditions of different materials [a].

Sample Code	Isotopic Labeling	Heating Conditions
GT250/10-CNT-GC	Uniform ^{13}C: glucose	Pyrolysis temperature of 250 °C Ramp of 1 °C from room temp to 120 °C,10 °C from 120 °C to 250 °C
GT250/1-GC	Uniform ^{13}C: glucose	Pyrolysis temperature of 250 °C Ramp of 1 °C throughout the temperature range
GT250/10-CNT-GC/TN	Uniform ^{13}C: glucose Uniform ^{15}N: taurine	Pyrolysis temperature of 250 °C Ramp of 1 °C from r.t. to 120 °C,10 °C from 120 °C to 250 °C
GT250/1-GC/TN	Uniform ^{13}C: glucose Uniform ^{15}N: taurine	Pyrolysis temperature of 250 °C Ramp of 1 °C throughout the temperature range
GT250/10-CNT-TC	Uniform ^{13}C: taurine	Pyrolysis temperature of 250 °C Ramp of 1 °C from r.t. to 120 °C,10 °C from 120 °C to 250 °C
GT250/1-TC	Uniform ^{13}C: taurine	Pyrolysis temperature of 250 °C Ramp of 1 °C throughout the temperature range
GT200/1-CNT-TC	Uniform ^{13}C: taurine	Pyrolysis temperature of 200 °C Ramp of 1 °C throughout the temperature range
GT250/1-CNT-TC	Uniform ^{13}C: taurine	Pyrolysis temperature of 250 °C Ramp of 1 °C throughout the temperature range

[a] Taking GT250/10-CNT-GC/TN as an example, the samples are denoted as follows: GT is short for glucose and taurine together; 250/10 represents a pyrolysis temperature of 250 °C with a ramping rate of 10 °C/min; CNT means the reaction was carried out in the presence of CNT as support; GC indicates the use of ^{13}C-labeled glucose and TN indicates ^{15}N-labeled taurine.

3.2. Materials Characterization

The catalysts were characterized using scanning electron microscopy (SEM, Tokyo, Japan), transmission electron microscopy (TEM, Hillsboro, OR, USA), X-ray photoelectron spectroscopy (XPS, Eden Prairie, MN, USA), N_2 adsorption/desorption, and thermogravimetric analysis (TGA, Eden Prairie, MN, USA). TEM images were taken on a Tecnai G2 F20. XPS testing was performed on a Phi 5500 Multitechnique system using an Al Kα X-ray source. A C1s location of 284.6 was used for peak corrections. TGA experiments were done on a Perkin-Elmer STD 600 under the flow of 50 mL/min air, and a temperature ramp of 10 °C/min, from 50 to 900 °C. N_2 adsorption/desorption testing occurred on a Micromeritics ASAP 2020 with the BET method used for surface area calculations (Norcross, GA, USA).

3.3. Solid-State NMR

Solid-state magic-angle-spinning NMR studies were carried out on a Bruker Avance 400 spectrometer (Billerica, MA, USA) using 4 mm and 2.5 mm magic-angle spinning (MAS) double-resonance probeheads. All ^{13}C experiments were carried out under MAS at 14 kHz.

^{13}C was observed at 100 MHz. The ^{13}C pulse lengths used were 4 μs. The samples were prepared according to the synthesis described above and in Table 1. To reduce the conductivity of the MWCNT-containing samples, two additional steps were added to the preparation of some samples: dilution with laponite clay and grinding in a cryomill (Spex SamplePrep). Both of these steps reduce the size of conductive domains within the sample. Conductivity in an ssNMR sample reduces the effectiveness of the radio-frequency pulses through power absorption. The existence of conductivity-induced power loss caused concern that the experiments conducted in the 4 mm probehead could suffer from insufficient ^{1}H decoupling, which could cause the suppression of alkyl ^{13}C peaks in the spectra. In order to confirm that the alkyl peaks were not systematically underrepresented due to poor decoupling, some experiments on the ^{13}C-taurine materials were repeated using the 2.5 mm probehead. A smaller rotor diameter is known to increase effective decoupling power, which reduces possible distortions from incomplete decoupling. The effective decoupling power in the 2.5 mm probehead was measured by the length of a ^{1}H 180° pulse.

Both direct-polarization (DP) and cross-polarization (CP) were used to provide the initial polarization in the MAS ^{13}C NMR experiments. DP experiments were performed with recycle delays ≥40 s, and included a Hahn echo. CP experiments used the multiCP pulse sequence [20] in order to obtain quantitative spectra, and also included a Hahn echo.

In order to determine the amount of non-protonated ^{13}C in each sample (aromatized carbon), recoupled dipolar dephasing was applied to both DP and CP experiments. In a recoupled dipolar dephasing experiment, ^{1}H decoupling is removed for a period of time, which causes the signal of protonated ^{13}C nuclei to decrease. The total dipolar dephasing period was made up of two 30 μs periods flanking a ^{1}H 180° pulse for all recoupled dipolar dephasing experiments.

To further probe the aromatized carbon, we applied ^{13}C REDOR [21] experiments. As the recoupling time in the REDOR experiment increases, ^{13}C nuclei farther and farther from ^{1}Hs are dephased, and their associated signal decreases. Since ^{13}C nuclei closer to ^{1}H nuclei will dephase more quickly, the REDOR experiment is sensitive to distance, and can be particularly useful for probing the size of aromatic domains.

3.4. Reaction Testing

Glucose and fructose were tested as model carbohydrate molecules for HMF production with the prepared CNT-based catalysts in a 10 mL thick-walled glass reactor (Alltech). In a typical experiment, 2.0 g of the selected organic solvents containing 2 wt % of glucose or fructose, and one of the solid catalysts, was added to the Alltech reactor, and a triangular stir bar (Fisher Scientific, Hampton, NH, USA) was also added to allow for adequate agitation. The reactor was then sealed with PTFE liner covered caps (Fisher Scientific) and immersed in the oil bath at desired temperatures and stirred at 400 rpm. A catalyst recyclability test was performed with fructose as the substrate in the THF/water solvent system. After each run, the organic solvent was decanted, and the catalyst was kept and dried, and washed using sugar-free solvent before fresh solvent containing fructose was added to run the reaction again. In order to test for possible Al leaching from the catalyst, the decanted supernatant from the fructose reaction was sampled and prepared for ICP analysis.

Reaction product analysis was performed using a Waters 1525 HPLC system equipped with a 2998 PDA UV detector (Milford, MA, USA) at 280 nm, and a 2414 refractive index (RI) detector maintained at 333 K. The aqueous phase samples were analyzed using a PL Hi-Plex H-form carbohydrate column at 353 K with 5 mM H_2SO_4 at a flow rate of 0.6 mL min^{-1} as the mobile phase. The organic phase samples were analyzed using a Zorbax SB-C18 reverse phase column (Agilent, Santa Clara, CA, USA) at 308 K, with a methanol/5 mM H_2SO_4 (8:2 v/v) binary solvent as the mobile phase at a flow rate of 0.7 mL min^{-1}. The conversion/selectivity being used in this manuscript are defined as follows: the conversion of glucose is defined as the moles of glucose reacted, divided by the moles of initial glucose; the selectivity of HMF is defined as the moles of HMF produced divided by the moles of glucose reacted; and the yield of HMF is defined as the glucose conversion multiplied by the HMF selectivity.

3.5. Inductively Coupled Plasma Mass Spectrometer (ICP-MS) Analysis of Al Leaching

The aluminum content in each sample was analyzed using an inductively coupled plasma mass spectrometer (ICP-MS) (Bruker Aurora Elite, Billerica, MA, USA). The instrument parameters (Table 1) were optimized for maximum sensitivity with low levels of metal oxide ($CeO^+/Ce^+ \leq 2\%$) and doubly charged ($Ba^{2+}/Ba^+ \leq 3\%$) ions. Calibration standards were prepared by serial dilutions of a commercially available aluminum ICP-MS standard (SPEX CertiPrep) with 5% aqueous hydrochloric acid (HCl). The samples were also prepared to have a matrix of 5% aqueous HCl. The solid MWCNT catalyst samples (~5 mg) were dissolved in concentrated HCl, and then diluted with deionized water to a final HCl concentration of 5% (final mass of approximately 10 g). The original reaction supernatant samples were diluted 50- to 100-fold with 5% aqueous HCl.

4. Conclusions

In summary, the synthesis of sulfonic group-tethered MWCNT solid catalysts has been successfully demonstrated. By manipulating synthesis conditions, catalysts with the desired acid density can be made. Functionalized MWCNTs can be used as a catalyst support to immobilize homogeneous metal cations for various catalytic applications. Solid-state NMR has been shown to be a powerful tool to provide quantitative information about the chemical nature of surface carbon content, in order to optimize those synthesis conditions and observe the changes due to the MWCNT support. The hybrid Al/CNT catalysts tested in this work demonstrated good catalyst stability and recyclability for both glucose and fructose dehydration to HMF. More study is required to design a better ligand field on the surface of the MWCNTs, to achieve higher selectivity and less catalyst leaching, and this will be the focus of future research directions.

Author Contributions: H.J. performed the experiments, data collection, and analysis; J.F. analyzed the results; T.W. devised the project. The manuscript was written jointly by all the authors.

Funding:: This research was funded by National Natural Science Foundation of China, grant number 21506246.

Conflicts of Interest: The authors declare no conflict of interest.

References

1. Kobayashi, H.; Fukuoka, A. Synthesis and utilisation of sugar compounds derived from lignocellulosic biomass. *Green Chem.* **2013**, *15*, 1740–1763. [CrossRef]

2. Nikolau, B.J.; Perera, M.A.D.; Brachova, L.; Shanks, B. Platform biochemicals for a biorenewable chemical industry. *Plant J.* **2008**, *54*, 536–545. [CrossRef] [PubMed]

3. Shanks, B.H. Conversion of Biorenewable Feedstocks: New Challenges in Heterogeneous Catalysis. *Ind. Eng. Chem. Res.* **2010**, *49*, 10212–10217. [CrossRef]

4. Ravenelle, R.M.; Schüßler, F.; D'Amico, A.; Danilina, N.; van Bokhoven, J.A.; Lercher, J.A.; Jones, C.W.; Sievers, C. Stability of Zeolites in Hot Liquid Water. *J. Phys. Chem. C* **2010**, *114*, 19582–19595. [CrossRef]

5. Tessonnier, J.-P.; Villa, A.; Majoulet, O.; Su, D.S.; Schlögl, R. Defect-Mediated Functionalization of Carbon Nanotubes as a Route to Design Single-Site Basic Heterogeneous Catalysts for Biomass Conversion. *Angew. Chem. Int. Ed.* **2009**, *48*, 6543–6546. [CrossRef] [PubMed]

6. Fukuoka, A.; Dhepe, P.L. Sustainable green catalysis by supported metal nanoparticles. *Chem. Rec.* **2009**, *9*, 224–235. [CrossRef] [PubMed]

7. Serp, P.; Castillejos, E. Catalysis in Carbon Nanotubes. *ChemCatChem* **2010**, *2*, 41–47. [CrossRef]

8. Anderson, J.M.; Johnson, R.L.; Schmidt-Rohr, K.; Shanks, B.H. Solid state NMR study of chemical structure and hydrothermal deactivation of moderate-temperature carbon materials with acidic SO_3H sites. *Carbon* **2014**, *74*, 333–345. [CrossRef]

9. Anderson, J.M.; Johnson, R.L.; Schmidt-Rohr, K.; Shanks, B.H. Hydrothermal degradation of model sulfonic acid compounds: Probing the relative sulfur-carbon bond strength in water. *Catal. Commun.* **2014**, *51*, 33–36. [CrossRef]

10. Johnson, R.L.; Anderson, J.M.; Shanks, B.H.; Schmidt-Rohr, K. Simple One-Step Synthesis of Aromatic-Rich Materials with High Concentrations of Hydrothermally Stable Catalytic Sites, Validated by NMR. *Chem. Mater.* **2014**, *26*, 5523–5532. [CrossRef]

11. Roman-Leshkov, Y.; Chheda, J.N.; Dumesic, J.A. Phase Modifiers Promote Efficient Production of Hydroxymethylfurfural from Fructose. *Science* **2006**, *312*, 1933–1937. [CrossRef] [PubMed]

12. Zhao, Q.; Wang, L.; Zhao, S.; Wang, X.; Wang, S. High selective production of 5-hydroymethylfurfural from fructose by a solid heteropolyacid catalyst. *Fuel* **2011**, *90*, 2289–2293. [CrossRef]

13. Nikolla, E.; Roman-Leshkov, Y.; Moliner, M.; Davis, M.E. "One-Pot" Synthesis of 5-(Hydroxymethyl)furfural from Carbohydrates using Tin-Beta Zeolite. *ACS Catal.* **2011**, *1*, 408–410. [CrossRef]

14. Xiong, H.; Wang, T.; Shanks, B.H.; Datye, A.K. Tuning the Location of Niobia/Carbon Composites in a Biphasic Reaction: Dehydration of D-Glucose to 5-Hydroxymethylfurfural. *Catal. Lett.* **2013**, *143*, 509–516. [CrossRef]

15. Huang, F.; Su, Y.; Tao, Y.; Sun, W.; Wang, W. Preparation of 5-hydroxymethylfurfural from glucose catalyzed by silica-supported phosphotungstic acid heterogeneous catalyst. *Fuel* **2018**, *226*, 417–422. [CrossRef]

16. Zuo, M.; Le, K.; Li, Z.; Jiang, Y.; Zeng, X.; Tang, X.; Sun, Y.; Lin, L. Green process for production of 5-hydroxymethylfurfural from carbohydrates with high purity in deep eutectic solvents. *Ind. Crops Prod.* **2017**, *99*, 1–6. [CrossRef]

17. De Souza, R.L.; Yu, H.; Rataboul, F.; Essayem, N. 5-Hydroxymethylfurfural (5-HMF) Production from Hexoses: Limits of Heterogeneous Catalysis in Hydrothermal Conditions and Potential of Concentrated Aqueous Organic Acids as Reactive Solvent System. *Challenges* **2012**, *3*, 212–232. [CrossRef]

18. Yang, Y.; Hu, C.-W.; Abu-Omar, M.M. Conversion of carbohydrates and lignocellulosic biomass into 5-hydroxymethylfurfural using $AlCl_3·6H_2O$ catalyst in a biphasic solvent system. *Green Chem.* **2012**, *14*, 509–513. [CrossRef]

19. Pagán-Torres, Y.J.; Wang, T.; Gallo, J.M.R.; Shanks, B.H.; Dumesic, J.A. Production of 5-Hydroxymethylfurfural from Glucose Using a Combination of Lewis and Brønsted Acid Catalysts in Water in a Biphasic Reactor with an Alkylphenol Solvent. *ACS Catal.* **2012**, *2*, 930–934. [CrossRef]

20. Zhang, Z.; Wang, Q.; Xie, H.; Liu, W.; Zhao, Z.K. Catalytic Conversion of Carbohydrates into 5-Hydroxymethylfurfural by Germanium(IV) Chloride in Ionic Liquids. *ChemSusChem* **2010**, *4*, 131–138. [CrossRef] [PubMed]

21. Wang, T.; Pagan-Torres, Y.J.; Combs, E.J.; Dumesic, J.A.; Shanks, B.H. Water-Compatible Lewis Acid-Catalyzed Conversion of Carbohydrates to 5-Hydroxymethylfurfural in a Biphasic Solvent System. *Top. Catal.* **2012**, *55*, 657–662. [CrossRef]

22. Johnson, R.L.; Schmidt-Rohr, K. Quantitative solid-state ^{13}C NMR with signal enhancement by multiple cross polarization. *J. Magn. Reson.* **2014**, *239*, 44–49. [CrossRef] [PubMed]

MDPI

St. Alban-Anlage 66

4052 Basel

Switzerland

Tel. +41 61 683 77 34

Fax +41 61 302 89 18

www.mdpi.com

Catalysts Editorial Office

E-mail: catalysts@mdpi.com

www.mdpi.com/journal/catalysts

www.ingramcontent.com/pod-product-compliance
Lightning Source LLC
Chambersburg PA
CBHW051840210326
41597CB00033B/5725